Surviving Climate Chaos
by Strengthening Communities and Ecosystems

Surviving climate chaos needs communities and ecosystems that are strong enough to cope with near-random environmental impacts. Their strength depends upon their integrity, so preserving and restoring this is essential. Total climate breakdown might be postponed by extreme efforts to conserve carbon and recapture pollutants, but climate chaos everywhere is now inevitable. Adaptation efforts by Paris Agreement countries are converging on community-based and ecosystem-based strategies, and case studies in Bolivia, Nepal and Tanzania confirm that these are the best ways forward. But success depends on local empowerment through forums, ecosystem tenure security and environmental education. When replicated, networked and nurtured by governments, they can strengthen societies against climate chaos while achieving sustainable development. These vital messages are highlighted for all those who seek a role in promoting adaptation: students, researchers and teachers; government officials and aid professionals; and everyone now living under threat of climate chaos.

JULIAN CALDECOTT is a Fellow of the Schumacher Institute for Sustainable Systems and Director of Creatura Ltd, and has a background in wildlife research and conservation in tropical rainforests. Since 2000 he has led evaluations of major aid investments for the European Commission, Denmark, Norway, Finland, Switzerland, the United Kingdom and the World Bank, focusing on climate change, biodiversity, ecosystem management, sustainability and institutional and community development. He has written seven books, including *Designing Conservation Projects* (Cambridge University Press, 1996).

Surviving Climate Chaos

by Strengthening Communities
and Ecosystems

JULIAN CALDECOTT
Schumacher Institute for Sustainable Systems

CAMBRIDGE
UNIVERSITY PRESS

CAMBRIDGE
UNIVERSITY PRESS

University Printing House, Cambridge CB2 8BS, United Kingdom

One Liberty Plaza, 20th Floor, New York, NY 10006, USA

477 Williamstown Road, Port Melbourne, VIC 3207, Australia

314–321, 3rd Floor, Plot 3, Splendor Forum, Jasola District Centre,
New Delhi – 110025, India

103 Penang Road, #05–06/07, Visioncrest Commercial, Singapore 238467

Cambridge University Press is part of the University of Cambridge.

It furthers the University's mission by disseminating knowledge in the pursuit of
education, learning, and research at the highest international levels of excellence.

www.cambridge.org
Information on this title: www.cambridge.org/9781108840125
DOI: 10.1017/9781108878982

First published 2021

Printed in the United Kingdom by TJ Books Limited, Padstow Cornwall

A catalogue record for this publication is available from the British Library.

ISBN 978-1-108-84012-5 Hardback
ISBN 978-1-108-79378-0 Paperback

This is for my son Ben, who is promoting climate action in realms that were beyond our dreams in 1996, when I last dedicated a book to him; and in loving memory of Dean, who was taken by the sea at the start of a brilliant design career, a reminder of human frailty in the face of nature.

Contents

Colour plates can be found between pages 208 and 209.

Preface

Surviving Climate Chaos by Strengthening Communities and Ecosystems explains why it is important to distinguish between the effects of climate change at large and small scale, and why and how to build survival strategies that are appropriate to each. At the continental and global scale climate change is directional and predictable, with rising temperatures and sea levels, melting ice, changing seasons and regional trends in droughts, floods and storminess. This is the level at which national and international responses are focused, and where the interests of governments tend to be concentrated. But the story is very different at the local and landscape level, where people actually live, since here the same effects are experienced as chaotic and unpredictable. Little is certain at this level except that risks will increase in novelty, frequency and intensity. Surviving climate chaos is therefore something that every community on Earth must do in its own way, each in their own circumstances and dependent upon their own local ecosystems.

Surviving climate chaos requires communities and ecosystems to be strong enough to cope with whatever a changing climate throws at them. This depends upon their resilience, resistance and flexibility – three dimensions of strength that are properties of all systems, and that depend on the integrity of those systems. Adapting to climate chaos is therefore the process of preserving and restoring the integrity of communities and ecosystems. Because climate change and chaos now reach into every corner of the world, this must be done everywhere if no one is to be left behind. The usefulness of this approach is highlighted here for all those with key roles in promoting adaptation, including staff of the UNFCCC Secretariat, national and local government officials in developing countries, aid professionals, students, researchers and teachers, and indeed all people who live under threat of climate chaos.

The global context is one of dire urgency. Climate and ecological emergencies have now been declared by thousands of nations and public institutions worldwide, and UN Secretary General António Guterres has called for all countries to join them, both in declaring an emergency and in seeking peace with nature. They are supported by the scientific community and tens of millions of concerned citizens, whose emotions are engaged and whose reasons are convincing: that the beauty and integrity of nature are being destroyed, that human progress is being undermined and reversed and that Arctic, equatorial and oceanic tipping points threaten a 'perfect storm' of runaway climate breakdown in mid-century. That is, the middle of *this* century. And once that happens, the process will be driven by its own internal feedbacks – one change automatically amplifying another change – so that any opportunity for further human influence will be lost.

This schedule has powerful implications for what humanity should now be doing. Since climate change is well underway, and climate chaos is already eroding and at times devastating localities and landscapes, we have to adapt and be prepared to adapt further. Nowhere will be safe for long, so this means everyone and everywhere. The sooner we start to strengthen our systems against chaos, the better for now, and the better for our prospects later. By doing so we will be building a world in which much of what we hold dear might survive a period of severe climate instability. But there are some extremes that we cannot survive, so we must try to avoid them, and this means we are in a race against the unknown calamity of mid-century climate breakdown. This we must delay if we can, while we strengthen our systems and bring the causes of climate change under control. This in turn requires net greenhouse gas emissions to be slashed immediately and quickly reversed, at any cost and with maximum effectiveness, while ecosystems and communities are strengthened against mounting chaos.

Finding ways to induce such major efforts at a global level is hard, and they cannot be imposed so must happen through the sharing of knowledge, voluntary compliance and common purpose. Local people and institutions are responding to their direct experience of climate chaos at the micro level, but to ramp up the overall effort quickly enough at the macro level requires governments and major financial institutions to help. Under pressure of evidence, reason and public alarm, however, these are also starting to react with a new degree of realism, offering hope that decisive breakthroughs are becoming possible. This story is also important, since the micro and macro levels are just as connected for the climate response as they are for climate change itself. So here I explore the role of the 2015 Paris Agreement in mobilising knowledge, political will and useful investment in mitigating and adapting to climate change.

The agreement is an experimentalist treaty that depends upon overarching goals, autonomous actors and iterative learning processes. To explore its influence I analyse official adaptation communications submitted in 2015–20 by 158 countries to the UNFCCC Secretariat – the knowledge hub of the agreement. Based on these I describe how governments see climate change and what to do about it, and how they are increasingly recognising diverse benefits from ecosystem-based and community-based adaptation, and the synergies between them. I highlight themes from countries in Europe, the small-islands group, Africa and the Americas. Seeking practical details, I explore these issues through case studies in Nepal, Bolivia and Tanzania, all involving the strengthening of local social and ecological systems. And I report on adaptation in urban environments, particularly the role of self-organising neighbourhood networks and their value to local governments and their members in making cities stronger and more sustainable.

These observations all support the case that complex systems can be strengthened in certain specific ways, based on how systems work and what harms them, and this can guide the design and evaluation of all adaptation investments. Entire aid portfolios can be designed to deliver progress both on adaptation and on all the Sustainable Development Goals, for these aims are now effectively the same. To join them fully it needs to be recognised that communities and ecosystems are mutually dependent, and that in the face of climate chaos a local empowerment and environmental education package is required everywhere. Where this is applied then systems will be strengthened, regardless of whatever else is done; where it is not, then systems will continue to weaken, and nothing else will compensate for its lack. Larger aid programmes can be built using this package, through replication, networking, technical additions and the shielding of local community–ecosystem units by higher authorities. Getting this mixture right for every locality on Earth is the essence of surviving climate chaos, and this book is part of the search for the best and cheapest ways to achieve it.

Acknowledgements

My special thanks go to Mary Monro for sharing our move to a new country, for keeping me well during the Covid lockdowns, and for many new ideas and perspectives. Also to Mike Speirs of Danida for the initial spark for the book, to Jenneth Parker and Ian Roderick of the Schumacher Institute for Sustainable Systems, to Dominic Lewis and Aleksandra Serocka of Cambridge University Press and to three anonymous peer reviewers.

The insights of many others enriched this book, whether they know it or not, and to whom thanks are due: Aklilu Amsalu, Kasper Thede Anderskov, Abubakar Diwani Bakar, Carmen Barragan, Govinda Basnet, Laxmi Kumari Basnet, Getachew Eshete Beyene, Ananta Bhandari, Neil Maclean Bird, Mary Bolingbroke, Ben Caldecott, Susan Canney, Molly Scott Cato, Muita Chacha, Henrik Chart, Bennett Collins, Elizabeth Colwell, Sue Cormack, Ingrid Dahl-Madsen, Thinh Quan Dang, Resham Bahadur Danghi, Abhoy Kumar Das, Sonya Dewi, Sindhu Prasad Dhugana, Martin Dickler, Jane Dunn, Anton Adriaan Eberhard, Aino Efraimsson, Andree Ekadinata, Anna Filipova, Shehana Gomez, Andy Green, Helene Rask Grøn, Helena Haakana, Sheha Hamdan, Finn Hansen, Minna Hares, Chris Jordan, Chudamani Joshi, Ali Amin Omar Juma, Mohammed Juma, Vuokko Jutila, Annika Kaipola, Minna Kallio, Jens Holm Kanstrup, Ganesh Bahadur Karki, Ville Karvinen, Mikhail Kavanagh, Niina Käyhkö, Manohara Khadka, Mgeni M. Khamis, Surya Khanal, Bernadeta Killian, Miriam Koenig, Rajan Kotru, Jakob Kronik, Yki Laine, Ram Prasad Lamsal, Juho Lappalainen, Kari Leppänen, Adam Ley-Lange, Olivia Lousada, Edmund Mabhuye, Machindranath, Alastair Macrae, Avi Mahaningtyas, Bustar Maitar, Makame Omar Makame, Ibrahim Khalid Mambo, Maulid Masud, Magnus Merkle, Sheha Mjaja, Musa Mkubwa, Hashim Muumin, William Nambiza,

Santosh Mani Nepal, Saroj Nepal, Tung Lam Nguyen, Thomas Nielsen, Pentti Niemistö, Mila Nuh, Jonathan Oates, Bishwa Nath Oli, Saida Omar, Nicholas Ostler, Bharati Pathak, Riikka Raatikainen, Aayush Rai, Andy Lee Robinson, Jack Ruitenbeek, Jenny Ruskin, Sadan of the Nepal Scouts, Omar Saif, Sanjaya Shah, Markku Siltanen, Lorna Slade, Ron Smit, Peter Birch Sørensen, Sam Staddon, Mauri Starckman, Keshar Man Sthapit, Suyanto, Olivia Tanujaya, Riitta Teiniranta, Ali Thani, Ida Theilade, Tea Törnroos, Sue Turner, Sauli Valkonen, Caroline VanderSluys, Gwen Vaughan, Markku Viitasalo, Mette Vinqvist, Elina Virtanen, Muriel Visser, John Waters, Pamela White, Melissa Wilson, Pius Yanda and Dan Zevin.

PART I Context, Tools and Systems

I Adaptation and the Paris Agreement

I.I THE GLOBAL CONTEXT

Around 1950 humanity began a phase of explosive growth in manufacture, trade, consumption, technology and the transformation of natural ecosystems and traditional societies. The speed of change ensured that weaknesses inherited from the past continued to deform our societies, including the exclusion and oppression of many people on grounds such as 'race', gender, caste, class and faith. Ignorance and greed also ensured that economic change had many negative side effects, notably the destruction of ecosystems and ecological services that sustain society, and the pollution of the air, food and water that sustain health.

Carbon dioxide (CO_2) from burning organic carbon in wood and coal was one pollutant that would soon come to have a particular significance. For by 1950 the biosphere – the global system comprising all life – had already absorbed almost as much extra CO_2 as it could without changing the composition of the air and the heat balance of the biosphere. As emission rates grew further the atmospheric concentration of CO_2 quickly rose to a level not seen for at least 800,000 years (Snyder, 2016; Our World in Data, 2020a). It has continued to rise ever since, with our annual carbon emissions soaring from a few billion tonnes in the 1960s to 40 or 50 billion tonnes in the 2010s (Ballantyne et al., 2012; Our World in Data, 2020b).

Because CO_2 is a greenhouse gas (GHG), this at once began the process of trapping abnormal amounts of solar radiation within the biosphere. Land use change and industry then added more and different GHGs, some of them far more potent than CO_2, including methane (CH_4), nitrous oxide (N_2O), sulphur hexafluoride (SF_6) and

3

compounds based on bonds between atoms of carbon and fluorine (such as the chlorofluorocarbons or CFCs). All have different heat-trapping (and other) effects, persist in the atmosphere for different lengths of time and react differently with other chemicals and under varied physical conditions in the biosphere.

The various sources (emission origins), sinks (absorption processes) and net rate of growth in GHG concentrations in the atmosphere are monitored and reported in detail for CO_2 (Le Quéré et al., 2015, 2018; Friedlingstein et al., 2019, 2020) and CH_4 (Saunois et al., 2016, 2020). These studies show not only an increasing understanding of the complex heat-trapping effect of GHGs over time but also a series of discoveries that call into question each level of understanding almost as soon as it is reached. These uncertainties have arisen, for example, from methane sources in melting permafrost, decaying peat and warming sea beds (Chapter 2), and from nitrous oxide released by the breakdown of fertilisers in farmland. These are capable of amplifying climate change and its impacts beyond the scope of previous models.

While GHG emissions were escalating, we were also changing ecosystems and extinguishing species. This was degrading the capacity of the biosphere to absorb GHGs and buffer their effects. The net result of all these processes came to be seen as an approaching crisis of global heating, mass extinction and ecological breakdown. Our first response was a false dawn in the early 1970s, when the United Nations Environment Programme (UNEP) was founded, followed by a pause when the political world was polarised by the Cold War. There was a more complete effort in the early 1990s, built around the United Nations Conference on Environment and Development in Rio de Janeiro, where two key environmental treaties were agreed: the Convention on Biological Diversity (CBD), which sought to head off mass extinction and ecological collapse, and the United Nations Framework Convention on Climate Change (UNFCCC).

The latter sketched out a path by which we would bring net GHG emissions under control (a process known as 'mitigation'), in order to head off the climatic effects of global heating ('climate change') and cope with their consequences ('adaptation'). The story since has been one of long pauses, scientific progress, political controversy, denial, distraction and occasional flurries of constructive thought and useful activity, notably in 2007 and 2015. In the process it came to be realised that the drivers of global heating and climate change are so foundational to our ways of life that mitigating them adequately would be very hard and expensive.

With public support, political will, leadership and cultural change this might not be impossible, but the difficulty of achieving adequate mitigation meant that adaptation came to be seen as an equal priority. This is partly an admission of defeat but mainly a pragmatic survival response. Besides which, many adaptation actions can contribute to mitigation and vice versa, as well as helping to reduce biodiversity loss and ecosystem breakdown. Thus, we have realised that all these problems are connected and can only be solved through systemic action based on holistic thinking.

1.2   THE CLIMATE CONVENTION

The UNFCCC entered into force in 1994 and provides the main framework for global discussions on mitigation, adaptation and 'means of implementation' aspects of the climate response (Kamphof, 2018a). Decisions are taken each year at a Conference of the Parties (CoP), the first of which, CoP 1/1995, was held in Berlin.[1] Some of these were game-changing: CoP 13/2007 in Bali, for example, coincided with and contributed to a sea change in governments' perceptions of climate change as a major economic threat, and hence their engagement with mitigation; while CoP 21/2015 in Paris yielded an agreement that set out new paths for mitigation and adaptation

---

[1] Other recent CoPs took place in Cancún (16/2010), Durban (17/2011), Doha (18/2012), Warsaw (19/2013), Lima (20/2014), Paris (21/2015), Marrakech (22/2016), Bonn (23/2017), Katowice (24/2018) and Madrid (25/2019). The next CoP (26) is planned for Glasgow in late 2021, having been rescheduled from 2020 due to the Covid-19 pandemic – see later in this chapter and Chapter 2.

efforts to follow, based on new ways for nations to cooperate (see Section 1.5). Decisions of special significance for adaptation had also previously been made at CoP 11/2005 in Nairobi, where the Nairobi Work Programme was agreed, and at CoP 16/2010 in Cancún. The latter authorised an Adaptation Committee at the UNFCCC Secretariat, and also issued the Cancún Adaptation Framework, which called for equal priority between mitigation and adaptation, while focusing adaptation on water, health, farms, food security, coastal zones and ecological and other systems.

Pre-dating, informing and later paralleling the UNFCCC process, the Intergovernmental Panel on Climate Change (IPCC) was set up in 1988 by UNEP and the World Meteorological Organisation. Its role is to analyse scientific findings on climate change and to inform the United Nations (UN) system about them, which it has done through a series of assessment reports (IPCC, 1992, 1995, 2001, 2007, 2014) and reviews on particular topics (most recently: IPCC, 2018, 2019a, 2019b). The sixth IPCC Assessment Report is due in 2022, and is expected to spell out: the certainty of human agency in driving climate change; the true dimensions and urgency of the emerging climate threat; and the transformative scale of global, economy-wide interventions needed to mount an adequate climate response. Many hopes are therefore pinned on the success of the CoPs in 2021–2023.

National and international laws have a common origin in top-down rule by governments, where leaders and apex forums make decisions that bind citizens and institutions to certain norms of behaviour. International law continued this tradition, and the CBD, which originated at the Rio Conference alongside the UNFCCC in 1992, reflects this top-down approach as a binding treaty imposed by all governments on all governments and the citizens and institutions over which they have jurisdiction. The UNFCCC could not be formulated in the same way, however, since even at the time (it became worse later) there was too much debate on the causes of climate change and what to do about it to agree upon anything more definite than a 'framework convention', with the details to be worked out later. These details would be provided by the CoPs, which were expected to produce leadership statements, technical guidance documents and specific binding protocols, which they did, for example, in the Kyoto Protocol at CoP 3/1997 (and its amendment at CoP 18/2012 in Doha) on reducing and reporting GHG emissions.

Meanwhile, three things happened. First, the climate response became embroiled in intense and extended debate, based partly on scientific uncertainties but mainly on the political exploitation of those uncertainties by groups with an interest in preventing binding GHG emission reductions (Chapter 2). Second, the subject of climate change became much more complex: 'mitigation' grew to embrace many different GHGs and their diverse and changing sources and sinks in all economic sectors in all countries; and 'adaptation' grew to cover an extraordinary range of factors as it was realised that vulnerability extended to every aspect of everyone's economic system and society, and they would all need to be strengthened in different ways against changing threats. Third, it became clear that this dynamic complexity, in the absence of an all-knowing 'hegemon with the power to impose a single set of rules' (Overdevest and Zeitlin, 2011: 2), meant that the top-down approach to organising the climate response would not work (Overdevest and Zeitlin, 2014). Opinion among European Commission (EC) and European Union (EU) member state stakeholders seemed to reach this conclusion after a humiliating failure of EU climate diplomacy at CoP 15/2009 in Copenhagen, and thereafter 'the EU moved away from its ambition of legally binding instruments towards more soft yet universal agreements' (Kamphof, 2018a: 3).

1.3 EXPERIMENTALIST GOVERNANCE

These three factors opened the way for a new approach based on 'experimentalist' governance, a form that is typically established by agreement among central, global or apex actors and local, national or subsidiary ones. It has three defining characteristics: (1) there are overarching but provisional goals and ways to assess progress; (2) there is broad discretion for subsidiary actors to pursue the goals in their own way, provided that they report regularly and transparently so that they can all learn from each other (e.g. through peer dialogue and periodic reviews); and (3) there are opportunities to revise the goals and ways of assessing progress, and the decision-making procedures themselves, in response to the results of the review process (Sabel and Zeitlin, 2012; Zeitlin and Sabel, 2013). Thus, it involves free actors in a common enterprise where progress is made iteratively, through repeated cycles of design, effort and learning, followed by redesign, renewed effort and new learning until the goal is reached or changed.

This kind of governance system emerged in large cultural domains where centralised rule was hard to sustain, yet all actors recognised their common interests and the need to cooperate in protecting those interests. This combination often occurs in large political entities, but not necessarily so. The ancient Roman Empire, for example, maintained centralised rule over a large area by means of professional legions, good roads, loyal colonies and intimidated client states (Luttwak, 1976), and its immediate successor, the Byzantine Empire, retained centralised control using its military and religious prestige (Rocker, 1937). For clearer cases of *experimentalist* govern-ance, we would have to look to vast cultural domains with weak central control, including the 1,000-year Holy Roman Empire of the German people (Wilson, 2016) and the EU (Sabel and Zeitlin, 2012).

Historically the aims of subsidiary actors were mainly collect-ive security and efficient trade, but more recent experimentalist regimes have been used in the domains of food, the nuclear power generation industry and air-traffic safety (Sabel and Zeitlin, 2011). The EU is a particularly rich source of experimentation in this model, owing to its Holy Roman Empire heritage via the Federal Republic of Germany, as a hands-off oversight and standard-setting body, and the creative tension between and among the EU institutions and member states. By 2000 it had already developed an experimentalist approach to internal problem-solving, an example being the Water

Framework Directive (WFD, Sabel and Zeitlin, 2012). In this process, tensions between the top-down regulatory and bottom-up experimentalist preferences of the various member states occurred in the 1990s, until the decisive shift in favour of experimentalism occurred by 2000 (Box 1.1).

BOX 1.1　**Experimentalist governance and the EU Water Framework Directive**

Years of negotiation among EU Member States produced a series of directives, including the Urban Waste Water Treatment Directive (1991) and the Nitrates Directive (1991). These aimed to tackle the problem of eutrophication, the accumulation of nitrate and phosphorus compounds from sewage and fertiliser pollution, which causes excessive algal growth that can suffocate aquatic life. They also targeted health issues such as microbial pollution in bathing water, and nitrates in drinking water. ... Realising that the world is complex, that local conditions vary, that member states all have different legal systems, priorities and capabilities, and that a 'one-size-fits-all' approach might not be the best way forward, the EU then developed its Water Framework Directive or WFD (2000). This requires integrated river basin management, and aims to ensure clean rivers, lakes, ground water and coastal beaches throughout its member states. It is a unique 'gold standard' in the management of water resources. It sets standards for river basin planning, and for the ecological quality and chemical purity of surface and ground waters. For river basins, the aims are general protection of aquatic ecology, and specific protection of unique and valuable habitats, drinking water resources, and bathing water, and all these objectives must be integrated for each river basin.

The central requirement of the WFD is that the environment must be protected to a high level, in its entirety. For ecological quality, water bodies are supposed to show no more than a slight departure from the biological community which would be expected with minimal human impact – the equivalent, say, of a Canadian lake

BOX I.I **(cont.)**

exposed only to summer campers and duck-hunters. ... As the member states tried to put the WFD into effect, they quickly developed a Common Implementation Strategy. In this, each country developed its own ideas of what good practice actually meant and how to measure progress, then applied them while studying the results, and compared notes so that they could all learn from each other. Every now and then the European Commission would study progress and lessons learned, and make proposals for everyone to think about. This kind of networked, exploratory peer learning, now called 'experimentalist governance' by academics, has proved to be an immensely powerful approach to managing systems that are too complex and dynamic for top-down rule-making to work very well.

Caldecott (2020): 163–165

1.4 EXPERIMENTALISM, SUSTAINABILITY AND SYSTEMS THINKING

The Sustainable Development Goals

Once the EU had abandoned a top-down approach around 2010 it began to exert a stronger influence on the UN, by supporting the UNEP and more generally being in favour of experimentalist solutions to major problems of environment and development. This approach contributed to the agreement in 2015 of the UN 2030 Agenda for Sustainable Development (UN, 2015) and the Sustainable Development Goals or SDGs (Kamphof, 2018a, 2018b; Table 1.1). The SDGs are overarching goals in an experimentalist sense, with autonomous actors and iterative learning processes, but each is related to the outputs of different complex systems. For example, SDG 6 (on water) depends upon the management of water resources, and those resources are themselves outputs of complex systems involving catchments, aquifers, farms, dams, pipes, treatment

Table 1.1 *The SDGs for 2015–2030*

SDG	Summary description
1	*No poverty*: End poverty in all its forms everywhere, through inclusive economic growth and equality.
2	*Zero hunger*: End hunger, achieve food security and improved nutrition, and promote sustainable agriculture.
3	*Good health and well-being*: Ensure healthy lives and promote well-being for all at all ages, as essential to sustainable development.
4	*Quality education*: Ensure inclusive and equitable quality education and promote lifelong learning opportunities for all, as the foundation for improving people's lives sustainably.
5	*Gender equality*: Promote gender equality and empowerment of all women and girls as a necessary foundation for a peaceful, prosperous and sustainable world.
6	*Clean water and sanitation*: Ensure availability and sustainable management of water and sanitation for all.
7	*Affordable and clean energy*: Ensure access to affordable, reliable, sustainable and modern energy for all, as this is central to nearly every major challenge and opportunity.
8	*Decent work and economic growth*: Promote sustained, inclusive and sustainable economic growth with full and productive employment and decent work for all.
9	*Industry, innovation and infrastructure*: Build resilient infrastructure, promote inclusive and sustainable industrialisation and foster innovation.
10	*Reduced inequalities*: Reduce inequality within and among countries, through policies that are universal in principle and pay attention to the needs of disadvantaged and marginalised populations.
11	*Sustainable cities and communities*: Make cities inclusive, safe, resilient and sustainable, with opportunities for all and access to basic services, energy, housing, transportation and more.
12	*Responsible consumption and production*: Ensure sustainable consumption and production in all sectors.
13	*Climate action*: Take urgent action to combat climate change and its impacts, as global challenges that affect everyone, everywhere.

Table 1.1 (*cont.*)

SDG	Summary description
14	*Life below water*: Carefully manage the oceans, seas and marine resources for sustainable development.
15	*Life on land*: Sustainably manage forests, combat desertification, halt and reverse land degradation, halt biodiversity loss.
16	*Peace, justice and strong institutions*: Ensure access to justice for all, and build effective and accountable institutions at all levels.
17	*Partnerships*: Revitalise the global partnership for sustainable development.

Sources: UNDESA (2018); UN (2020).

facilities, etc. Moreover, every such system and every output depends upon or affects one or more of the others. For example, SDG 3 (on health) depends on the outcomes of the systems behind SDG 1 (on poverty), SDG 2 (on hunger), SDG 4 (on education) and others.

Systems Thinking

Because of their interlinkages, to make sense of the SDGs and to plan for or monitor their achievement requires systems thinking (e.g. Bateson, 1972; Meadows, 2008). This makes sense of complex phenomena using ideas such as *interconnectedness* (everything is connected to everything else), *synthesis* (understanding the whole and its parts at the same time), *emergence* (new phenomena arise from interactions among other phenomena), *feedback loops* (outputs of phenomena are inputs to other phenomena and affect their behaviour), *causality* (one thing leads to another) and *systems mapping* (tracing all the connections and effects among the parts of the system). For these reasons, experimentalism, sustainability and systems thinking are deeply connected (e.g. Sanneh, 2018), and together they provide the pervasive approach of this book. But to return to the immediate story, 2015 was also the year of CoP 21/2015 in Paris, and by then the EU had had several years after Copenhagen to encourage experimentalist thinking as an alternative to making a fresh attempt to agree to a top-down treaty.

1.5   THE PARIS AGREEMENT

Experimentalist Features

The Paris Agreement's role is to enhance implementation of its parent convention (UNFCCC, 2016a). Thus it shares with the UNFCCC most of its purposes (such as capacity building and reducing GHG emissions), methods (such as transparency and rules of procedure), principles (such as equity and common but differentiated responsibilities) and mechanisms, including its financial and technology mechanisms, its Secretariat, its CoPs[2] and its technical subsidiary bodies. In experimentalist terms, it has overarching goals for mitigation and adaptation (which it repeatedly states are to be given equal priority), and both a reliance on and freedom for its parties to choose, explain and report transparently on their own paths towards those goals, thus supporting both a continuous peer-learning process and periodic reviews. The latter are described as 'global stocktakes', the first of which is to be at the CoP in 2023 (Article 14), with others at five-year intervals unless the scheduling is changed.[3] As the supreme decision-making forums of the convention, the CoPs have the power to redefine goals, methods and anything else in the light of experience and lessons learned.

Global Mitigation Goal

In Article 2 the parties accept the 'temperature goal' of holding 'the increase in the global average temperature to well below 2°C above

---

[2] The UNFCCC CoPs since 2015 have had an overlapping role in governing the Paris Agreement through 'CMAs' (from 'Conference of the Parties Serving as the Meeting of the Parties to the Paris Agreement').

[3] CoP 26 has been delayed by the Covid-19 pandemic from 2020 to 2021, and it is unclear whether or how this will affect the timing of the first global stocktake.

Temperature rise since 1850

Global mean temperature change from pre-industrial levels, °C

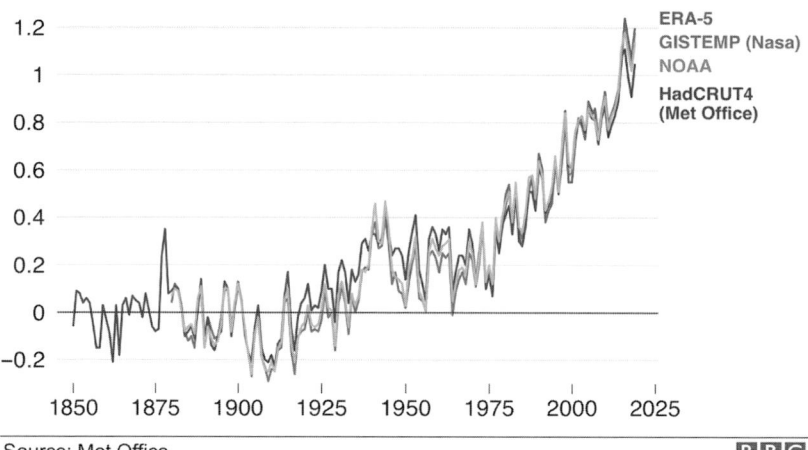

Source: Met Office   B B C

FIGURE 1.1 **The heating biosphere**
Notes: Data processed by the Met Office. Graphic from BBC News at
www.bbc.co.uk/news. Reproduced with the permission of BBC News and
the Met Office. (A black and white version of this figure will appear in
some formats. For the colour version, please refer to the plate section.)

pre-industrial levels and pursuing efforts to limit the temperature
increase to 1.5°C above pre-industrial levels'. This seems straightforward, since rising temperatures can be measured and depicted
(Figure 1.1), and 'everyone knows' both that they can be felt and that
they are the central issue in 'global warming'. But it is not so simple,
because (1) measuring global average temperature in real time requires
the continuous collection and analysis of enormous numbers of temperature records over the whole planetary surface and (2) its action
significance is based on projected heating effects from GHGs in the air
(as tonnes of carbon dioxide equivalent, tCO_2e), but the relationship
between the two, although roughly linear, contains significant uncertainties (Matthews et al., 2009). Moreover, non-linear emissions can
occur from drastic and unexpected ecological change, making projections based on linear models unrealistically reassuring, and GHGs

exert their heating effects over long periods so there are time lags that confuse the links between actions and responses.

Only supercomputers and satellites make this indirect approach possible, and it would have been much simpler to have adopted instead the overall goal of reducing to zero, and then reversing, the rate of increase of the concentration of GHGs in the atmosphere. This is much easier to measure, and in the case of CO_2 it has been measured continuously since 1958 (Keeling, 1986; GML, 2020a, 2020b). It is also more directly relevant to the defining purpose of mitigation, which Article 4 makes clear is to reduce net GHG emissions and accumulations in the air. But the temperature goal does not obscure the key task, which is for all parties to reduce their net GHG emissions as much and as quickly as possible.

Global Adaptation Goal

In Article 7 the parties adopt 'the global goal on adaptation of enhancing adaptive capacity, strengthening resilience and reducing vulnerability to climate change'. The adaptation goal is therefore expressed in process terms. This is amplified in Article 2, where it is stated that the agreement aims to strengthen the global climate response by 'increasing the ability to adapt to the adverse impacts of climate change and foster climate resilience'. The key ideas are to increase resilience and reduce vulnerability, and build capacity to do both. Meanwhile the articles add the requirements that whatever is done to adapt should contribute to sustainable, low-carbon development (both articles), while not threatening food production (Article 2), and being seen in the context of the temperature goal (Article 7).

These articles together describe a domain in which adaptation makes societies stronger and better able to resist climate impacts, while also improving well-being sustainably without undermining other goals. Other articles add nuance to these ideas: Article 11 refers to the need for capacity building 'at the national, subnational and local levels'; Article 7 reaffirms the multilevel nature of adaptation challenges and efforts, and calls for 'a country-driven, gender-responsive, participatory and fully transparent approach, taking into consideration vulnerable groups, communities and ecosystems', guided by science and 'traditional knowledge, knowledge of indigenous peoples and local knowledge systems' and promoted through knowledge sharing; and Article 12 stresses the priorities of education and public participation in whatever is done. Box 1.2 gives a summary of the UNFCCC Secretariat's recent thinking on what adaptation is.

BOX 1.2   **What is adaptation?**

Adaptation refers to adjustments in ecological, social, or economic systems in response to actual or expected climatic stimuli and their effects or impacts. It refers to changes in processes, practices, and structures to moderate potential damages or to benefit from opportunities associated with climate change. In simple terms, countries and communities need to develop adaptation solution[s] and implement action[s] to respond to the impacts of climate change that are already happening, as well as prepare for future impacts.

Adaptation solutions take many shapes and forms, depending on the unique context of a community, business, organization, country or region. There is no 'one-size-fits-all-solution' – adaptation can range from building flood defences, setting up early warning systems for cyclones and switching to drought-resistant crops, to redesigning communication systems, business operations and government policies. Many nations and communities are already taking steps to build resilient societies and economies, but considerably greater action and ambition will be needed to cost-effectively manage the risks, both now and in the future.

Successful adaptation not only depends on governments but also on the active and sustained engagement of stakeholders including national, regional, multilateral and international organizations, the public and private sectors, civil society and other relevant stakeholders, as well as effective management of knowledge. Adaptation to the impacts of climate change may be undertaken across various regions, and sectors, and at various levels.

BOX I.2 **(cont.)**

Parties to the UNFCCC and its Paris Agreement recognize that adaptation is a global challenge faced by all with local, subnational, national, regional and international dimensions. It is a key component of the long-term global response to climate change to protect people, livelihoods and ecosystems. Parties acknowledge that adaptation action should follow a country-driven, gender-responsive, participatory and fully transparent approach, considering vulnerable groups, communities and ecosystems, and should be based on and guided by the best available science and, as appropriate, traditional knowledge, knowledge of indigenous peoples and local knowledge systems, with a view to integrating adaptation into relevant socioeconomic and environmental policies and actions.

Source: UNFCCC (2020a).

Adaptation Reporting and Learning

Parties to the UNFCCC agreed at CoP 19/2013 in Warsaw that each would submit a report to the Secretariat, to be known as its Intended Nationally Determined Contribution (INDC), on its circumstances and willingness to contribute to the global climate response. The call was reiterated at CoP 20/2014 in Lima, along with further non-binding guidance on the expected content of each INDC and special provisions for least developed countries (LDCs) and small island developing states (SIDS; Holdaway et al., 2015). Most did so, and the INDCs were synthesised to provide a reference level of ambition and commitment for use in negotiations when CoP 21/2015 convened in Paris (UNFCCC, 2015), with the synthesis being updated immediately afterwards (UNFCCC, 2016b). Parties to the Paris Agreement then agreed in Article 4 to submit the same kind of report, now known as a Nationally Determined Contribution (NDC), as soon as possible after formally joining the agreement and every five years thereafter (UNFCCC, 2020b; WRI, 2020). Since most of the first generation of NDCs were dated 2016, the majority of updates are expected in 2021.

In Article 7 the parties further agreed that they would 'submit and update periodically an adaptation communication', which could be part of or separate from the other reports that the UNFCCC requires of its parties, including Biennial Update Reports, National Communications, the NDCs themselves, and also the National Adaptation Plans (NAPs) envisioned under Article 7 of the Paris Agreement. The intention of allowing this overlapping reporting was to avoid 'creating any additional burden for developing country Parties'. The result is that there are a number of routes through which parties can convey descriptions of their circumstances, vulnerabilities and actions, and articulate their adaptation needs and priorities. To these can be added the 'adaptation scorecard' reports that EU member states have prepared, which are not part of the UNFCCC/Paris Agreement arrangements, but which relate to the EU-wide NDC (Chapter 11).

The adaptation communications are the collective mechanism by which the parties inform each other and the entire global community of interest, through the Secretariat, of their intentions regarding, and progress towards, the adaptation goal in its 'local, subnational, national, regional and international dimensions' (Article 7). Apart from being a source of insights, experience and lessons learned for use and sharing by the Adaptation Committee of the Secretariat, the adaptation communications are raw materials for the global stocktake process. In Article 14 the parties agree to assess collective progress 'in a comprehensive and facilitative manner, considering mitigation, adaptation and the means of implementation and support, and in the light of equity and the best available science'. In Article 7 the parties agree that the stocktake will use the adaptation communications to review 'the adequacy and effectiveness of adaptation and support provided for adaptation' and 'the overall progress made in achieving the global goal on adaptation', with a view to enhancing 'the implementation of adaptation action'. Informed by the stocktake, the relevant CoP will then make decisions as appropriate to amend goals, strategies and anything else where changes are needed.

The Talanoa Dialogue

One of the decisions of CoP 21/2015 in Paris was to encourage dialogue among countries, through which they could share insights and experiences as they prepared to implement the Paris Agreement (UNFCCC, 2020c). An official process of this kind was launched at CoP 23/2017 under the presidency of Fiji, and named the 'Talanoa Dialogue' from a Pacific (Fijian-Tongan-Samoan) word meaning 'talk' or 'discussion' (Robinson and Robinson, 2005), or less prosaically, 'storytelling without concealment' (Moorhead, 2019). It was a one-year process designed to promote understanding and mutual aid among countries in thinking and talking through the implications of their Paris Agreement commitments, and, it was hoped, would be reflected in more ambitious mitigation goals being announced at CoP 24/2018 in Katowice. The resulting *Talanoa Call for Action* was short on specifics, but the *talanoa* approach offers a way to promote adaptive thinking, where sharing ideas and knowledge is critical to progress (Chapter 13).

Rethinking the CoPs

Before reaching the global stocktake, new issues have arisen over how CoPs work in practice. These started as large meetings and grew larger over time: the mean number of participants was 5,040 in CoPs 1–10, 8,875 in CoPs 11–14, 16,482 in CoPs 15–22 and 22,733 in CoPs 23–25. In recent CoPs, nearly two-thirds of the participants were from states (parties and observers), more than a quarter were from observer organisations (mainly non-governmental organisations (NGOs)), and the rest were from the media (UNFCCC, 2020d). These numbers reflect intense and growing public interest in climate change, and also the fact that countries and organisations feel they have something to gain or lose from the outcomes.

The Covid-19 pandemic has called the model into question, given the potential of a physical meeting to result in the infection and potential disablement or death among specialist officials, journalists and activists. The main argument in favour of retaining a physical format is that deals are done and influence exerted through persuasion and consensus-building among people interacting directly, often in informal settings on the conference fringes. Also, the annual conference means that at least once a year there is significant media coverage of climate change issues. These effects would be hard to reproduce through remote digital/virtual conferencing, but the urgent need for decisions on climate change has driven discussion of alternatives (Calliari et al., 2020; Mori, 2020). In the present context, CoP 26 is due to make decisions on Article 6 of the Paris Agreement, and hence to agree a specific mechanism and operating guidelines for the supervision and coordination of international carbon emission offsetting and trading. This is an important and divisive topic, and there are many others.

There is also concern that the structure of the CoPs raises issues of representation, inclusiveness and influence in relation to gender and ethnicity, including limited opportunities for indigenous peoples' delegates to bring about change (e.g. Suseeya and Zanotti, 2019). The whole issue is complex, however, since only the CoP can legally make binding decisions (and must therefore meet in order to decide not to meet), and attending a physical meeting is a poor way to obtain the insights and influences of all the world's peoples and interest groups. A solution may lie in something like an upward cascade of virtual citizens' assemblies and other national consultations to provide guidance to each country as it prepares and then submits its proposals for specific actions and decisions to the UNFCCC Secretariat. The latter would collate and circulate these proposals, in advance of a parties-only meeting to decide what to do through consensus or voting. A scaled-down and careful physical CoP in 2021 may be the most likely scenario for now, with a more limited, regulated and therefore controversial NGO and media presence, while other arrangements are gradually devised.

1.6 DESIGN OF THE BOOK

There are many ways to consider and describe the problem of growing instability of the world's climate, and many points of view upon which to base potential solutions. It is often described as a 'wicked' problem (Chapter 9) because it has no common meaning for everyone, and almost every aspect of it is worrying and debatable. This book targets a topic at one extreme of this uncertainty – adaptation – and tries to make it make sense in terms of the behaviour of complex systems. The aim is to consider climate systems and the ecological and social systems with which they interact, and to develop some simple ideas for how best to adapt to chaotic system change.

Evidence and case studies are therefore deployed to explain how and why to strengthen ecological and social systems as a key way to promote adaptation at the local and landscape levels of all countries. This responds to the adaptation goal noted in Section 1.5, since it assumes that systems must be made stronger in various ways, including resilience, to reduce their vulnerability to near-random climate stresses, while the capacity of stakeholders to build such strength is increased, and the ability of the systems to meet human needs sustainably is preserved or enhanced.

The approach is inherently bottom up, so would benefit from the experimentalist dimension of knowledge sharing and networking, among adapting communities and also with governments that have an essential role in enabling and supporting local stakeholders in their efforts to adapt. This alliance between local and central stakeholders is complementary to other government roles in orchestrating sustainable national development in cooperation with other governments that are also faced by climate change. The key point is that local people face microclimatic chaos, rather than macroclimatic change, and this can only be adapted to through local actions that strengthen local systems.

If this is done effectively, then each country will grow stronger at its 'grassroots', and a large part of the climate problem for each country will be made less severe. The problem itself cannot be solved without major progress on mitigation, but adaptation can buy time and limit casualties and costs while this is achieved. The design and performance of recent aid projects are considered in light of this approach, in the hope of guiding future adaptation investments to perform better in future and in practice. Along the way it is also hoped to establish that much of what many good aid projects do already is in fact helpful to adaptation, and that the need is for *more of the same but better, plus systems thinking.*

Chapter 2 describes the scale and urgency of the emergency that we are facing, and some ways to think about what we are trying to do about it and why. It ends with a brief review of some changes that offer grounds for hope that peoples, governments and major institutions are reacting more realistically to the overall challenge than they have ever done, but also that this is just a beginning relative to the likely climate system breakdown in mid-century. A sea change in our collective attitude is needed, but this may already be underway. In this hope, Chapter 3 offers the conceptual tools that are needed to support a new and localised approach to adaptation, introducing systems and systems thinking, chaos and its relevance to climate, and the nature of ecological and social systems. Chapter 4 explores the sources of strength and weakness in these systems, and how this knowledge can be applied to help them cope with chaotic stresses.

Chapters 5–7 then relate how these principles were applied – if not deliberately then at least in practice – to aid investments in Nepal, Bolivia and Zanzibar, and with what effect. These chapters therefore also explain how to identify and plan for the telltale signs of good design and high performance in real-life projects that directly or indirectly affect the strength of systems in the face of climate chaos. They are followed in Chapter 8 by a brief discussion of some principles for adaptation in cities. The book then returns to the global perspective, with Chapter 9 on evolving ideas, priorities and choices of researchers, aid professionals and governments since the Paris Agreement, followed by a review in Chapter 10 of changing patterns in adaptation

action and adaptive thinking, as revealed by an analysis of the adaptation communications submitted in 2015–2020. Chapter 11 then
details the adaptation challenges and responses in Europe, small
islands, Africa and the Americas, before Chapter 12 explores the
question of how to design and evaluate adaptation investments in
light of all this.

As new approaches to adaptation are tried out and understood,
points of consensus start to be visible. This is a complex and dynamic
process, however, as there has been more innovation, and more progress, on the climate response in the last five years than in the
previous five decades. But the direction of travel is in line with
everything that had previously been discovered about the importance
of ecosystem-based and community-based sustainability. For
example, it has long been known that community-based ecosystem
management involving secure tenure, forums, environmental education and intercommunity networking tend to enable sustainable and
equitable outcomes, and that similar arrangements work similarly
well in African, Asian and American forest, savannah, wetland and
coastal marine ecosystems (observations and references in Caldecott,
1988, 1996, 2005, 2015, 2017a, 2020; Caldecott et al., 2013; Lutz and
Caldecott, 1996). Moreover, that this is so regardless of the kind of
renewable resource concerned, from ecotourism and bioprospecting
revenues to fish, wild meat, rattan cane and timber harvests.

This knowledge leads towards the conclusion that climate
chaos must be addressed primarily at the local and landscape levels,
where its impacts are most severe yet can also be resisted by strong
ecological and social systems. Hence it is possible to sketch out a
framework for designing and evaluating aid investments to make
them more effective in promoting such an approach. The aim is for
these findings to contribute to discussion in the years leading up to
the first global stocktake required by the Paris Agreement, and
beyond.

Chapter 13 concludes the book by considering some of the
distinctive issues involved in thinking about mitigation and

adaptation, and their implications for our understanding of the emergency that faces us. It emphasises that our collective responses do, and must, go much deeper than anything we have yet attempted as a global system of peoples, cities, countries and ecosystems. To keep focused we will need hope, for which there is at least some good reason, and also a sense of purpose based on a commitment to take the kinds of collective action for which humans are best equipped, with the aim of building 'Peace with Nature'. The final section offers specific messages for the UNFCCC Secretariat, national and local governments, aid institutions, students, researchers and teachers, and for the citizens of localities and landscapes everywhere.

2 Chaos and Climate Emergency

2.1 BIRTH OF THE ANTHROPOCENE

Adaptation is a potent idea in biology where it describes outcomes of natural selection among lineages that are more or less fitted to changing environmental conditions at any given time. The term is also used in the faster-moving social realms of culture, technology and economics, often with the value-laden implication that old ways must be replaced by new ones. An extraordinarily high capacity to adapt is a key feature of humanity, having arisen during our evolutionary origin in the ever-changing, ice-age ridden Pleistocene era (2.58 million years ago to 11,700 years ago).

This experience of radical change in temperature, rainfall and sea level shaped us and promoted the cognitive skills that allowed us to exploit more stable environments in the subsequent Holocene era, which allowed the diversification and competitive growth of urban-agricultural civilisations in parts of Eurasia. With our growing understanding of history and evolution, the idea of adaptation has become ever more central to modern thought, prompted especially by the process of global transformation that accelerated from around 1950. This has involved cultural homogenisation, economic growth and environmental impacts of great scale and speed, all of which needed to be adapted to and made sense of by people everywhere.

The year 1950 corresponds approximately to the end of the Holocene era and the beginning of the Anthropocene ('the age of people'), a change that is expected to be clearly visible in future sedimentary deposits, which among other things will be rich in plastic residues and poor in fossil species diversity (Waters et al., 2016). The Holocene–Anthropocene Transition is imagined here to have started

25

sometime around 1750, and chaotic transitional conditions are expected to persist until at least 2150. Human activity since the start of the transition has driven an enhanced greenhouse effect, leading to biosphere heating, polar melting and a prospective methane surge from the Arctic (Wadhams, 2016), and therefore incipient runaway system change.

Recognising this has brought adaptation into new focus, now in the sense of 'adapting to climate change'. In this context the central issue is whether and how we can adjust our ways of life to survive and prosper in a new environment. The answer will depend on the extent and rate of imposed change, but the assumption here is that many changes in many places will be slow enough to be manageable if our considerable adaptive skills are applied effectively.

2.2   'WAR' WITH NATURE

The 'Dying Planet Index'

Since the early 1990s, scientists in the IPCC, Millennium Ecosystem Assessment, UNEP, Intergovernmental Science-Policy Platform on Biodiversity and Ecosystem Services (IPBES), the secretariats of the climate and biodiversity conventions, and other networks, have been reporting that humanity is violating multiple boundaries of biosphere integrity. These warnings are reaching a crescendo (IPCC, 2014, 2019a, 2019b; Steffen et al., 2015, 2018; UNEP, 2019a; CBD, 2020; Dasgupta and HMT, 2020; Xu et al., 2020).

The multi-taxon Living Planet Index (LPI) has meanwhile revealed an accelerating decline in global wildlife abundance (WWF, 2012, 2014, 2016, 2018), and has now fallen to a 68 per cent loss since 1970 (WWF, 2020). It would be demoralising but accurate to rename the LPI the *'Dying* Planet Index'. Moreover, IPBES reported that close to a million species are threatened by human actions (IPBES, 2019), while other analyses imply that up to a million species are now becoming committed to extinction each year due to 'web of life' failures such as trophic shifts and the loss of co-evolved species (Caldecott, 2017a).

As climate zones shift, even the protected areas set aside for ecosystems and wild species become unviable in new conditions, further contributing to mass extinction (Hoffmann et al., 2019). In short, the ecosystems that sustain water supplies, environmental security, pollination of crops, fisheries and soil fertility are deteriorating fast, exposing people, farmlands and settlements to severe risks and costs. Perceptions of the dire consequences have spread gradually into the economic mainstream, from forward-looking analysts at the World Bank (e.g. Burton and van Aalst, 1999) to those in major private corporations (e.g. Swiss Re, 2020; WEF, 2020a, 2020b).

The Leaders' Pledge for Nature

The overall message is that the living systems that provide food, water and security for people and businesses everywhere are failing, and the failures are starting to join up. Each part of the pattern is a spreading desert, drought, wildfire, flood, storm, mudslide, epidemic, extinction, famine or social crisis induced by them. The preamble of the Leaders' Pledge for Nature (2020), signed on behalf of 77 countries and the EU, puts it thus:

> We are in a state of planetary emergency: the interdependent crises of biodiversity loss and ecosystem degradation and climate change – driven in large part by unsustainable production and consumption – require urgent and immediate global action. Science clearly shows that biodiversity loss, land and ocean degradation, pollution, resource depletion and climate change are accelerating at an unprecedented rate. This acceleration is causing irreversible harm to our life support systems and aggravating poverty and inequalities as well as hunger and malnutrition. Unless halted and reversed with immediate effect, it will cause significant damage to global economic, social and political resilience and stability and will render achieving the Sustainable Development Goals impossible.

Considering the human-caused realities of a heating and chaotic climate, biodiversity loss at mass extinction levels and the widespread breakdown of natural ecosystems, to say that we are 'at war' with nature seems at times like a reasonable metaphor. It was used, for example, in a powerful speech by UN secretary general in December 2020 (Guterres, 2020). But it would be an odd kind of war, between one side that relies on conscious, high-speed activity and great destructive capability, and another which works slowly and reactively, with unconscious inventiveness and with infinite power in the long term.

Strategy and Artistry

Odd though the 'war with nature' metaphor may be, it can be useful in posing the question of what 'peace with nature' might look like (Chapter 13), while also offering a connection to the idea of strategy. This term originated in military thinking, with the original (Greek) meaning of 'generalship', but is now universal in development planning, including in the climate response. Trying to clarify its meaning after generations of misuse by soldiers and politicians confused over their relative roles, one military historian concluded that strategy is 'the art of distributing and applying military means to fulfil the ends of policy' (Liddell Hart, 1954: 321).

Replacing 'military means' with 'public investment' would yield a definition that satisfactorily suggests the need for artistry in designing projects, and that investments are simply ways to put policies of sustainable development – as drawn up by governments – into effect. The same author uses the historical record of warfare since antiquity to derive a number of firm conclusions, including the value of flexibility and mobility, and of campaigns having multiple potential objectives, as well as two key principles: that *indirect approaches are best*, and that *surprising or confounding the enemy is vital*.

These are relevant here for two reasons. The first is that natural dangers, aggravated by climate change, are often indirect, threaten different but equally important human interests in different ways at the same time, and often surprise us because of our own ignorance of ecology. The expectation of being 'crept up on and surprised' is itself a form of defence, however, and here it is considered central to adapting to climate chaos. The second reason is that these principles apply as much to non-violent contests as to violent ones, so can be used in competitions over the direction of policies, priorities and budgets, and between central governments and local societies seeking greater autonomy and empowerment.

2.3   GOALS OF THE PEACE

Grand Strategy

The 'war' metaphor also has a higher level of relevance, in the sense of what Liddell Hart calls 'grand strategy'. This includes but transcends policy, and seeks 'to co-ordinate and direct all the resources of a nation, or band of nations, towards the attainment of the political object of the war – the goal defined by fundamental policy [while also looking] beyond the war to the subsequent peace' (Liddell Hart, 1954: 322). Grand strategy should thus calculate and mobilise all the economic, physical and moral resources of the nations concerned, and organise their application towards their ultimate goal of a satisfactory post-war settlement.

This is the level at which the climate change response necessarily operates. It is anchored in the ultimate goals of international environmental treaties, especially the UNFCCC and CBD, and is bound to proceed through grand-strategic ideas like the transformation of entire economic systems to new, low-carbon, climate-resilient and biodiversity-friendly forms. Our grand-strategic aim should be to achieve these changes while also laying the foundations of benign relationships among people and between people and the biosphere, doing nothing in the short term that prevents our learning the lessons that we will need in the long term.

This is why, for example, the UN stresses that no one should be 'left behind' in squalor and oppression (Chapter 3), and why the Extinction Rebellion is as much focused on improving the quality of society through inclusive participatory democracy and tolerant sharing as it is on the emergency measures needed to save society from climate change (see Section 2.9). Thus, a grand-strategic climate response might envision severe regulation to save our collective future, but with such powers being exercised temporarily and under constitutional restraint 'to avoid damage to the future state of peace' (Liddell Hart, 1954: 322).

Entitlement Myths

One consequence of our species' tendency to form competitive groups and strata is the need to rationalise the distribution of success and failure that prevails in each society at each time. Such rationalisations are embedded within entitlement myths, stories that explain why those who have a certain education or first language, or belong to a certain ethnicity, nationality, gender, age, religion or some other category of human, deserve more or better than anyone else. These myths are persuasive, especially when taught from birth as they often are, as well as pervasive and potentially harmful. They distort decisions, maintain competitive inequity and make it hard to agree inclusive solutions to common problems as they deny the existence of common interests. Even when brought to consciousness and denied, they can still exert unconscious influence through habits of individual thought (thus affecting the quality of leadership) and 'group think' (thus affecting the outcomes of elections), so they are a formidable constraint on progress. This is a key warning to bear in mind as we respond to calls to reduce GHG emissions on an emergency basis, and as we try to adapt to a world fraught by increasing levels of ecological risk.

2.4   ECOLOGICAL RISK

An aim of citing military history is to introduce some key principles that apply to social and ecological systems in the context of adapting to climate chaos. Here a drawback compared with military history is that outcomes are uncertain, being harder to measure in the absence of historical judgements, and because the story is still unfolding and outcomes lie in the future. Thus, we must be guided by what we think will work, based on reason, expectation and directions of travel, and such evidence as we can find, rather than by certainty based solely on the experience of outcomes.

Adapting to climate chaos is ultimately about coping with ecological risks. These may be enormously uncertain at the local level, but at the global level the risk landscape is fairly clear as well as alarming. Here the definition of a 'global risk' used as in the *Global Risks Report* is as good a starting point as any: 'an uncertain event or condition that, if it occurs, can cause significant negative impact for several countries or industries within the next 10 years' (WEF, 2020c: 88). To this can be added a sense of system change that creates transformative jeopardy for *all* human interests, and some of the major global, ecological risks are summarised in Box 2.1.

2.5   TIPPING POINTS

The Precautionary Approach

Among the global risks in Box 2.1, only tipping points are uncertain in the sense of being inferred rather than demonstrable. The 'tipping point' idea occurs often in the social sciences (Gladwell, 2000), where it is relevant to fashion, marketing, political consensus and the concept of *Zeitgeist* (the 'spirit of the times'), in the natural sciences of physical, chemical and biological systems, and especially in relation to ecological transformations (Dakos et al., 2019). In the climate context it raises doubts over whether we can manage major environmental change at all, other than by using the precautionary approach of doing only things that are known to be harmless. This would avoid negative but unpredictable consequences, and is an alternative to starting by doing easy and profitable things and then trying to manage their consequences. Precaution is slow and cautious, the opposite of a 'get-rich-quick' scheme, and often appeals to ecologists and pessimists, rather than to optimists and those inclined to expect technological fixes to solve complex problems without a clear understanding of how complex systems behave.

BOX 2.1   **Severe global ecological risks**

Things that *might* go wrong:

- *Climate chaos*: dramatic change to climatic conditions that have prevailed over thousands of years, to which our expectations, and farming systems, settlements, businesses, etc., are adapted.
- *Ecosystem breakdown*: the near-simultaneous failure of ecosystems across the biosphere, so that they can no longer provide essential services at local, national or continental scale.
- *Mass extinction*: the extinction of millions of lineages, wild species and higher taxa, including many of great antiquity, and those that pollinate crops or maintain soil fertility.

Reasons to *expect* them to go wrong:

- *Climate systems*: assessments using knowledge of system turbulence, past behaviour, greenhouse effect and physical chemistry, observed changes and their human causes.
- *Ecological systems*: assessments using knowledge of evolutionary history, ecological rules, responses of known ecosystems, observed changes and their human causes.
- *Extinction processes*: assessments using knowledge of past mass extinctions, needs of wild species, patchy distribution of wild species, observed changes and their human causes.

Reasons to expect them to go wrong *soon*:

- *Worsening trends*, such as polar and glacial ice melt, and drying and burning trends in the Amazon Basin, southern Europe and South East Asia.
- *Recent changes*, such as deforestation in large parts of the equatorial tropics, desertification, oceanic acidification and declining wildlife populations.
- *Approaching 'tipping points'*, in which systems change suddenly from one stable state to another, through a brief (milliseconds to millennia) chaotic transition.

Sources: MA (2005a, 2005b); UNEP (2012, 2019b); WWF (2014, 2016); IPCC (2019a, 2019b); IPBES (2019).

Complexity and Transformation

Complexity is familiar in the worlds of finance, business and ecology, but many journalists and politicians are more used to simplicity. It is hard to reduce complexity to simple stories or political messages, and this tends to obscure the need for change with the effect of preserving 'business as usual'. Yet several large-scale ecological tipping points are now coming into view to challenge this default position. Deforestation in the Amazon Basin, for example, which is now at about 20 per cent of its land area, is thought to be very close to the point where there will be insufficient rainforest to maintain the region's moist climate (Lovejoy and Nobre, 2018; Bolle, 2019). Sustained and repeated drought would then permit the rapid replacement of most or all Amazonian forests by fire-maintained grassland. A similar scenario is in prospect in Borneo and Sumatra (e.g. Gaveau et al., 2014; Marlier et al., 2015; Voiland, 2019). In all three cases, forest and land fires are underway and consistent with tipping point predictions, with catastrophic implications for tropical biodiversity, environments and livelihoods.

The Arctic 'Death Spiral'

The case of the Arctic further illustrates the tipping point principle, but with even greater intimations of global danger. Thus, the Arctic 'death spiral' displays the volume of sea ice in the Arctic ocean each month since 1979 (Figure 2.1), its depth having been measured by military submarines and its area by satellite imagery (Wadhams, 2016, 2017; B. Horton, 2020). Sea ice volume has declined steadily, driven by sustained heating well in excess of 1.5°C (Figure 2.2), but before 1980 there was little seasonal variation as so much of it was in the form of deep, multi-year ice. Then, in the 1980s and 1990s, multi-year ice declined and seasonal effects became more marked, and after 1997 most of the ice became single-year and began dramatically expanding in the winter and contracting in the summer months. The declining minimum ice volume in September each year is the

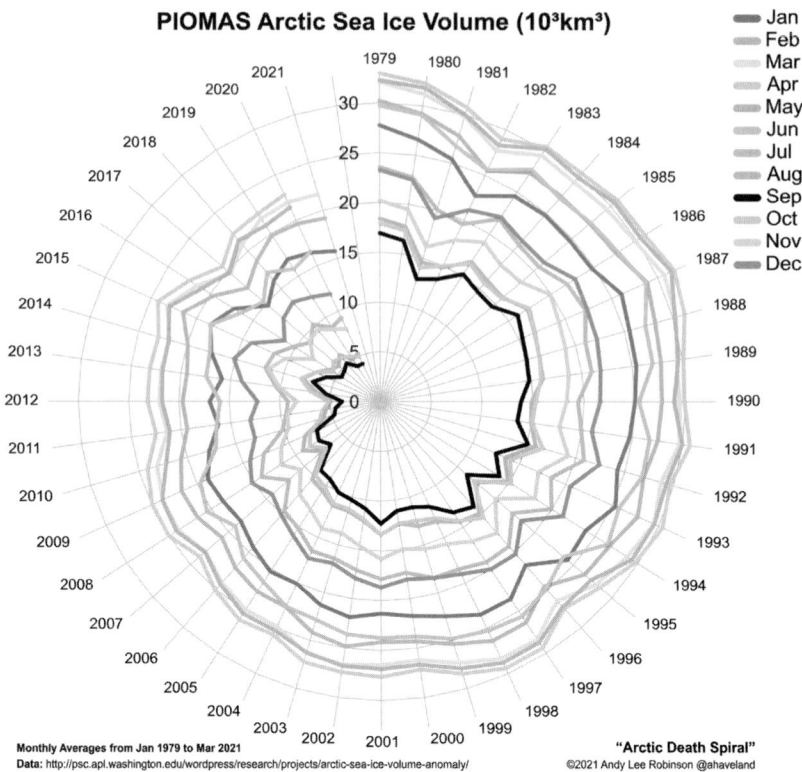

FIGURE 2.1 **The melting Arctic, 1979–2021**
Notes: (a) PIOMAS is the Pan-Arctic Ice Ocean Modelling and
Assimilation System developed by the Polar Science Center at the
University of Washington (PSC, 2020a, 2020b); (b) the Arctic Sea Ice
visualisation is by Andy Lee Robinson, and is used with permission.
(A black and white version of this figure will appear in some formats.
For the colour version, please refer to the plate section.)

key point, since from the trends this is likely to approach zero in the
early 2030s. To appreciate the significance of this, it is necessary to
understand several processes that are at work simultaneously in the
Arctic system: melting ice, melting peat, methane release, and fire
(Box 2.2).

The melting and burning in the Arctic since 1979 is from the
small amount of global heating so far, as a result of carbon emissions

2020 temperature anomalies

Based on HadCRUT5 (Met Office) data set

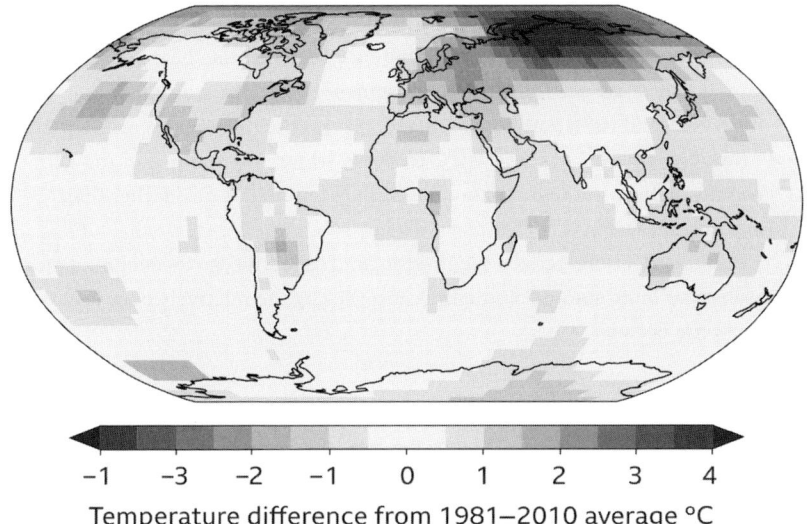

Temperature difference from 1981–2010 average °C

Source: Met Office ⒷⒷⒸ

FIGURE 2.2 **The heating Arctic, 2020**
Notes: Data processed by the Met Office. Graphic from BBC News at bbc
.co.uk/news, reproduced with permission. Contains public sector
information licensed under the Open Government Licence v3.0, © Crown
copyright, Met Office. (A black and white version of this figure will appear
in some formats. For the colour version, please refer to the plate section.)

from industry and deforestation since about 1950 when the carbon
balance tipping point was reached for the biosphere as a whole
(Chapter 1). The implication here is that much of the excess heat
trapped on Earth by the greenhouse effect and not absorbed by the
deep ocean (Figure 2.3) has so far been going into melting ice, but from
the 2030s this will no longer be the case. With no ice to absorb extra
greenhouse heat in the 2030s, Arctic water will warm much faster
than before, accelerating the melting and decay of permafrost peat,
and release of methane. Methane is much more potent as a GHG than

BOX 2.2   Ecological risk processes in the Arctic

Melting sea ice:

- The 'mass-melting' effect. It takes 80 times more heat to melt ice than to warm liquid water, because of the difference between the 'heat capacity' of water (i.e. it takes one calorie or 4.186 joules to heat one gramme of water by one degree) and its 'latent heat of fusion' (i.e. it takes 80 calories or 334 joules to turn one gramme of ice at $-1°C$ to liquid water); thus, much of the extra greenhouse heat in the biosphere has so far been melting ice rather than warming sea water.
- The 'area-reflectivity' effect. Ice is white and reflective, so its declining area means that more summer sunlight is absorbed by the relatively dark waters of the Arctic ocean.

Melting peat:

- The Arctic system comprises large areas of land in Siberia, Canada, Alaska, northern Finland and Scandinavia and elsewhere within the Arctic Circle, including several trillion tonnes of carbon-rich organic peat that has existed for millennia as wet but frozen permafrost.
- Arctic warming is two or three times faster than the global average, leading to the large-scale melting of permafrosts and the decay of peat.

Methane release:

- Decaying Arctic peat is releasing significant amounts of methane annually (possibly in the billion tonne range), adding to the 50 billion or so tCO_2e released worldwide from all other sources.
- Very large quantities (possibly in excess of a trillion tonnes) of methane exist in frozen, water-bonded form in shallow sea beds around the Arctic Ocean, and these molecules become unstable when warmed even slightly.
- Methane degassing around the Arctic is already underway, and would be expected to increase dramatically with heating of the Arctic Ocean.

Fire:

- As permafrosts melt, water is released and soils dry out, allowing land and forest fires to occur which are now underway throughout the Arctic Circle territories.

BOX 2.2   **(cont.)**

- These fires release CO_2 and smoke, soot from which darkens the remaining ice and speeds its melting under sunlight, contributing to the observed rapid melting of the Greenland ice cap.

Sources: Wadhams (2016); Biskaborn et al. (2019); Joosten (2019); Pendleton et al. (2019); UNEP (2019b); Welch (2019).

CO_2, so the sudden release of very large amounts during the 2030s and 2040s will greatly amplify the worldwide greenhouse effect. Humanity has not yet taken any effective action to head off such an Arctic tipping point. Moreover, we currently have no way to mitigate or undo such a rapid and massive methane release, should it occur.

2.6   LOCAL RISKS

Global risks manifest themselves locally in extremely diverse ways, and are perceived by local people from their own points of view. An example of local risk perception is given in Box 2.3, which summarises what local stakeholders said when they were asked about environmental risks by those preparing Dominica's NDC report in 2015. Dominica is a small, isolated and hurricane-prone tropical island, however, and specific local risks vary greatly between countries (Chapters 10 and 11).

The list of perceived threats in Box 2.3 draws attention to the fact that climate chaos at the local level is largely about damage inflicted by an agitated climate system, the form of which is unpredictable in detail, even though constrained by general circumstances – such as being a small Caribbean island. Other places have other characteristic envelopes of calamity associated with being in mountain, subdesert or inland rainforest locations, and where problems come when events exceed their 'normal' intensities and timings.

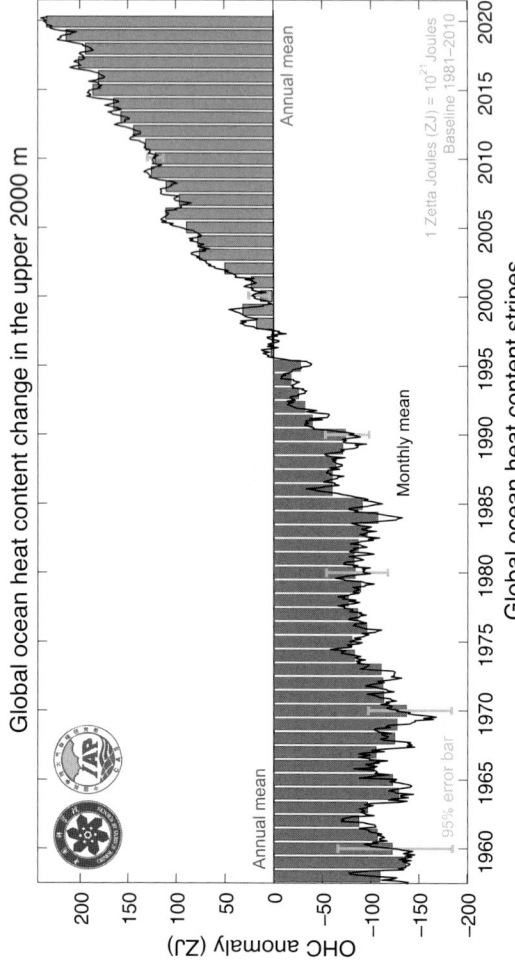

FIGURE 2.3 **The heating ocean, 1956–2020**

Notes: (a) Ocean heat is measured in zettajoules (ZJ), with 1.0 ZJ being equal to approximately double humanity's annual world energy use in 2020; (b) the graphic is from Cheng et al. (2021). Upper ocean temperatures hit record high in 2020. *Advances in Atmospheric Sciences*, reprinted by permission from Springer Nature. (A black and white version of this figure will appear in some formats. For the colour version, please refer to the plate section.)

BOX 2.3   **Extreme, severe and moderate environmental risks perceived at a local level**

A compilation of stakeholder opinions in the Commonwealth of Dominica, a 750 km^2 island nation in the far-eastern Caribbean, identified the following and scored them at 6–10 on a 10-point scale of risk.

- *Extreme risks (level 10).* Cumulative effects of increased extreme events and climate variability: physical damage to crops and access roads, reduced farm and fisheries productivity and food security, more pests and diseases and compromised livelihoods. Other expected or observed impacts include more frequent economic setbacks, prolonged recovery periods, stress on economy (including increased loss of life, impact on tourism, impact on agricultural production, food security, forest cover, human health and social capital), and less attractive environment for foreign investment due the destruction of critical infrastructure for tourism, manufacturing, agriculture and trade. In addition, a specific severe risk was identified from increased intensity of hurricanes, with flooding and landslides damaging houses, human settlements, critical infrastructure, forest resources and business and other properties.
- *Severe risks (levels 8–9).* Sea level rise combined with storm surges damages coastal infrastructure (roads, ports, jetties, storage, processing, packing, landing sites) that are used for agricultural trade and access to markets, while also damaging coastal tourism facilities (beaches, hotels, airports, sea ports and cruise ship/ferry terminals), and impacting on coral reefs and other marine ecosystems with associated effects on livelihoods and food security, and harm to tourism (hotels, dive industry, yachting), as well as significant cultural loss in Carib Territory and loss of beaches for recreation. In addition, changes in rainfall patterns bring water shortages and crop damage due to increased drought and storms (including impacts on Kalinago people and lost income to farmers), and increased incidents of landslides affecting houses, human settlements and infrastructure, and forest resources, in addition to costs for insurance, reinsurance and building loans. Finally, more intense rainfall causes damage to corals and fisheries, and increased climate variability leads to changes in fish and marine mammal migration patterns affecting food security and tourism.

BOX 2.3   **(cont.)**

- *Moderate risks (levels 6–7).* To an increased frequency and/or intensity of extreme events is also attributed the risks of damage to water resources/ infrastructure and impact on water quality and costs for water supply, increased costs of coastal resources management and (indirectly) reduced availability of international donor funding due to increased demand for emergency assistance from vulnerable countries. Increased climate variability and increased mean temperature are seen as risk factors in land degradation (impacting on food production, water quality, health and nutrition), increased damage to buildings and water cisterns (from warping, cracking and subsidence in dry conditions) and (indirectly) a decline in tourism visitor arrivals due to milder conditions in their home countries affecting the winter tourism market. Finally, changes in rainfall patterns are seen as impacting on water quality/supply and costs of water treatment/delivery and damage to water/communication infrastructure, with hotels and restaurants in particular being at a tipping point where loss of income due to lack of water could put them out of business.

Source: Government of Dominica (2015).

2.7   THE CLIMATE EMERGENCY

Climate science emerged from weather forecasting, meteorological research, palaeoclimatology and computer modelling of climate systems. Its growth included changes in vocabulary, and 'chaos' featured early, in the mathematical sense of great system complexity, unexpected consequences of small variations in starting conditions and a resulting difficulty in predicting outcomes. Chaos implied that 'the interactions among winds, oceans, ice sheets, forests, and so forth might be so unstable that scientists would never be able to predict a future climate for sure' (Weart, 2004: 180).

Meanwhile, the idea and terminology of 'climate change' had emerged from the study of past ice ages and combined with 'global warming', while both had become attributed in the modern world mainly to burning fossil fuels, and were widely reported and discussed in these terms. The two themes of unpredictable versus directional climatic change clashed in the 1980s and 1990s when uncertainty in the climate models over what was understood and could be predicted limited their influence on policy. Despite this, public opinion and political leadership were such that in 1992 it was possible for most governments to sign the UNFCCC.

The convention was put into effect only slowly, however, as the debate about climate science dragged on and the uncertainties involved were exploited politically to inhibit global action on limiting GHG emissions. Even so,

> by the start of the twenty-first century the modelers could confidently declare what was *reasonably likely* to happen. The protracted research efforts of a dozen teams had converged on answers. Even the most prominent scientist critics quietly admitted that sooner or later the greenhouse effect must be felt. The old predictions were solid. Doubling the CO_2 level was almost certain to raise the average temperature, and most likely by around 3°, give or take a degree or so. (Weart, 2004: 181)

The palaeoclimatology and modelling then suggested what a 3°C rise in average global temperature implied (Lynas, 2007):

> The Amazon rainforest would dry and burn completely, and be replaced by desert. The further drying and heating of drought-prone areas such as southern Africa, north-western and central America, and most of Australia, would render them uninhabitable and largely covered by shifting sand dunes. Remaining glacier ice would melt completely, almost halting the flow of rivers that rely on glacial melting for much of their water, such as the Indus, Ganges/ Brahmaputra, Mekong, Yangtze and Yellow, which together sustain half the world's current population. Storms and storm-surges would begin to engulf low-lying cities and countries. The Sahara Desert would enter southern Europe. Most of those species not killed by

land use changes in the late twentieth and early twenty-first centuries would die out, and 'the age of loneliness' would have begun. (Caldecott, 2020: 33)

As public alarm steadily increased in the early 2000s, the idea of climate *change* began to shift to climate *chaos*, with 'chaos' being used to describe the unpredictable damage inflicted on human interests by randomised behaviour of the climate system, rather than to explain the difficulties of predicting that behaviour. Thus, by 2007 a UN report could say that 'Powerful new coalitions for change are emerging. In the United States, the Climate Change Coalition has brought together non-government organizations (NGOs), business leaders and bipartisan research institutions. Across Europe, NGOs and church-based groups are building powerful campaigns for urgent action. "Stop Climate Chaos" has become a statement of intent and a rallying point for mobilization' (UNDP, 2007a: 65). Similarly, commentators were by then in the habit of writing phrases like 'the negative changes we make to our environment are contributing to mass extinction, local water crises, and further climate chaos' (Goldsmith, 2007: x).

2.8   THE POLITICS OF CLIMATE CHAOS

The word 'chaos' has presentational as well as scientific value, chosen along with 'breakdown' and 'collapse' to convey a feeling of impending crisis. And since at least the mid-2000s the sense of crisis has indeed increased with every scientific report on the state of the living world. Over the following decade or so it hardened further into the notion that we are living in a 'climate and ecological emergency', a concept central to the Climate Strike and Fridays for the Future movements inspired by Greta Thunberg, and also to the Extinction Rebellion, all of which took off internationally in 2018 and then grew in impact and influence (see Section 2.9).

Accepting such an emergency would, it is felt, justify measures that are too disruptive and too expensive to be contemplated in normal times. But many stakeholders are reluctant to do this, perhaps because they see no safe, cheap, profitable or comfortable way forward if they did. This resistance to voluntary change might be expected of those with a lot to lose from changes to ownership, business and fiscal regimes. That these can have disproportionate influence in their own national societies may explain some denial actions by countries and businesses: the USA withdrawing from the Paris Agreement during the Trump presidency, for example, and nearly half of 1,000 sampled energy companies still intending to expand their investments in thermal coal (Ambrose, 2020).

Yet many countries and companies have taken a different path, by committing themselves to policies and investments that lead towards systematic economic decarbonisation. Section 2.9 describes initiatives that respond to the need for urgent change, often explained in terms of meeting the Paris Agreement's temperature goal. But most have the potential weakness that they are based on modifying 'business as usual' at a pace that is comfortably affordable by almost everyone, and that offers opportunities for jobs and fortunes to be created through investment in new technologies and markets. This is understandable, but making soft and easy progress against a vague temperature goal is not at all the same as accepting hard, time-bound biophysical limits on major economic activities. The latter implies that the schedule of change must be set primarily by the needs of the biosphere and those who depend upon healthy ecosystems, and not by the expectations of capital and those who benefit from unsustainable economic growth. This is a much more challenging prospect, and requires a rethinking of human priorities if it is to be undertaken.

2.9 SIGNS OF A *ZEITGEIST* SHIFT

A change on the needed scale could be described as a shift in the *Zeitgeist* – the 'spirit of the times' that reflects prevailing global opinion on values and priorities at any given moment of history. In the context of climate change this would involve the alignment of effective leadership, collective understanding of the climate and eco-logical emergency, and a willingness to pay whatever it takes to extri-cate ourselves from it. As there are countervailing forces at work on the climate response, neither the timing nor the impact of a possible *Zeitgeist* shift can be predicted. But it is assumed that leadership and public willingness to pay are the key constraints on reducing GHG emissions, that these will improve and that an increasing share of climate response investment in the aid sector will be directed to adaptation. This book is therefore written in the hope that there is still sufficient time in which to use effectively the increased resources for adaptation, and that more time can be obtained through our efforts to reduce GHG emissions.

The Paris Agreement

Something that happened in 2015 may seem distant, but it was a moment long in gestation and one with a long shadow in a vastly complex system that can react only slowly. To become possible it required the abandonment by a critical mass of stakeholders of miti-gation as their only real purpose in favour of equal treatment for adaptation, and the abandonment by the EU of a top-down 'hege-monic' model in favour of an 'experimentalist' one. These new pos-itions were much more closely aligned with the thinking of most UN members (including the LDCs, SIDS, most EU member states and many of their long-term developing country partners), and they freed up the whole system to explore and discuss their experiences and priorities. The Talanoa Dialogue (Chapter 1) and Placencia Ambition Forum (Chapter 11) were only two of the resulting channels that contributed to network dialogue surrounding the NDCs and other adaptation communications that began pouring in before, during and after Paris. The result has been a system-shift towards shared know-ledge and competitive ambition in the climate response, 'algorithms' that will continue operating for decades, and that are changing everything.

The Elysée Palace Climate Initiative

In 2018, France convened a dialogue on climate risks and decarbon-isation investment among the sovereign wealth funds of Abu Dhabi, Kuwait, Saudi Arabia, Qatar, Norway and New Zealand (Rose, 2018). In 2019 the process was extended to include the major financial asset managers Blackrock, Goldman Sachs, BNP Paribas, HSBC, Natixis, Amundi, State Street and Northern Trust (Farand, 2019), thereby influencing the management of some US$17 trillion in assets. This approach combines with central bank insistence on the reporting of climate risks in exerting major influence on corporate behaviour.

Corporate Reporting of Climate Risks

One way to describe an aspect of the *Zeitgeist* is in terms of 'techno-economic paradigms', and how these shift from time to time (Perez, 1985). The current shift may be more potent than any yet seen, simultaneously involving physical climate impacts, technological changes, new resource landscapes, unprecedented litigation and liabil-ity exposures, and quickly evolving social norms that will produce new policies, laws, judicial decisions and employment rules (Caldecott, 2019).

Many of the risks are non-linear and exposures to them are often systemic and permeate the entire economic and financial system, with every future scenario featuring serious risks and significant stranded assets (Caldecott, 2018). There are also a number of factors – including those known as 'optimism bias', 'short-termism', 'the fal-lacy of sunk costs' and 'loss aversion' – that all discourage accurate accounting of stranded assets, and encourage irrational commitments to vulnerable assets and investments.

Efforts to mobilise shareholders to insist on greater environ-mental responsibility by corporate executives have been underway for decades (e.g. Sparkes, 2002), but these calls are now being ampli-fied by regulators interested in the risks posed by stranded assets. Thus, in 2017 the Bank of England facilitated the establishment of the Network for Greening the Financial System, the members of which have jurisdictions covering companies and populations responsible for nearly half the world's GHG emissions (NGFS, 2019), and whose purpose is 'to help strengthen the global response required to meet the goals of the Paris Agreement and enhance the role of the financial system to manage climate-related and environmental risks' (NGFS, 2020: 8).

The same direction of travel led to the Prudential Regulation Authority Supervisory Statement that took effect in October 2019 (Bank of England, 2019). This requires, for example, all businesses to appoint people who are responsible and liable for climate risk, to allocate sufficient resources to deal with climate risk, to report transparently and accountably on the management of climate risk and to provide evidence of compliance. These enforceable transparency requirements can exert immense leverage on corporate behaviour.

The Extinction Rebellion

The Extinction Rebellion was launched on 31 October 2018, with the reading of a Declaration of Rebellion (XR, 2019) to a thousand or so people (including me) in Parliament Square in London. The declaration asserted that governments had failed in their duty to protect humanity and the biosphere from ecological breakdown, and served notice that Extinction Rebellion would oppose directly and non-violently, without fear of prosecution (Box 2.4), the causes of further environmental, social and spiritual harm. The movement would thereafter agitate for citizens' assemblies to formulate emergency responses and influence laws, for declarations of climate and ecological emergency, and especially for urgent action by public and private institutions (Taylor, 2020).

The Extinction Rebellion launch was addressed among others by Greta Thunberg, whose 'Fridays for the Future' Climate Strike, which began in August 2018 in Sweden, had already begun to spread (for her messages at the time, see Thunberg, 2019). During 2019, Extinction Rebellion and the Climate Strike movements mobilised

BOX 2.4 **The experience of an Extinction Rebel in an English court**

You go in alone now. I scribbled some notes about what I wanted to say while waiting, so my talk was not very polished – it's kind of raw and lot of emotion behind it though, made myself cry reading it. The Magistrate said 'I don't want to hear people's political views, I just want to hear the circumstances about the arrest'. But I said 'I need to tell you why I did it because it's a serious thing and there are powerful reasons that made me break the law'.

I was pretty quick, probably under 2 minutes, but got to say a lot about the government, about being angry that I feel forced to do things to get arrested due to the government's inaction. I said that the government and big business are the real criminals not us, that we need system change and cannot do that without [government] action. They talk but no action, too concerned with keeping power and votes. I said they and the billionaires who have raped and destroyed this world are the real criminals.

I spoke about Extinction Rebellion, that they might think it's silly or wrong but I don't know any other group that are putting themselves on the line like we do. And what IS amazing is that everyone isn't out there demanding change. The planet is dying and we carry on as usual. I said I don't know what else to do and brought up my granddaughter Eva and said I would rather break the law than be ashamed when she is old enough to ask me what I did to stop this. Its horrific and disgusting the burden we are leaving to the young. An impossible mess.

I got a small fine and the contribution to court costs and I can't remember what the other fee is called but all comes to £159 – not too bad. I don't have a conditional discharge, *yay!* Which is great [as] it means I can get into trouble again soon. But I think I'd better get a job first.

They did really seem to listen and be engaged and I think it's very powerful that every day these magistrates have to listen to a stream of us giving our heartfelt reasons. And I think it must surely sometimes really resonate with them and touch their hearts. It felt very good to be

BOX 2.4   **(cont.)**

able to share with them how damn angry I am and just how wrong and crazy it is to do nothing. Keep on wonderful Rebels! You are on the right side. Jenny xxx.

Source: Bath Extinction Rebellion chat message from Jenny Ruskin, 8 November 2020 (quoted with permission).

millions of people in actions and street demonstrations in many countries. In late 2019, 11,264 scientists from 153 countries signed a declaration of climate emergency in a paper that spelled out their reasons in detail (Ripple et al., 2020), with a further 2,502 adding their names before it was closed (Alliance of World Scientists, 2020). By December 2020, binding climate emergency declarations had been issued in 1,859 jurisdictions by governments responsible for over 820 million citizens in 33 countries (CED, 2020), and the UN secretary general was calling for all remaining countries to do likewise (Guterres, 2020). And in reaction to all of this, there was urgent discussion of how to mainstream the climate response into public budgets and development plans at all levels, up to and including the EU.

In 2019 it was clear that EU institutions now expect an ambitious climate response from staff and consultants. As one EU study concluded:

> It makes sense to learn from the past but also to acknowledge that each year humanity still releases tens of billions of tonnes of GHGs while conserving only tens of millions, and that the public is worried, frustrated by denial of overwhelming evidence, and is now being radicalised into direct action. Hence there is an immediate need for the EU and Member States to communicate with the public, and particularly with the millions of young Europeans who are now mobilised, to explain how their governments can and will solve the climate problem. This will require major concessions to

the public mood, including uplifting declarations of Peace with Nature (following Costa Rica) or the Rights of Mother Earth (following Bolivia), along with plausible targets and reform processes, and convincing evidence that emergency action is underway. To head off the climate change threat itself will require much more even than this, and adaptation to inevitable change remains a poorly-defined goal that requires far more attention than hitherto. (Caldecott, Clark et al., 2019: §Climate, 6)

Citizens' Assembly Climate Initiatives

In January 2020 the French government appointed 150 randomly selected individuals to a Citizens' Assembly tasked with identifying additional elements for the national climate response (Chrisafis, 2020). It reported in June with 149 recommendations (Convention Citoyenne, 2020), coinciding with French local and regional elections in which the green vote was greatly enhanced. Almost all the assembly's recommendations were promptly accepted by the French government which also promised both an additional €15 billion for the climate response and a referendum on whether to create the crime of 'ecocide' (Willsher, 2020), the latter being the deliberate violation of biosphere boundaries. Comparable processes ran during the first half of 2020 in the UK (Climate Assembly UK, 2020), and the second half in Scotland (Scotland's Climate Assembly, 2020a). These are all complex processes in rapidly changing political circumstances, but as with the Talanoa Dialogue the inclusive format and influx of new ideas can only be helpful and might, in unexpected ways, prove transformative.

Awareness of Mitigation Deadlines

Several countries, including the UK in 2008 and Denmark in 2019, have established legally binding deadlines (2050 in both cases) for decarbonising their territorial economies to net zero. Since 2019 the EC has been encouraging other countries to do the same, calling for net zero carbon emissions by 2050 and a halving or more of emissions by 2030. This is in the context of a 'European Green Deal' which also addresses the need for incentives to encourage private investment and action plans for halting species loss, cutting waste and making better use of natural resources (EC, 2019; Blasetti and Williams, 2020; Harvey and Rankin, 2020). Similar goals were also announced in late 2020 by China, Japan and South Korea (McCurry, 2020), and the US Biden administration is expected to do likewise.

Deadlines that are legally binding change the way in which the economic consequences of different mitigation strategies are calculated (DECC, 2009; CarbonBrief, 2017). This has often been based the 'social cost of carbon' (SCC), an idea that focuses on the marginal social utility of saving each tonne of carbon emissions relative to other potential investments, using a range of discount rates to represent social preferences for early, delayed or intergenerational benefits. The choice of discount rate strongly affects the SCC, and can become politicised when the SCC is used to guide public investments. Some individuals value the interests of future people, or those living in other countries, differently than others, so there is ample scope for debate and potential polarisation. The resulting uncertainty over basic assumptions, combined with remoteness from events in the real biosphere, undermined the value of the SCC approach.

Instead, the Paris Agreement imposes a limit – the temperature goal – on what can be allowed to happen. Time-bound limits on net emissions in national legislation, based on an estimate of what is needed for the temperature goal to be reached, require planners to work backwards from their legal deadline, setting carbon prices and other incentives and rules that are consistent with the time-bound emissions limit. This 'target-consistent' approach should make it much more likely that the limit will not be exceeded, and hopefully therefore that the temperature goal will be met as well. This was a breakthrough in thinking, because for the first time it introduced an agreed boundary to human activities and intentions based partly on biophysical knowledge.

The next generation of mitigation evaluators and planners are exploring further in the same direction, taking a 'deadline-aware' approach in assuming that by 2050, at the latest, and unless circumstances change, further investments in mitigation will become nearly pointless (Caldecott et al., 2021). This is not because of legal constraints, but because of the high likelihood of irreversible/runaway climate breakdown by mid-century, induced by the activation of Arctic, equatorial and oceanic tipping points. After that, climate change will be driven by feedbacks within the biosphere system, rather than by human actions. This same distinction can be made when someone chooses to walk to the edge of a high cliff and step off, since with the last step gravity takes over and second thoughts become irrelevant.

As biophysical returns on mitigation investments are delivered closer and closer to the deadline, therefore, they become less and less valuable for mitigation, simply because the system as a whole is becoming more and more committed to breaking down. The real deadline cannot be dated precisely, since there are far too many variables involved and many of them are unknown – for all we know it could already be in the past. But a precautionary deadline can be used, if only because a dated schedule rather than a temperature goal offers greater clarity in comparing real mitigation usefulness between early emission gains from fast-acting projects versus later gains from slow-acting ones. This approach fits into the middle of the following three-step process for comparing proposed mitigation investments.

- *Step 1* is to estimate in tCO_2e the net physical emission savings expected in each year, after taking account of all increased emissions from construction, transport, operation and indirect social and economic effects of the investment. This allows actions that deliver relatively powerful results soon to be distinguished from those that deliver weaker ones later.
- *Step 2* is to correct the annual physical tCO_2e savings according to when they occur, by multiplying the net tCO_2e saving expected in each year (from Step 1) by a factor that declines exponentially from 1.0 in the starting year to almost zero 30 years later, to yield a dated mitigation value (dmv) for each. For example, if 2020 is the year from which the mitigation effects of an investment are counted, then 1.00 $tCO_2edmv2020$ would become 0.37 $tCO_2edmv2030$, 0.14 $tCO_2edmv2040$, and 0.05 $tCO_2edmv2050$ (Table 2.1).

Table 2.1 *Exponential decay in dated mitigation value for correcting tCO_2e to tCO_2edmv at exp(0.1)*

Year (0-31)	Exp (0.1)
0	1.00
1	0.90
2	0.82
3	0.74
4	0.67
5	0.61
6	0.55
7	0.50
8	0.45
9	0.41
10	0.37
11	0.33
12	0.30
13	0.27
14	0.25
15	0.22
16	0.20
17	0.18
18	0.17
19	0.15
20	0.14
21	0.12
22	0.11
23	0.10
24	0.09
25	0.08
26	0.07
27	0.07
28	0.06
29	0.06
30	0.05
31	0.05

All the dated mitigation values would then be added to yield a total mitigation value ($\Sigma tCO_2 edmv$) over a standard period, say 20 years. This will amplify contrasts between investment returns identified in Step 1, making alternatives clearer, while adding a sense of urgency and comparability among portfolios. It could also be useful to divide the $\Sigma tCO_2 edmv$ by total investment cost ($\Sigma \euro$) over the period to yield a mitigation benefit per unit cost ($tCO_2 edmv/\euro$), which can be compared with the $tCO_2 edmv/\euro$ of other potential investments over the same period.

- *Step 3* is to recognise the value of co-benefits in each case, using proxies and policy preferences to list and weight all those considered important, including adaptation, water, biodiversity, local cultural and amenity values and contributions to the SDGs. This will highlight actions that yield abundant co-benefits for many sectors or interests, rather than benefiting only one or a few of them.

With this approach, mitigation actions that deliver results early and cheaply, and that yield abundant co-benefits for many, will tend to be favoured consistently over those that are late, expensive and benefit only one sector or interest. It also has other implications for mitigation strategies, for example by giving high potential value to precautionary investments in avoiding deforestation (Box 2.5), which can be used to offset residual emissions by actors that are otherwise committed to zero net emissions (Allen et al., 2020). It further implies that even temporary success in preventing emissions is worth seeking, since this will buy time either to make the measures permanent or to find ways to offset or recapture the emissions that will eventually occur. This in turn means that 'deadline-aware' mitigation planners should care less than designers and evaluators usually do about whether impacts are sustainable, since what really matters is only that they happen *soon*.

The Well-being Economy Alliance

This is one of many networks that promote ecological and social sustainability, along with an economic theory that values equity, well-being, sustainability and creative diversity. It has more traction

BOX 2.5 **Precautionary mitigation investment**

If climate change runs away from human control, in a 'perfect storm' of activated tipping points, then mitigation will have failed and all we can do is try to survive. But until that point it makes sense to invest as effectively as possible in mitigation, in the hope of buying time and taking pressure off the biosphere. Renewable energy and energy efficiency, regulation and capacity building, planting forests and changing public expectations and consumption patterns can all help as part of an integrated strategy. But all these are slow-acting measures from the point of view of a mid-century precautionary deadline, and it is also now vital to conserve all remaining high-carbon-density ecosystems. These include tropical forests, which contain hundreds of tonnes of carbon (sometimes a thousand tonnes or more) in each hectare, as well as whole webs of life and the people who depend on them. Not destroying such ecosystems would be wise, as the impact of released carbon from dying forests will have to be coped with by everyone, or offset or recaptured later at great cost.

For these reasons, the idea of precautionary investment in avoided deforestation is fast gaining ground. In September 2020, for example, the British NGO Fauna & Flora International delivered an open letter to the UN General Assembly on behalf of some 180 affiliated conservation organisations, calling for an initial US$500 billion annual funding commitment 'to reverse ecosystem degradation and protect the natural world' (FFI, 2020: 9; Knight, 2020). A share of this for saving the remaining tropical forests would be an excellent investment, yielding enormous gains at costs that could be very low if spent in partnership with local communities who want to protect their own environments. Costs of a dollar per hectare per year are reported for community protection of 6,100 forest hectares in the Monteverde indigenous territory of Bolivia (Jens Holm Kanstrup, personal communication, November 2020), for example, and a quarter of that for setting up a community monitoring system for the 500,000 ha Prey Lang Wildlife Sanctuary in Cambodia (Theilade et al., 2021).

BOX 2.5 **(cont.)**

This kind of investment often cannot create official 'carbon credits' for sale, since it cannot be proven that the work adds to what the government concerned claims to want to do anyway. A legally defined protected area ought not to be at risk, after all, but deforestation experience tells a different story. Adequate funding for local partnership-based protection over millions of hectares can add certainty that billions of tonnes of carbon will not be released quickly by accident, and as a precaution during a life-or-death gamble with the biosphere this assurance is well worth having.

Nor does the value of forest protection rely only on preserving *intact* forests. Even damaged forests contain lots of carbon, and they will regrow under protection, absorbing several tonnes of carbon per hectare per year as they do so, and meanwhile offering more and more ecological goods and services to reward their local guardians. Several Danish-funded projects (smaller than those in Nepal and Bolivia described in Chapters 5 and 6) have been based on local community stewardship to save and restore damaged but regenerating forests, including in Sumatra and Flores in Indonesia with net gains in carbon emissions, biodiversity and environmental security for local people (Caldecott et al., 2021).

than most because it aligns a growing number of smaller countries where innovative social, environmental and economic thinking is more easily put into practice than in bigger states. This is a similar process to the networking of cities and local governments (e.g. through United Cities and Local Governments and C40 Cities – Chapter 8), being also driven by the divergence of priorities between larger- and smaller-scale societies. Thus, a 'well-being economy' is one that gives special attention to environmental protection, common-good resources, community empowerment, intergenerational justice, eradicating poverty and inequality, materials recycling, mental health, natural capital accounting and social care (WEAll, 2020).

This approach received new energy through the alliance of Iceland, New Zealand and Scotland (Sturgeon, 2019), which have agreed to work together to put their ideals into practice through specific policies and regulations, against defined goals and success indicators. Membership of the Wellbeing Economy Alliance by organisations in countries that include Bhutan, Trinidad and Tobago, Costa Rica and Namibia promises further inclusive growth especially among smaller progressive nations. A *Zeitgeist* shift such as the one imagined here is likely to involve the sudden consolidation of all such relatively bottom-up initiatives into a common purpose with widespread public support.

The 2020 'Stillpoint'

The novel zoonotic coronavirus (SARS-Cov-2) responsible for the 2019 outbreak of coronavirus disease (Covid-19) was first noticed in China, quickly spread to other countries, and was recognised as a global pandemic by the World Health Organization in early 2020. It later became clear that the virus held particular dangers for those with comorbidities such as obesity and diabetes. Since these are well known as non-communicable diseases of poverty, Covid-19 came to be recognised as a *syndemic* disease, being 'characterised by biological and social interactions between conditions and states, interactions that increase a person's susceptibility to harm or worsen their health outcomes' (R. Horton, 2020: 874). That being the case, it was clear that Covid-19 could not be addressed fully without also correcting the underlying causes of poverty and deprivation in the areas of public health education, nutrition and economic equality.

In this analysis there are parallels with the climate emergency, which also fits into a syndemic pattern to which multiple forms of environmental abuse and their consequences all contribute and synergise. In both syndemics, many elements can be traced to an underlying cause: for Covid-19, an imbalance between of the interests of the rich and the poor; for climate, an imbalance between the interests of humanity and nature. Both kinds of imbalance also synergise with each other, to make poor people singularly and immediately vulnerable to the malign effects of instability in the whole system (Swinburn et al., 2019).

The experience of Covid-19 induced a steep global public and political learning curve which can be seen as a model and stimulant for aspects of the climate response. The head of the UNFCCC Adaptation Committee, for example, observed that Covid-19 'is giving us the opportunity to rethink our perception of a lot of things including vulnerability, early action, rapid response and what timescales mean when confronted with a crisis' (Youssef Nassef, in Adaptation Committee, 2020). In order to reduce infection rates and minimise loss of life, many governments took extraordinary measures to suspend travel, social interchange and much economic activity (e.g. Portes, 2020), initially in early 2020 and later, when the virus resurged, towards year's end.

The sudden relaxation of relentless human pressure on nature has been called the 'anthropause' (or properly, anthropopause) by scientists studying its effects on wildlife behaviour (Rutz et al., 2020). This literally means 'human ending', so 'anthropic hiatus' would be more appropriate, but I prefer 'stillpoint' which in osteopathy means the moment of balance that signals transformation in a dynamic and complex system (Brooks, 1997). In any case, one of its effects was a sharp decline in worldwide GHG emissions (IEA, 2020). Against the expected bounce-back with a return to 'business as usual', some governments took the opportunity to consider how to redesign their economies by investing more in low-carbon sectors than high-carbon ones during the lockdown and recovery (e.g. Harvey, 2020a, 2020b; Hepburn et al., 2020). Thus the first commitment in the Leaders' Pledge for Nature (2020) stated an obligation to put 'biodiversity, climate and the environment as a whole at the heart both of our COVID-19 recovery strategies and investments and of our pursuit of national and international development and cooperation'.

The Danish ambassador to Singapore summed up the position when she wrote that:

Clearer skies, disappearing smog and thriving eco-systems have given us a sight of how the environment may come to life when consumptions are balanced and emissions reduced. ... The pandemic has taught us that national solutions won't solve global challenges. The pandemic has reminded us of the importance of international collaboration in combatting global risks, and in many ways, it gives us a clearer vision of what it takes to save the planet. Multilateralism is essential. Covid-19 has shown us that people are ready to fundamentally change behaviour if they see that the threat is imminent and serious enough. (Landi, 2020)

As well as reaffirming the value of fiscal intervention, central planning, public health services and international cooperation, the whole response to the Covid-19 pandemic showed that urgent, expensive and far-reaching emergency measures are quite feasible if they are thought necessary by enough influential people.

PART II Understanding Climate Chaos

3 Systems, Climate and Ecology

3.1 SYSTEMS AND CHAOS

A system is a collection of living or inanimate things that relate to one another by exchanging influence, energy, information or materials in some way. Those involved are distinct entities, the relationships among them are governed by rules, and the system exists because of the entities and the rules. Information is what distinguishes entities and describes relationships (Bateson, 1979), so information is inherent to all systems. Life on Earth can be thought of as one self-organising system which has access to thermonuclear fusion energy from the Sun (and in some cases from subterranean nuclear fission), and that uses its own information content to guide material and energetic relationships between molecular structures and other entities, thereby maintaining itself as a complex system (Capra and Luisi, 2014). The information concerned has been accumulated through evolution and is stored in heritable molecules and elsewhere throughout the system. The transactions that maintain all life occur at scales ranging from the molecular to the biospheric, and over time periods ranging from nanoseconds to billions of years – adding up to a long, complex and dynamic story that as far as we know is unique to Earth.

Entropy is the name for all transformations in which distinctiveness and information are destroyed, and it is in the nature of isolated systems for entropy to increase over time, a principle stated in the second law of thermodynamics. If allowed to proceed unopposed, the end point of entropy is chaos, an equilibrium state of complete disorder and confusion. This is the fate of all systems that lack both access to an external source of energy and a way to use it to maintain themselves. The word 'chaos' has two senses, however, beyond its direct meaning as the end point of entropy. One is philosophical or mystical, as it has been used to describe the condition of the universe before creation, as well as that of the self after death. The other is physical or mathematical, as it has been used to describe complex systems whose behaviour is so unpredictable as to appear random. In this formal sense, chaos is a boundary beyond which order changes to something which is impossible to describe using the mathematical models that work up to that boundary but no further.

Efforts to extend human understanding into the realm of chaos have helped to describe the high-energy plasma that was the universe in its first three minutes (Weinberg, 1993), and to explore the creative tension that is thought to exist between order and chaos (e.g. Thom, 1975; Davies, 1987; Kauffman, 1995; Ball, 1999). The main relevance here is that climate systems have been called 'chaotic' because they are extremely complex and very sensitive to starting conditions, making their behaviour historically unpredictable. The way in which 'chaos' is used in relation to the climate has changed, however, from a description of the climate system itself (and why its behaviour is so hard to predict) towards a description of the effects of an unstable climate on human interests (and what can be done about them). In the last sense it is most relevant to the health of our ecological and social systems, and to any action we might take to heal or strengthen them before they are degraded to the point of collapse.

3.2   CLIMATE AND CLIMATE CHANGE

The idea of 'climate' implies a recognisable regularity of pattern within a weather system over the long term (i.e. at least decades, but often centuries and sometimes millennia). To mention climate is to expect certain weather conditions in particular times and places, ones that result from a combination of factors that include latitude and altitude, slope, exposure and season, annual and multi-annual cycles and oscillations in energy, water (as ice, liquid or vapour), prevailing winds, ocean currents, sinks and up-wellings, and variation in net inputs of solar energy to the biosphere as modified by sunspots and greenhouse gases, among other things. Thus, climate is what we have learned to expect by observing and understanding the behaviour of an extremely complex – but not chaotic – system on Earth, one that is strongly coupled with what goes on in the Sun.

For most people, most of the time, the climate where they live strongly influences their farming systems and food supplies, the design of their houses, the contents of their clothes cupboards, the plants in their gardens and much else. These influences decline when people live in larger-scale trading systems, and have access to foods grown and items made elsewhere. The global economy of recent times involves the progressive insulation of people from local influences, including climate, native species and cultivars, soils and traditions. It favours instead a more homogenous, cosmopolitan lifestyle based on a global average level of productivity and human creativity, mediated by something approximating free trade.

In this context, 'climate change' implies a rearrangement of parts of this global system in ways, and at a speed, that are predictable enough for investments to be redistributed in a timely way, and/or that allow the effects to be smoothly absorbed by the averaging effect of free trade. In this relaxed conservative or laissez-faire view, most local losers from climate change will become winners eventually, and those who own the capital that is moved around as the system adapts will also mostly turn out to be winners. Thus, climate change does no more than stir the sea of opportunity, and responding to it demands only the flexibility with which the capitalist system has responded to previous changes.

And capitalism does indeed have a strong record of adapting to changes of circumstance. The family, company or national identities of the owners may change, sometimes dramatically and with blood-shed, and millions of people may have to move jobs or countries, or suffer and die early as others prosper, but the system itself has sur-vived centuries of paradigm shifts, revolutions and wars (e.g. Hobsbawm, 1962a, 1962b, 1987, 1994; Piketty, 2014). Whether this can continue depends on the true dimensions of current, cumulative environmental threats, including climate change and climate chaos, and our capacity to mitigate and adapt to them.

3.3   CLIMATE CHANGE AND REFERENCE LANGUAGES

Physics

The subject of climate change is confused and confusing because it includes three distinct issues with different scientific contexts and social implications. First there is the phenomenon itself. Here the reference language is that of physics – GHGs, radiations, energies and temperatures – with measurements yielding hard data and computer algorithms manipulating those data according to assumptions rooted in the worldview of the physicist. We have known about the phenomena involved for generations, and the physical models have grown very convincing. As a result, we are certain that climate change has occurred in the pre-human past, that there is now a human influence driving extra warming that has melted polar and mountain ice, and that this is creating new conditions and mechanisms that both accelerate ice melt and start other processes (like methane release) which could make the heating process intensify very quickly, leading to catastrophe. Although people can see seasons change and glaciers retreat, none of the pattern itself would make sense without physics.

Then there are two questions about what to do about human-caused climate change. The first is: how do we make it less severe by changing what we are doing to cause it? Here physics is of much less use, but still some as it allows us to measure and compare our actions in a common 'currency' of tCO_2e to decide if we are doing enough, quickly enough, to change the outcomes predicted by physicists. The latter may be able to say that a certain quantity of tCO_2e must be taken out of the atmosphere by a certain date, but decisions on how to do so can only come from elsewhere. They involve answering questions like: what must change, and how will collective motivation be organised? How much is worth paying to achieve it, when and by whom? And what technology can we invent that can do the job, and with what intended and unintended consequences? The answers change in a kaleidoscopic way as new knowledge pours into the system, along with new feelings, new political compromises and new technologies, and as these are debated, voted on, put into effect and rendered obsolete by yet more knowledge.

Beyond Physics

The second question is the subject of this book: how do we adapt if climate change cannot be stopped? Here physics has little to offer and we must respond to the fact that climate change transcends all cultural and national boundaries, yet it is also active and must be anticipated and responded to in different ways at the local level everywhere. It therefore has to be understood in a way that allows for seamless coherence between the very large and the very small, and between all locations regardless of whether, who or how people live there. This leads to all sorts of cognitive, behavioural and social (economic, political, institutional, etc.) issues and challenges, starting with that of describing, understanding and discussing the phenomena themselves. This calls for a language that can make sense of complex systems regardless of scale and place.

Since the challenge of climate change is unprecedented in modern human experience, and we may have only one chance to respond to it correctly, we should be omnivorous in seeking a suitable reference language, and try to learn from past efforts to make sense of things at the local to global range of scales. Several fields of knowledge could be relevant, which are summarised in the following paragraphs. These apply only to the major cosmopolitan knowledge systems, however – ones that come from the great common intellectual experiences of multinational history: the administration of large empires, the revelations of prophets, the insights of enlightenment thought and the practices of modernity, all of them widespread and influentially accepted as 'true'. But the list therefore misses out the overwhelming majority of traditional knowledge systems that are possessed by individual peoples and transacted in thousands of languages. It cannot be ruled out that one or other of these is in fact the ideal vehicle for discussing all aspects of climate change, and it is in fact certain that important local insights, and highly likely that vital global ones, also exist among these other systems of knowledge and thought.

Legalism

When Roman, French and British systems of law were being widely imposed through imperial rule they were believed to reflect universal principles of justice, but were in fact just 'projections of aspects of our own [premises] onto the world stage' (Geertz, 1993: 221). As they spread and took root they directly extinguished many local and alternative laws (Rocker, 1937) and contributed to the loss of many languages (Ostler, 2005). Much military and missionary energy was invested in spreading these ideas, and a pastiche from the main systems derived from them resulted in international treaties that offer legalism as a way to understand climate change.

These arrangements, however, were not inclusive because of the effects of post-sixteenth-century European expansion, which suppressed the influence of other major traditions, such as Shariah, Indic and Adat law, and eclipsed the distinctive thousand-year heritage of the 'Holy Roman Empire' (in fact a Germanic, non-Roman system of governance). The latter collapsed in 1806, and its themes of strong local autonomy under diffuse rule continued in parts of Europe that had never been Roman, until its resurgence through the EU after 1950 and its influence on later treaties such as the Paris Agreement (Chapter 1).

Apart from these cultural biases, the main drawback here is that law is devised to consolidate and codify existing arrangements, so cannot be used to explore 'why' or 'vision' questions, or explain how things are or should be in terms of points of view other than those from which the laws themselves were formulated. Individual lawyers can induce legal changes through principled activism – the work of Polly Higgins and Valérie Cabanes on ecocide and of Granville Sharp on slavery are examples – but as a whole approach legalism is too retrospective and too elitist to support a complete discourse on climate change, although it can have a useful guiding and constraining function in certain contexts.

Socialism

The series of socialist 'internationals' since 1864 were a major source of universalist ideas and vocabulary, but none had much to say about climate change which was only recognised much later, and they anyway focused on other issues of class struggle, social equity and governance. On the other hand, all forms of socialism enabled global thinking which, with increasing awareness of the roles of ecosystems and biodiversity in ensuring the safety of society, inevitably produced some new ideas. At its best, socialism inspires people and can encourage great artistic and collective creativity, but there is a tension between community and state orientations, the latter often leading to inhuman and anti-environmental outcomes (e.g. forced collectivisation and pollution). Parts of the socialist tradition led however to the broad idea of 'community-based resource management' (CBRM), in which communities own and receive benefits (of all kinds, including spiritual, cooperative and self-esteem rewards that have no cash equivalents) from ecosystems, and therefore have good reasons to look after them. The CBRM approach can work well, especially if combined with supportive governance arrangements, and is a key part of the environmental conservation 'toolkit'.

Capitalism

Liberalism, conservatism and neoclassical economics are theories of why capital should be free, and have much to say about how economies should work, how wealth is created and how it should most efficiently be distributed, but tend to discount future, long-term or sustainable benefits and disregard equity and environmental issues. On the other hand, capitalism enabled global thinking which, with increasing awareness of the roles of ecosystems and biodiversity in ensuring the safety of capital, inevitably produced some new ideas. At its best, capitalism rewards innovation and the invention and spread of new ideas and technologies, but there is a tension between individual and corporate orientations, with the latter often leading to inhuman and anti-environmental outcomes (e.g. wage-slavery and pollution).

Parts of the capitalist tradition led, however, to the broad idea of 'resource economics', in which some of the things that ecosystems do are identified as goods and services that can be valued in monetary terms. These values are used to make calculated decisions about what to preserve, what to sacrifice and who and how people should be compensated for monetised losses. This led to environmental impact assessment (EIA) and related ideas of development planning, which recognise the claims of society in addition to those of capital. These ideas led in turn to some useful hybrid ideas, notably that of payment for ecosystem services (PES). Here the owners of an ecosystem (who might be a community, a business or an individual) are paid (in cash or kind) for the economic value of the services provided by the ecosystem to society (e.g. in the form of water supplies, flood protection, biodiversity conservation or carbon storage), or else capture a willingness to pay more directly, such as through ecotourism, in either case giving them a reward for protecting the ecosystem itself. The EIA, planning and PES approaches can work well, especially if combined with supportive governance arrangements, and all are key parts of the environmental conservation 'toolkit'.

Economics

Considered more broadly than its neoclassical school, which has been discredited in recent years (e.g. Chang, 2011; Stiglitz, 2013, 2017; Piketty, 2014; Anon. 2017), it is mainly the EIA and PES traditions that support the claim by economics to provide the 'language of climate change' (e.g. Stern, 2007; TEEB, 2010; Dasgupta and HMT, 2020). The central idea is that economic values reflect human reality, so can be used to discuss all aspects of that reality. Economic thinking and economists also possess a disproportionate influence politically, owing to their fluency and credibility in the matter of money – the universal solvent of many human values.

The main drawback here is that what humans value is only part of the full picture, which also includes the whole biosphere and the long-term future. This is problematic since in conventional economics future benefits must be discounted or it would be too expensive to do anything in the long term, and it is also incomprehensible to do anything for spiritual or other purposes that cannot be monetised. This leaves economics with immense influence from both its socialist and capitalist inheritances, but also with serious weaknesses in making sense of climate change. A convincing economic theory is still awaited that unifies the nature of humanity as a social animal, the communities that humans live in, the ecosystems that support those communities at all scales, from the very local to the global, and the long-term future of all people and non-human species.

Religion

The notions of shared creation and compassion for all beings in a universe permeated by divine consciousness offered enough common ground for people who are individually aligned with many faiths to talk and agree with one another (e.g. Palmer and Finlay, 2003). In practical terms, driven by fear of mass extinction, which is interpreted as a human assault on divine purpose, the basic idea of interfaith dialogue is twofold. First, it is to encourage leaders in each faith group to remind their followers that their particular faith requires them to treat the environment and other living beings with respect (e.g. Pope Francis, 2015; Islamic Declaration on Global Climate Change, 2015). And second, it is to amplify the same message for all believers while also reaching out to freethinking people with a common message of spiritual blessing for nature from multiple religious leaders (e.g. Interfaith Statement on Climate Change, 2015).

Any such process that lowers barriers between peoples can support cooperation, which is an inherently positive effect, and focusing religious minds on environmental matters may also help shape laws, treaties and individual behaviour. But the essential problem for interfaith dialogue, as a way to develop a reference language for climate change, is that few people care or think deeply about the nature of the divine, and what divinity may or may not want. Most simply absorb their religion from birth, as part of a cultural package that binds their society together and distinguishes it from other societies. Moreover, compared with the potency of faith in the past, expressed through the creation of sacred sites that were treated as off limits to ordinary human use, the quest continues to convey the idea to everyone that protected wildernesses and biodiversity refuges are still, or also, sacred. A convincing religious philosophy is still awaited that unifies the infinite universe with the whole living world, the whole history of life on Earth and the details and diversity of ecosystems and human cultures, and that also compels collective action to care for the whole fabric of life.

3.4 ANTHROPOLOGY AND SOCIAL SYSTEMS

In anthropology, humanity is recognised as one natural species whose distinguishing feature is an ability to create different cultures, each with its own myths, traditions, language, etc., and at least partly shaped by the environment in which it arose. Each culture can be seen as a phenomenon in its own right and understood in its own terms while using standard methods of description, analysis and comparison. The challenge then is to reconcile 'the massive fact of cultural and historical particularity ... with the equally massive fact of cross-cultural and cross-historical accessibility – how the deeply different can be deeply known without becoming any less different; the enormously distant enormously close without becoming any less far away' (Geertz, 1993: 48). This is done through

> a continuous dialectical tacking between the most local of local detail and the most global of global structure in such a way as to bring them into simultaneous view. Hopping back and forth

between the whole conceived through the parts that actualize it and the parts conceived through the whole that motivates them, we seek to turn them ... into explications of one another. (Geertz, 1993: 69)

This form of anthropological systems thinking allows for additional universalist approaches, including mythology, archaeology and a Jungian psychology that emphasises the common structures and processes of the human psyche and its connectedness far back in individual, cultural and evolutionary time (e.g. Stevens, 1993). Thus anthropology should be a rich source of universalist language and, since each culture reflects adaptation to local conditions, it also offers an obvious link to climate change and other crises of our time. As a coalition of indigenous peoples' organisations put it:

Overcoming dualism, separation and imbalances in relationships between humans and nature is central to addressing the biodiversity and health crises, including the rise of zoonotic diseases and pandemics. Sustained interactions and partnerships between sciences and indigenous and local knowledge systems – inclusive of women, men, elders and youth – are enriching contemporary problem-solving with holism and reciprocity. Indigenous ways of knowing and being evoke and inspire new narratives and visions of culture and nature working together within a living and sacred Earth. (FPP et al., 2020: 22)

But one of the few universalist conclusions of anthropology is that

each people knows their own kind of happiness: the culture that is the legacy of their ancestral tradition, transmitted in the distinctive concepts of their language, and adapted to their specific life conditions. It is by means of this tradition, endowed also with the morality of the community and the emotions of the family, that experience is organized, since people do not simply discover the world, they are taught it. They come to it not simply as cognitions but as values. (Sahlins, 1995: 12)

This is important, since the ability of a group to express its own culture in its own way is a key requirement of freedom and well-being. But it does create difficulties for any attempt to talk or think about, much less do something about, climate change at a global level. Thus anthropology both contributes to, yet also constrains, the quest for a suitable universal reference language for climate change.

3.5   ECOLOGY AND ECOLOGICAL SYSTEMS

Ecology is the science of living systems. As a science it relies on the use of common standards of observation, analysis and collective criticism to build up reliable descriptions and explanations of reality. As biology it assumes the engineering of organisms through evolutionary responses to design challenges imposed and opportunities offered by the real world (Dennett, 1995). This makes biological thought intensely practical, and often to do with budgets, investments, costs and benefits that may be expressed and accounted for in terms of energy, nutrients and surviving offspring, but which is no different in principle to economic thought. But being concerned with systems, ecology focuses on describing the parts, their relationships, the things that connect them and the properties that result from all the parts, relationships and connections being active within or upon each system.

Since every living system is connected to every other, an ecologist distinguishes them only as a matter of convenience, and has to remain alert to possible influences from abroad at all times. Here, 'abroad' means any kind of distance – spatial, but also sensory (i.e. what can be detected using our human senses, as constrained by the expectations of our culture and the abilities shaped by our inherent aptitudes and training, and what other organisms can detect using their own senses, which may differ from ours) and instrumental (i.e. what can be detected by the instruments we use, as constrained by the expectations of existing knowledge that are designed into the equipment, and which may simply be unable to detect something important). So while proceeding within the boundaries of consilient science, this alertness to 'abroad' makes ecologists open to the phenomeno-logical diversity that is inherent to anthropology, mythology, psychology and religion, and also to potential connections within and between systems that are invisible to human senses and current instruments.

At the same time the systems approach requires an ecologist to think in terms of every system being part of a bigger system, in a connected sequence from the molecular to the global level, and with the characteristics of every level both influenced by and influencing every other level, over every imaginable scale of time and space. This is a way of thinking reminiscent of anthropology, since both have the same challenge of reconciling detail and pattern at all scales. But unlike anthropology, ecology has the weakness that it tends to see humanity as 'just another organism' in an ecological system, a point of view not widely appreciated among fellow humans. This can perhaps be corrected by infusing into ecology the humanistic and open-minded relativism of the anthropologist.

Living systems include our own selves, families, communities, farms and dwellings, as well as the atmosphere, oceans, coasts, swamps, grasslands, soils, drylands and forests and all their non-human inhabitants, which make up the living world and every part of it. These are all systems, and climate change is a system-wide phenomenon, so we have no choice but to address it at a system-wide level. In doing so, since every part of the ecological story has to be supported by evidence and reason, the result in ecology is an edifice of systems knowledge that is reliable, vast and inclusive enough to make sense of climate change and help with the challenge of adaptation in all parts of the world and at all scales of human society and the ecosystems that sustain it.

So the conclusion of this brief review is that it is ecology, in alliance for many purposes with anthropology, that holds by far the most useful set of keys to the adaptation part of the climate response discourse. As the science of living systems it is uniquely placed to address issues of living systems, which is what we are, where we live and which are threatened by climate chaos. The other disciplines contribute potent ideas and experiences that are useful in choosing practical adaptation investments: international environmental law from 'legalism'; CBRM from 'socialism'; EIA, planning and PES from 'capitalism'; political influence from 'economics'; and interfaith dialogue and occasional flashes of leadership from 'religion'. All have weaknesses as well as strengths, but all have necessary roles in the climate response, as do the many local variants and wholly different traditions of thought embedded in less cosmopolitan cultures.

3.6   UNDERSTANDING ADAPTATION

Six Principles of Adaptation

The definition of adaptation used here is the process of understanding, anticipating, planning for and responding to changes in ecological and social systems during a transition from one set of environmental conditions to another. Breaking this process down into its component parts and necessary actions, while also seeking ecologically and socially sustainable outcomes, leads to the conclusion that to adapt effectively requires:

- understanding how the systems work and what is important about them;
- understanding the changes that are underway and what they mean;
- exchanging ideas and knowledge about systems, changes and coping strategies;
- envisioning an outcome that is tolerable, stable and safe for all;
- recognising and choosing wise, credible and accountable leaders; and
- leaving no one behind.

Understanding and Sharing Knowledge

While it is possible to imagine trying to adapt without understanding, in the absence of an inspiring vision, learning, leadership or equity this is unlikely to work very well. The first three of the six principles, on understanding and sharing knowledge, are particularly important, since the contemplation of climate change adaptation by any group or institution, at any scale, will have to cover a lot of ground, including the nature of the social and ecological systems that support them, what features are particularly important (in terms of livelihoods, amenity or sacredness) to the people concerned, and how changes in those can be detected and responded to. The various intellectual traditions and reference languages mentioned in Sections 3.3–3.5 can contribute useful questions here, including the following:

- From law: what did past elites prescribe and proscribe, and why?
- From socialism: how can collective action help solve problems?
- From capitalism: how can innovation be rewarded?
- From economics: where is it useful to be guided by prices, and what things are priceless?
- From religion: what is truly important, and why?
- From anthropology: what do events and actions mean within our various imaginative universes?
- From ecology: more system-wide questions on how to safeguard water supplies, soil fertility, farm productivity, wild harvests and environmental security.

A synthesis of all of them can then allow people to decide whether some of the strategies that often work (CBRM, PES, EIA, participatory planning, etc.) can be applied in their own cases, and how this can be done given the system of governance in that society – the system which defines who owns what, and how decisions are made. The ways used to find answers to these questions, the kinds of social and ecological relationships that influence those answers, and their implications for how people live or could live more sustainably, are all shareable, and if shared will increase the efficiency with which new groups begin to ask similar questions of their own circumstances. Networking and knowledge sharing are therefore important, but the other three of the six principles, on vision, leadership and equity, while just as important are much less straightforward, and far harder to address through investment projects. Key principles are noted in the following paragraphs.

Envisioning the Future

This is a topic of great significance but it is also one in which detailed prescriptions are of little use. Some kind of inclusive dialogue around 'where do we want to go' is integral to stakeholder participation in development planning, although instances vary in quality between being based on locally generated ideas and free prior informed consent, and those involving solutions proposed by outsiders. Such vision statements are constrained by the constant needs of people, but can still be very diverse since each culture possesses its own distinctive view of what is good for its members, and hence what an ideal future would look like. Thus the most useful role for outsiders is to offer reliable knowledge (e.g. about ecology) freely, but advice (e.g. on what to do) only sparingly, while accepting, encouraging, enabling and validating (as appropriate to the political context) the indigenous processes of dialogue that can lead to wise decisions.

Group visioning processes contain dangers as well as opportunities, however, since a group can fragment around different visions when new information, ideologies, incentives and alliances are introduced. Thus challenged, discord and polarisation become possible if the capacity of a society for peaceful consensus building is exceeded, offering opportunities for factions and outsiders to exploit differences. Box 3.1 provides two examples from the anthropological record. As is typical of information from this source, the cases are highly specific yet can also draw attention to more universal phenomena. For example, all societies have mythologies and factions, and the capacity to decide, consciously or otherwise, whether to react with violence or peaceful arbitration when challenged, and quite different processes are recorded in these two cases.

Conservative Tendencies

The cases in Box 3.1 unfolded over one to three centuries, but there are hints of another phenomenon over the longer term. The influence of traditional Polynesian social values – already encountered in the

BOX 3.1 **Alternative responses to new challenges in traditional societies**

Fragmentation in Hawaii, 1778–1898

Traditional Hawaiian society was disturbed by the arrival of Captain James Cook of the Royal Navy in 1778–1779. Priests of the major god Lono recognised Cook as a manifestation of Lono, and ritually inducted him as such. But rival priestly groups and their secular allies then killed Cook and treated his corpse as an enemy sacrificed to royalty – the body was flensed, and the durable parts distributed among chiefs and high-status warriors, with some to HMS *Resolution* as a courtesy. The subsequent unification of the whole archipelago under Kamehameha from 1795, the downgrading of the Lono cult and the folding of Cook into a cosmological story less threatening to the elite, the overthrow of the whole system through missionary activity after 1820 and annexation by the United States in 1898, can all be traced to polarisation and chaos among conflicting worldviews that started with Cook's arrival.

Christianisation in Iceland, 1000–1262

A pagan smallholding society with strong participatory and egalitarian traditions inherited from settlers in the 870s considered the propositions of Christian missionaries carefully in their annual parliament, the *Althing*, in 1000 AD. They decided collectively to delegate the decision to a respected shaman who concluded, after meditating and dreaming, that they should convert, and they then peacefully did so. In a series of completely unforeseen consequences, the decision allowed the acquisition of lands by the church which created a new class of tenant farmers, which broke down the traditional governance system based on small-group loyalties and led to centuries of civil war, which eventually opened the way for annexation by the Kingdom of Norway in 1262.

Sources: For Hawaii, see Sahlins (1995); for Iceland, see Magnusson (1960).

concept of *talanoa* – might be reflected in Hawaii's consistently left-democratic political culture in the modern United States, and Icelandic society eventually (after seven centuries) returned to its independent, participatory and egalitarian roots. The fact that Iceland, as 'a haunted and uncanny land' (Holland, 2008: 212), retains a strongly pagan atmosphere is also a reminder that there are conservative forces in human cultures that preserve something of their original spirit and can exert a consistent influence for extremely long periods.

The same phenomenon can perhaps be seen in modern Europe, where there seem to be persistent fault lines that map on to ancient divisions, for example between areas of Viking influence and the rest of England, marked by place names north and south of the 'Watford Gap' (Alberge, 2017; Adams, 2018). Less well documented, but possibly just as significant, are the sharp political differences between northern and southern Italy, which might map on to the 'Rubicon' boundary between the ancient Roman provinces of Cisapline Gaul and Italia; and perhaps some cultural differences between Portugal, pacified as the ancient Roman province of Lusitania but otherwise little touched by Roman rule, and Spain, governed ruthlessly by Rome and extensively mined and farmed by slaves. More obvious is the division between the ancient Roman world and the world of the Germanic peoples. This can be interpreted as directly relevant to the evolution of the climate treaty, since as noted in Sections 1.3 and 3.3 the Germanic world retained its decentralised system of governance and law, and this gave rise eventually to the Holy Roman Empire (famously described by Voltaire as 'in no way holy, nor Roman, nor an empire').

The Germanic cultural area produced two other examples of loosely federal systems: the Hanseatic League and the Federal Republic of Germany. And the modern EU is clearly (though perhaps unconsciously) modelled on the Holy Roman Empire, with a commission replacing the emperor as the entity responsible for setting standards for trade and human rights, and the member states replacing the hundreds of autonomous territories of the empire, which were governed variously by kings, dukes, counts, bishops and abbots. Even the flag is similar in concept, with the EU's circle of equal stars on a royal blue field replacing the many equal territorial crests on the empire's eagle's wings.

The EU includes territories that were once parts of the ancient Roman Empire, however, where wholly different traditions of centralised, top-down governance prevailed for centuries. This could be the ultimate source of the tensions between member states already noted in the development of the WFD, with an expectation of top-down regulation competing with an expectation of experimentalism. This contest was apparently settled in favour of experimentalism internally by about 2000, and externally 10 years later, after the failure of top-down regulation at CoP 15/2009 in Copenhagen.

It is therefore suggestive that political dissatisfaction with the EU began growing at about the same time in England, the dominant country of the UK, which is by tradition a centralised unitary state and little influenced by the historical development of post-Holy Roman Empire governance on the European mainland. This tendency, aggravated and exploited politically, eventually led to England (and Wales, arguably the least independent-minded of the four UK countries) voting to leave the EU in 2016, and taking the whole of the UK into a very difficult and costly departure process that continues to date. It may therefore not be an accident that England (but not Scotland, arguably the most independent-minded and pro-EU of the UK countries) was consistently among the least successful of EU regions in cleaning up its waters in line with the WFD (EEA, 2018, 2019).

These consistent differences between cultural zones in the enthusiasm of their peoples for complying with externally agreed standards of behaviour may therefore be legacies of deep historical experience. This is all relevant to aid programming, since it implies that taking the time and making the effort to build and educate in the context of the local culture, and creating the space for local people to find their own answers at their own pace, may well yield more robust and durable solutions than programming against imposed deadlines can possibly achieve.

Choosing Good Leaders

Humans are a travelling species, and the verb 'lead' is linked to words that mean 'journey', 'way' or 'course', and also with 'load' (things that you carry on a journey) and 'lode' (as in lodestar and lodestone, things that guide you on a journey). Even settled peoples can never afford to forget how to travel as there is always the risk that where they live will be affected by drought, sea level rise or invasion. The idea of the person within a group who is responsible for starting and steering a journey must therefore be utterly primal. But a group must be ready to travel before a leader can shape a vague feeling that it is time to move into enthusiasm for a journey in a particular direction, with all its dangers and labours. This also applies to other meanings of leadership, which is no longer limited to physical travel (if it ever was) and now also concern responsibility for other kinds of journey, ones that involve change and progress in relationships between people, and between people and their environments. All require skills in managing conflict by dispensing justice, managing relations with other groups, understanding and articulating the needs and desires of groups, and choosing directions and destinations.

Leadership lies in the artistry with which all these tasks are done. It brings together the usefulness of many other mental capacities, since it requires all signs in the environment to be seen and understood, many of which are subtle and require intimate familiarity with the natural and social world. Particular indicators must be chosen and their significance marshalled into a story that can help the group's thoughts and wishes take form. An important factor is that leadership confers high status so there is competition for it, and leadership stories must therefore be able to persuade or outmanoeuvre rivals – those who want to stay, or make peace with the invaders, or try something else, or head in a different direction.

A particular risk arises where important environmental and social signals cannot be appreciated without a great deal of knowledge and attention to detail, so they are recognised only by a small minority. This applies often in large, complex or fragmented societies in which there are many distractions, and especially where slowly deteriorating environmental conditions are involved, or where there are slowly growing social threats such as inequality, corruption and polarisation. Here, if the danger is severe and solutions are needed urgently but there is little public appreciation of the need for action, an essential quality of leadership is a willingness to act decisively to safeguard the group but in advance of public opinion. An accountable leader must then explain what happened in the emergency and why, and be willing to accept responsibility for all the consequences. That said, it is still only possible to define 'good leadership' in generic terms, as comprising:

- the competence to identify the most important challenges;
- the attention given to diverse signals about those factors and how they are likely to affect the group;
- the intelligence needed to seek, discriminate and absorb sound advice on what to do about them;
- the articulacy to explain and build support for a collective course of action that will minimise harm and maximise benefit for most people in the long run; and
- the flexibility to maintain alliances while adapting to events.

So a good leader must at least be competent, attentive, intelligent, articulate and flexible. Other lists can be composed that stress more interpersonal skills, for example: 'Integrity, competence, kindness, compassion, empathy, vision and hard work: these seem to be the core traits of the leaders who have won the trust of their populations and led their countries effectively through this [Covid-19] crisis' (Sridhar, 2020: 2). But all candidates for leadership are likely to claim all these attributes, so the choice will be up to the electorate. The latter might be expected to be on the look out for stupidity, cupidity, criminality, extremism, narcissism or a tendency to lie, since these often-concealed attributes can endanger everyone.

With climate change, it would also make sense for voters to prefer leaders who understand ecology, at least to the extent that they can be trusted to grasp the basic ideas of where food, water and environmental security come from, and what climate change might mean for these necessities. Such knowledge can easily be tested at interview or by questioning during campaigns, but ecology is still seen as a minority interest rather than something that every citizen and journalist should know about. It is not feasible for leaders who are competent on climate change to be chosen consistently while this lack of interest in ecology persists. This is why the universal early teaching and learning of ecology is necessary for adaptation leadership, and therefore for adaptation itself.

Leaving No One Behind

The principle of inclusive development is summarised under the UN's call to 'Leave No One Behind' (UN, 2016). It is straightforward in the sense that people can be shown to have been 'left behind' using indicators of inclusion and equity (school attendance, poverty, literacy, health, wealth, participation, etc.) that are easily detected and assessed, while people are often aware, or can be reminded, that they have been left behind themselves or that others have been left behind. The last is much easier among neighbours but much harder if the people concerned live far away, or belong to a category of 'other' (by ethnicity, caste, class, gender, etc.) which the attitude of their society encourages them to ignore. Since making sure that no one is left behind requires other people to care if they are, key action points include:

- *equity* – challenging, educating, organising or using legal, legislative, administrative or direct action means to break down the perception of 'otherness', the social assumptions that sustain it and the different treatment of 'others' that results;

- *charity* – raising awareness of the plights of distant people who are suffering, and making it easier to do something to help them; and
- *cooperation* – making it easier for communities of neighbours to detect and respond helpfully to hardship among people who live among them or nearby.

The first action point encourages fairness, inclusion and human rights, for example through gender equity and social inclusion efforts to promote equal treatment of women and other groups who may be excluded, marginalised or weak. The second point motivates charitable giving, and some overseas aid and social welfare spending by governments. Institutional interventions motivated by equity and charity are both vital in reducing large-scale suffering, but the third point may be the key to no one being left behind at other scales. The case for this starts by seeing populations as aggregations of small groups – the 'Dunbar groups' of 50–150 people described in Chapter 4 – rather than as classes united and divided by economic interests. The issue then becomes how individuals behave in groups, rather than how classes behave in certain conditions of production and ownership.

Thus the nation, class or any other large group can be seen as a mass of small groups, all expressing the same social skills, all with the same limitations but high levels of effectiveness within those limitations, and with all the groups overlapping to create one system with its own system-wide behaviour. This becomes important when we consider that the macro-scale phenomena associated with climate change – such as fires, droughts, storms, floods and heat waves – are actually made up of immense numbers of tiny events that impact small groups, who are left to help each other as best they can until disaster relief arrives. These events only attract media coverage and high-level political interest when they affect large areas and populations, but at all smaller and intermediate scales of space and time it is the density and effectiveness of small groups that defines the strength of the society as a whole.

This has significant implications here, since it suggests that the most reliable, cheap and equitable way to ensure than no one is left behind is for individual hardship to be detected by other people, neighbours rather than machines, companies or functionaries of the state, and acted on by them alerting public services and monitoring the response. This offers a way to implement reliably policies of equity that require neither states nor corporations for their primary effectiveness. The key issue then becomes how to ensure that nations and corporations support and enable the operations of small groups, rather than replacing them with larger-scale systems that conflict with the ordinary sociability of our species. This choice, of whether to work with organic society or replace it with a machine state, is fast becoming urgent as the technology needed for universal surveillance is developed by corporations and bought by governments.

3.7 ECOLOGICAL SYSTEMS AND BIODIVERSITY

For nature to build a 'natural' ecological system (such as a forest or coral reef), or for people to build an 'artificial' one (such as a farm or plantation), bearing in mind the key role of information in describing (through human consciousness) and guiding (through biophysical processes) the construction and maintenance of *all* systems, at least three things are required:

- *a physical location*, where nutrients and energy can be sourced in forms that are useful to the organisms concerned;
- *'information-1'*, in the form of DNA that defines the species of organisms that are able to live there, and that have reached the place through colonisation or introduction because they are pre-adapted to local conditions; and
- *'information-2'*, in the form of genetic, epigenetic and metabolic and/or behavioural adaptation among the varieties, lineages and individuals of the species that have become accustomed to live there (i.e. for domesticated 'land-races', 'breeds', 'cultivars', etc., and their equivalents for populations of wild organisms, from local variants to subspecies).

There is always, in addition, a need for time and continuity of physical conditions, with a sufficiently long duration and a sufficiently predictable pattern of variability that adaptation can occur, whether naturally (over centuries or millennia) or artificially (over years or decades). If these conditions are fulfilled, an ecological system will arise that meets all the system-criteria of possessing distinct entities (animals, plants, fungi, microbes, soil particles, water, etc.) that have rule-bound relationships with each another (pollination, predation, commensalism, symbiosis, parasitism, digestion, competition, etc.), and that meet their needs through exchanges (of nutrients and energy) in line with those relationships.

This, in principle and in practice, is what an ecological system *is*. That I have used 'information-1' to describe species-level, and 'information-2' for lineage-level features of groups of organisms, is a reminder that *information* is inherent to all systems because they comprise different things and information is 'news of difference'. Since 'diversity' concerns differences between entities, the term biological diversity (or 'biodiversity') is a synonym for the information contained in living systems. Thus, in a slightly different way, the 1992 CBD categorises biodiversity into the three levels of:

- *genetic diversity*, roughly equivalent to my 'information-2', though with more emphasis on genetic resources that are or could be used by people, and less on natural epigenetics, behaviour and metabolism;
- *species diversity*, roughly equivalent to my 'information-1', which allows for *species richness* (the total number of species in a place) and *endemism* (the proportion of species that occur nowhere else) to have a strong role in guiding nature conservation investments, where the intention is usually to maximise both; and
- *ecosystem diversity*, which includes all 'information-1', all 'information-2' and all information contained in (i.e. that can be used to describe unambiguously) the physical location, as well as whatever emergent properties (e.g. overall structure, colour scheme, apparent texture, meteorological and hydrological functions) the ecosystem as a whole might have that distinguishes it from all other ecosystems (i.e. the diversity of *ecosystem types*, which is also useful in guiding nature conservation since it allows representatives of all the types that occur in a country or region to be included within its system of conservation areas).

3.8   SOCIAL SYSTEMS AND CULTURE

At the cost of introducing a few specialised terms above, a degree of consilience (consistency and mutual intelligibility) is possible between the theories of information systems, international law and conservation biology. This allows some parallels to be drawn with human social systems. These are distinguished from the social systems of non-human organisms, which are seen here as ecological systems with social behaviour prominent in their 'information-2'. This simplification allows a focus on the social systems of our own species. So, considering only these, to build a *human* social system also requires at least three things:

- *ecosystem tenure*, a physical location from which the members of the society are able reliably to draw the resources of food, water and raw materials that they need to sustain themselves, and also to support the production, invention and value-adding needed to generate surpluses and trade with other social systems;
- *ecological knowledge*, a body of wisdom and practical instruction on how to use the resources of the inhabited ecosystem productively and sustainably to meet all the diverse needs of all members of the community, and that also maintains and transmits that knowledge down the generations (e.g. through spoken, sung, danced, painted or written rules related to the landscape, its biota and their proper uses); and
- *social knowledge*, a body of wisdom and practical instruction on how to regulate or perform functions concerning leadership, ownership, use rights, ritual obligations, conflict management, status, gender, etc. within the society, and diplomacy, raiding, trading, warfare, etc. with other societies, and that also maintains and transmits that knowledge down the generations (e.g. through language, initiation, mythology, dance, story, art and artefact).

There is always, in addition, a need for time and continuity of social conditions, with a sufficiently long duration and a sufficiently pre-dictable pattern of variability that traditions can be established and act to stabilise the society, usually over years or decades, exceptionally over centuries or even millennia. If these conditions are fulfilled, a social system will arise that meets all the system-criteria of possessing distinct entities (individuals, classes, clans, age-sex groups, families, ritual sects, farmland, hunting and gathering grounds and waters, etc.) that have rule-bound relationships with each another (debts, closed seasons, sacrifices, taboo species, sacred sites, spirit groves, ancestral access restrictions, burial grounds, etc.) and that meet their needs through exchanges (of prestige items, sumptuary expenditure on gifts and parties, obligations, marriage partners, trade goods, etc.) in line with those relationships. This, in principle and in practice, is what a human social system *is*.

If 'biodiversity' is synonymous with the information contained in ecological systems, the equivalent word for social systems would be 'culture'. In both cases, the information is disseminated among all the entities and relationships that constitute the system and exists only because of the system. In both cases, fragments can be excised and put in collections (museums, zoos or botanical gardens), but the system itself remains behind in its place of origin. The implied view of culture as a system property, like biodiversity but emerging from a human system rather than an ecological one, is consistent with the thinking of philosophers on peoples and their distinctive cultures back to Johann Gottfried von Herder and throughout the tradition of cultural anthropology from Franz Boas to Clifford Geertz and beyond (e.g. King, 2019: 140, who describes the characteristic view as being that 'Cultures might look like things, solid and tangible, but they were more like systems: the way particular ideas and habits fit together').

The parallel cannot be exact, however, given the major differences in medium, mechanism, purpose and flexibility at a group level, and the ways of insight and invention at an individual level (even though it is hard to imagine an invention, or the motivation to make one, without some kind of social purpose or consequence). In any case, these differences reflect the extraordinary adaptive capabilities of our species, the details of which are as engrossing to anthropologists as the diversity of life is to biologists. But here the details are much less important than the similarities between ecological and social systems, which suggest that chaos may have a similar relationship to both as an obliterator of distinctiveness, relationships, patterns, rules and predictability, without which systems automatically cease to exist.

3.9   ADAPTATION, RESILIENCE, RESISTANCE AND FLEXIBILITY

Adaptation is a *process*, conveying the notion that all the entities and relationships that comprise a system are changing in order to make the whole system conform to new circumstances. The latter may involve the nature, intensity or direction of pressures on the system, or the deletion of elements within the system. These mean different things in social and ecological systems, but parallels are easy to draw. For example, in social systems new *pressures* might involve the arrival of powerful outsiders bearing new concepts of value, law and ownership, while in ecological systems they might involve rising or falling water tables or changes in seasonal mean temperature and rainfall. Similarly, in social systems *deletions* might mean the deaths of individuals and the forgetting of cultural elements such as words, stories and skills, while in ecological systems it may mean the hunting or felling of unusual numbers of particular organisms or the extinctions of species. In any case, the system adapts by changing itself to fit.

Such changes may be short or long term, minor or major, depending on the pressures and deletions in each case. For climate change at the macro level there is a steady directional alteration in climatic patterns and their consequences (such as rising seas) over the long term, to which people will adapt by changing their farming systems, livelihoods, locations of residence, etc., and wildlife will adapt by feeding and breeding less successfully in one area and more successfully in another, and by altering dispersion and migration accordingly, slowly for trees and more quickly for mammals and birds. In a chaotic transition from one macroclimatic regime to another, which is what we must deal with in this century and at least the next, a succession of near-random, brief, but often intense shocks is being superimposed on the underlying directional process.

Surviving these shocks so that the system can adapt to the change is the aim, and this depends on the strength of systems, which in turn depends on the qualities of the relationships among its components. Will people and wildlife be able to live on their food reserves or take shelter, and be able to wake up and become active again when the shock has passed? If so, they are *resilient* – a quality that might be called 'bounciness'. Will they be able to withstand cold or thirst, or build higher dams or deeper wells, or thicker walls and blankets? If so, they are *resistant* – a quality that might be called 'hardness'. Will they be able to migrate, or move their fields and livestock up-hill or up-latitude, down-hill or down-latitude, or change their crops and diets and still make a living until the shock passes. If so, they are *flexible* – a quality that might be called 'bendiness'.

Thus, in system terms, resilience, resistance and flexibility are aspects of the *quality* of a system known as its strength. The IPCC makes this distinction clear, by defining adaptation as 'the process of adjustment to actual or expected climate and its effects' (IPCC, 2014: 118) and resilience as 'the capacity of social, economic and environmental systems to cope with a hazardous event or trend or disturbance, responding or reorganizing in ways that maintain their essential function, identity and structure, while also maintaining the capacity for adaptation, learning and transformation' (IPCC, 2014: 127). Here, though, resilience is being made to serve as a synonym for strength, as well as some of the sources of resilience, resistance and flexibility, which may have been good enough in the past but will not serve if the matter of strength and strengthening of systems is looked at in more detail. This requires some unpacking, but in general, loose terms, it is understood that 'resilience' will continue to stand for a bouncy, hard and bendy form of 'strength' in the climate change literature.

4 Making Systems Stronger

If the unpredictable damage inflicted on human interests by climate chaos at the local and landscape levels is the threat that should be driving much of the adaptive response, then how does it work, what are the interests at risk, where do they reside and what can be done to protect them? This is of critical relevance for investment strategies of all kinds, including aid, since investment security – whether measured as value for money, sustainability performance, inclusive justice or long-term profitability – requires a predictive understanding of the systems where the investment occurs, and chaos makes this hard to have or to keep.

Answers to such questions must be sought through a focus on the nature of the systems themselves: ecological ones because they comprise the fabric of the living world, and so are responsible for human food supplies and the security of our homes and settlements; and social ones because almost everything we do is done with, for, by, to or in the context of our relationships with other people, especially the small groups of kin and acquaintances – 'Dunbar groups' – with whom we share our lives most directly.

From analysis of the ethnographic record, businesses and religious and other institutions and internet-based networks, it was found empirically that people tend to maintain social networks of up to about 150 individuals but often fewer (Hill and Dunbar, 2003; Dunbar, 2010; Dossey, 2016). This finding can be interpreted to mean that people are not able to remember the complete details of more than a few people at one time. That this could be a constraint is made clear by considering the level of detail that is relevant to social cooperation, which includes who is related to whom through multiple generations, marriages and adoptions, the quality and content of past interactions with individuals, and the reciprocal obligations created during them, all going far beyond remembering names and categories but normal in all human societies.

It is thought that when people know each other well they are better able to cooperate effectively, implying that people tend to work best in small groups. This is consistent with the idea that humanity evolved through small groups cooperating internally while competing externally. It is also consistent with the ease with which small, effective groups and teams can be established for particular purposes (to roof a barn, perhaps, or to lobby local government for better services), or just for social enjoyment (Chapter 8). This line of thinking has major implications for how investments can be made most effective, including those in disaster preparedness and other aspects of the climate response.

4.2   THE STRENGTH AND DISSOLUTION OF SYSTEMS

Theoretical points, for instance about chaos as the end point of entropy (Chapter 3), can quickly become very real should there be a wildfire in an ecological system or a genocide in a social one. In these cases, a wave of entropy sweeps through complex order, with no concern for the identities, relationships, rules or values of any entity within it, and reduces it all to uniform ash, or else to vacant land with bones and barbed wire. These extreme cases are reminders that all complex systems arise against the pressure of entropy. This is driven back in ecological systems by evolutionary processes that favour appropriate design and the segmentation of life-strategies among lineages, and in social systems by conscious design effort, differentiation and cooperation among distinctive individuals, lineages and groups.

The resulting accretion of complexity in either case means that these systems must have at least some inherent strength to resist entropic change. This arises because each relationship involves mutual adaptation to a range of behaviours by the other partner, as well as to a physical environment which is never absolutely constant. The resilience in each partnership should add up across many relationships to create the capacity to absorb change, to bend rather than break, and to adjust expectations or behaviour. It is this collective tolerance to change and challenge that defines the strength of systems.

Whatever a system's inherent strength, more often than wildfire or genocide its erosion involves the gradual dissolution of mutual dependencies ecologically, or of mutual trust socially, thus undermining the rule-bound relationships among the system's entities that made the system possible in the first place. A system can be degraded to some extent, for example by logging, disease, hunting or slave-raiding, without losing its essential character, but if the pressure continues to increase it can reach a point when vital relationships break down suddenly and all together, and the culture, or ecosystem, simply evaporates.

Until that point the system is recognisable: birds can still find mates and nesting sites, and people can still sing songs and gather medicinal plants in their old ways. But afterwards, nothing remains but weed species and demoralised survivors, drinking too much and dying young. Still mysterious though these transformations are, they seem characteristic of any complex systems under stress. For example, 'health and vitality' can be understood as integral to a living organism, from which blood samples, for example, can be taken without harm, but which can be damaged by poisons, pathogens, cancers or injuries that progressively break down the relationships among its parts, until suddenly its 'life energy' evaporates and the organism dies.

The death of a single organism and the departure of its metabolism and consciousness is a definitive kind of system transformation. More generally, systems comprising numerous diverse organisms and persons are self-generating as well as self-organising, and have enormous power to grow into new forms under pressure, to regenerate once a pressure is relieved and to resettle into new equilibrium arrangements among the survivors after a bout of serious damage.

The 'wildfire' and 'genocide' cases are rare, and seldom leave no survivors at all, in the form of fire-resistant seeds, stumps and roots, or fragments of language, art and memory, from which regrowth can occur in some form.

Even the most hacked forest or urban landscape will contain hidden relics of the former ecosystem, as well as newcomer weeds, that can regrow to form part of a new ecosystem if people were to depart, as the Ukrainian city of Pripyat after the Chernobyl Nuclear Power Plant disaster in 1986 illustrates. And even the desert in Maine in the United States and the Mega Rice project area in Central Kalimantan in Indonesian Borneo – the results of deforestation over infertile white sandy soils in the nineteenth century and 1990s respectively – are new equilibrium ecosystems with new species complements, even if they are extremely poor relative to their starting points.

In any case, it is proposed here that anything that contributes to the dissolution of relationships within a system will tend to weaken it. To the extent that connections confer strength, their loss at a rate faster than they can regrow or be replaced by new ones will make the system vulnerable to further stress, and less able to resist the further advance of chaos. This is now a world in which most ecological and social systems have been changed, and in many cases weakened:

- for ecological systems, by some combination of direct exploitation (hunting, fishing, logging, grazing, etc.), conversion (drainage, clearance, mining, dam flooding, etc.), fragmentation (by infrastructure, settlements, farmlands, etc.), pollution (by industrial and domestic wastes, agrochemicals, pharmaceuticals, chemical breakdown products, etc., whether waterborne or airborne) and the effects of climate change (salt intrusion, desertification, ground water depletion, etc.); and
- for social systems, by national and international cultural patterns abruptly replacing local ones, devaluing traditional systems of resource tenure, accountable leadership, language and the mythological sources of meaning and value, without putting in place substitutes that are entirely adequate to meet all human social needs without major adjustment.

4.3   COMPLEXITY AND STRENGTH

The relationship between complexity and fragility is ambiguous, since complex systems can dissolve as fast as, or faster than, simple ones if the right stresses are applied. For example, tropical moist forests have many species with narrow niche tolerances, and many depend on a continuously moist microclimate in large patch sizes. Thus, the system may simultaneously be very robust to disturbance in small patches (e.g. traditional shifting cultivation with small fields and long fallows, which allows for rapid regeneration by the vegetation, re-establishment of moist microclimates and recolonisation by fragile biota) but very vulnerable to disturbance in large ones (e.g. rapid selective logging and a damaged canopy and drying conditions over large areas).

Such non-linear responses to disturbance make it hard to attribute strength directly to complexity, but it does seem certain that ecosystems and societies that have arisen in stable environments over a long period are themselves stable as long as the conditions to which they are adapted prevail within their tolerances. The latter may be narrow or broad overall, or may be constrained by the specific tolerances of one or more 'keystone' species that are responsible for maintaining the environmental conditions that allow many others to exist. In this case it matters little how robust many individual species and their interrelationships may be, if they are all dependent upon a small number of keystone species. The latter are defined in Box 4.1, along with a number of other factors, features or qualities that are relevant to the strength of three kinds of system that might be affected by climate chaos: ecological systems, social systems and human-ecological or production systems.

Box 4.1 identifies parts of each kind of system where important entities and relationships are concentrated, and where a change in conditions might induce other changes that could destabilise the system as a whole, though without necessarily any certainty on what

BOX 4.1 **Factors, features or qualities that are relevant to the strength of systems**

Ecological Systems

- *Trophic structure.* Named from *trophē,* the Greek word for 'nourishment', trophic structure describes the pattern of relationships in an ecosystem between populations of (often diverse) organisms grouped according to their role in the flow of energy and nutrients, starting with plants that capture energy from sunlight and nutrients from the soil, whose tissues are consumed by animals that feed on plants and each other, and bacteria and fungi that decompose the wastes and tissues of plants and animals. The result is a complex and dynamic pyramid or network of feeding relationships, with characteristic amounts of resources available to support populations of different species at each major trophic level, each making a living in their own way in competition with others (i.e. each occupies its own ecological niche). Changes affecting one part of any level can affect other parts or even the whole structure.

- *Keystone species.* Named for the 'keystone' of an architectural structure, one that holds it together and imparts its special mechanical strength, a keystone species is one that has a particularly important role in defining and maintaining the physical or trophic structure of an ecosystem. Examples include elephants in African woodland savannah, largely because they push over so many trees and make space for new generations of trees to grow up (and process so much vegetation into dung, which they distribute very widely along with seeds that later germinate), and beavers in high-latitude northern river valleys and floodplains, largely because they build dams that create new ponds while slowing down the water flow, so providing life-opportunities for many other species.

- *Top predators.* Named for being close to the top of a food chain or apex of a trophic pyramid, top predators like big sharks in tropical marine ecosystems and wolves in northern taiga and tundra (or much smaller predators in much smaller ecosystems) influence the behaviour and populations of their prey which in turn affects all other populations in the ecosystem, thus maintaining the trophic structure overall. Removing a top predator can have dramatic and unexpected effects on the whole ecosystem as the populations and ranging behaviours of prey animals change, with new intensities of grazing, browsing,

BOX 4.1   **(cont.)**

trampling and dung deposition affecting the vegetation and soil overall and in new places.

- *Pollinators, symbionts, parasites, commensals,* etc. These are all names for different kinds of dependency that have evolved or co-evolved between different organisms: pollinators for those that plants attract to their flowers by offering nectar in exchange for carrying pollen from one to another; symbionts for those that live together with mutual advantage to both partners; parasites for where the relationship is less reciprocal and one partner suffers loss; commensals for where the relationship is advantageous to one side and neutral to the other. These relationships (and others) can be extremely diverse and complex, but the key point is that the loss of one species from an ecosystem can cause the loss of others that had co-evolved to depend upon it, causing cascades of extinctions and lost relationships throughout the system concerned.
- *Physical structure.* Ecosystems may have characteristic structures because of the large organisms that grow there, such as trees in a terrestrial forest or coastal mangrove, or giant kelp or reef-building corals in marine ecosystems, which through their bulk, rigidity and shape offer niches for the other organisms that live there. Removing or damaging them causes the whole ecosystem to change. Most ecosystems are rooted in soils, which are complex systems of rock fragments, plant residues, water, air and small organisms or the roots and hyphae of larger ones, and are supported by soils physically and in the supply of nutrients and water, so any change to the soils – such as by erosion or compression of water-bearing layers – will cause the ecosystem to change.
- *Water.* All life on Earth has always depended on water, and it is an essential part of all ecosystems. It carries dissolved chemicals which are captured as nutrients by plants and microbes, and within the body by animal cells, but also itself provides a necessary chemical environment for all the molecular processes of life. The same features that make water essential also mean that anything affecting the water in an ecosystem – in terms of its quantity, timing, quality and dissolved salts, nutrients and oxygen – can have major effects throughout the system, as species adapted to narrow tolerances for water conditions die off and are replaced by others that have different preferences or broader tolerances. If poisons (cyanides, pesticides, etc.) or

BOX 4.1 **(cont.)**

artificial fertilisers (phosphates, nitrates, etc.) find their way into water, mass death can result from direct toxicity or from algal blooms that consume all available oxygen and suffocate everything else, while too much water standing in a soil ecosystem can drown roots and kill trees.

Social Systems

- *Peace and war.* Factors in the resolution of social tensions and conflict include: leadership roles, judicial processes, priestly, shamanistic or divine oracles, etc. (an 'etcetera' is always needed when describing human social systems).
- *Resources and livelihoods.* Factors in the ownership and sustainable management of resources include: land, water and wildlife tenure and use rights, roles and responsibilities, collective defence arrangements (against invasion, ill-health, dangerous enchantments), diplomacy, rules of warfare, etc.
- *Meaning and demoralisation.* Factors that make life meaningful to all members of a society, without which demoralisation will become a threat, include: sacred cosmologies, myths and legends which give meaning to the landscape, wildlife and society, including reassuring explanations concerning existence (ontology), experience (phenomenology) and the afterlife (eschatology).
- *Roles and alienation.* Factors that define social roles and prevent alienation of the individual from the society include: collective memory systems (secret societies, club houses), initiation processes, communal rituals and other shared activities (trances, dances, worship, songs, parties).
- *Language.* Language is the key factor that permits human levels of social complexity, so language is therefore to culture what water is to ecology: *sine qua non*. Here a distinction is made between a language indigenous to a culture, of which some 6,000 are extant (but many others have gone extinct in recent centuries), which maps 'an entire cosmos of meaning' peculiar and specific to the culture concerned (King, 2019: 242), and the standardised international languages which are often used only to handle day-to-day transactions and cannot replace indigenous languages in traditional societies (but they do evolve local variants in parallel with the development of new local cultures).

BOX 4.1 **(cont.)**

Production (Human-Ecological) Systems

- *Profitability of production and sale.* Relevant factors include: affordable inputs (costs of labour, agrochemicals, equipment, maintenance, fuel/energy), secure, dry and refrigerated storage and transport to market (roads, rail, vehicles, etc.), sale venues (market facilities, shops), acceptable arrangements with middlemen, protection against organised crime and corrupt officials, moderate taxes and other market distortions, fashions that encourage or discourage marketability (demand for organics, hostility to GHG emissions, Fair Trade criteria, etc.), government policies and central purchasing authorities.
- *Social sustainability of production.* Relevant factors include: protection against theft (e.g. livestock rustling, squatting), fires (surveillance and intervention), labour relations (strikes), conflicting land use (e.g. mining, settlement, logging, dam lakes), official and unofficial land expropriation, war, conflict, terrorism and health of workers (e.g. pesticide poisoning).
- *Ecological sustainability of production.* Relevant factors include: protection against fires (drought, rainfall, microclimates), pests (local and invasive), diseases (of crops, livestock, prawns and fish), storm damage, floods, heat waves, cold snaps, soil depletion (deterioration and erosion, soil ecosystem health, soil acidity, fertility), genetic diversity depletion (pest resistance, populations of predatory insects and birds, populations of pollinators) and biodiversity health (e.g. pesticide poisoning).

---

the result would be in each case. Thus, for example, removing a top predator or keystone species would certainly impact an ecological system, while demolishing long-established resource tenure, leadership or religious arrangements would impact a social system, and destroying roads, union agreements or the health of crops or livestock would impact a production system. Where these changes occur suddenly and with little foresight or compensatory planning, the consequences may be dire among the components of each system, indicated in rates of mortality among wildlife, social distress among people and bankruptcies among businesses.

If these changes truly weaken many of the systems concerned, then climate chaos is a major source of further threat because it reduces predictability and increases agitation in climate systems. These take the form of more frequent and more violent or extreme incidents of storms, floods, cold snaps, heat waves and droughts. And these occur on top of progressive changes such as the spread of new diseases and salinisation of groundwater, with all of these and other phenomena being thrust upon weakened systems with little if any warning, at unexpected times as regular seasons vanish, and from unexpected directions as winds cease to 'prevail' from particular points of the compass.

Under these conditions, if many threats and impacts cannot be predicted then adapting to climate chaos must mean strengthening ecological and social systems against *any* threat or impact, from *any* direction, in *any* dimension. This in turn means that the design and performance of adaptation investments must be judged by their role in preserving or restoring the integrity, and therefore the strength, of systems. This they may do by opposing things that weaken systems, by supporting things that strengthen them and by encouraging the regrowth of new networks of relationship within them, if they have already been damaged.

4.4 INFLUENCES ON STRENGTH AND FRAGILITY

It is open to question whether it will ever be possible to measure the tolerance to change precisely for all the relationships within any complex ecological system, especially as the relationships themselves are so dynamic. But we do have an intuitive grasp of how to assess much the same thing for human social systems. Thus, it is easy (at least for outsiders with a common language and a degree of sensitivity) to gain an impression, for example, that a society is conservative and brittle, its members closed-minded and intolerant of new ideas and unfamiliar behaviours, or perhaps xenophobic and racist.

These features are accessible to comparative assessment, whether by intuition or psychometry, and the findings are relevant if it is true that brittle and conservative societies are less robust to imposed changes than flexible, open-minded ones. This seems likely, and indeed it is central to the justification of all 'democratic' social arrangements that they explicitly offer freedom to differ, and that they are therefore more adaptable and stronger in the long term than any less inclusive or less free form of governance.

A similar argument has long been made for free markets, based on the notion that consumer preferences expressed through purchasing choices can shape the supply of goods and services more effectively than is possible in any more centrally planned approach. A specific version of the same concept, known as 'experimentalist governance', has relevance to the climate change issue since it underlies the design of the Paris Agreement (Chapter 1). All these lines of thinking converge on the question of whether the strength of social systems can best be harnessed to the cause of adapting to climate chaos through local empowerment, or through central planning, or more likely through some hybrid that involves the horizontal networking of locally empowered communities that are informed, financed and protected by central or top-down arrangements.

Amid so many potential criteria of system health, and so many potential indicators of system strength and weakness, the challenge is to select the few that are as meaningful and easy to identify as possible, and for which evidence is as accessible and unambiguous as possible. If the management aims, or the objectives of an aid project (or other investment), include promoting adaptation to climate chaos, these criteria and indicators should feature in the documents that record the life of each project: the environmental profile, concept note or pre-appraisal report for the situation analysis; the detailed proposal, project document or appraisal report for the project design; and the mid-term review, final evaluation and/or *ex post* evaluation for the evidence-based judgements made regarding its performance.

These documents will ordinarily be prepared by professional staff or consultants on behalf of the proponent (e.g. a governmental or non-governmental institution) or potential investor (e.g. an aid agency, development bank, corporation or ministry), who may not be familiar with the principles of system behaviour, or have access to much detailed information about the ecological and social systems concerned. These uncertainties suggest that what is needed is a simple and robust approach that can draw attention only to those major factors that *must* be addressed in design, or that *must* be assessed in evaluation, if major errors are to be avoided. It is signs of these features that one would be looking for in the written records and interview notes that contain most of the data used in project design and evaluation.

Two questions therefore arise: what are these major factors and, bearing in mind that they can take many forms, how can they be detected? Chapter 5 examines a simple case in Nepal, in which the following key criteria of system health are used: physical structure of natural forests as an ecological system indicator (and as a proxy for regeneration, microclimate, species richness, water catchment and carbon storage functions, etc.); and local responsibility for forest management as a social system indicator (and as a proxy for short accountability pathways, decentralisation through governance reform, interest in environmental education, local capture of economic benefits from forestry, etc.).

4.5 ORIGINS OF SYSTEM STRENGTH

Based on the foregoing reasoning, the strategic principles in Box 4.2 concern the weakening and strengthening of social and ecological systems. The key design principles that flow from this analysis are contained under the 'active intervention' bullets in each case. Production systems such as commercial plantations are not considered here because of the great diversity of factors involved that originate far outside the systems themselves (Box 4.1), including technologies, market and macroeconomic conditions, and political influences at all scales. These are properly addressed within sectoral and national system analyses, but very complex causal loop diagrams would be needed to depict them and this is beyond the present level of discussion.

BOX 4.2   **Principles of strength and weakness in social and ecological systems**

Social Systems
- *These are strong* to the extent that their members (as individual people) agree on how to classify each other, how to relate to each other, how to resolve conflicts between each other and how to hold each other to account through a common understanding of values, and the roles, rights and responsibilities of leadership, all based on new arrivals being comprehensively inducted through upbringing, education and initiation.
- *They are weakened* and rendered vulnerable to further stresses by breakdowns in classification (as new categories of people arrive), relationship rules (as new laws are imposed), conflict management (as new sources of unlawful violence arise), values (as unfamiliar ideas and behaviours are introduced) and leadership (as accountability is reduced, stretched and fragmented), making induction difficult and error-prone, and also by changes to the underlying system of resource tenure (i.e. ownership and use rights).
- *These weaknesses may be offset,* by *conserving* the original system, by *reaffirming* aspects of the original system or else by *building a new system* through restoring clarity of identity and consistency of convention, values, roles, rights and induction processes.
- *Active intervention* focuses on *shielding* (which is seldom possible for long, and may be seen as unethical to the extent that it appears to treat peoples like 'zoo animals'), *validation* (e.g. of language and sacred myths, usually by endogenous activism), or *targeted protection* (often by outsiders) of universal rights, human security and the means of potential livelihoods (i.e. healthy ecosystems under clear ownership and accountable management), so that people can seek their own new equilibrium.

Ecological Systems
- *These are strong* to the extent that their members (as individual kinds of organism) have established and maintain stable (predictable, routine, constant) relationships with each other, based on being able to recognise each other (by sense or automatic response) and react appropriately to their relationship (by calculation or necessity).
- *They are weakened* and rendered vulnerable to further stresses by breakdowns in composition (as new mixtures of organisms arise or arrive),

BOX 4.2 **(cont.)**

making recognition difficult and error-prone, and also by changes to the underlying system of ecological resources (including energy, nutrient and water availability, structure and function).

- *These weaknesses may be offset* by *conserving* the original system, or by *regenerating* it, or else by *safeguarding* the underlying system of ecological resources and allowing a new system to grow up.
- *Active intervention* focuses on *shielding* (e.g. within protected areas), *validation* (e.g. using flagship species) or *targeted protection* (e.g. of soils, water, structure and microclimate and against alien invasives, etc.), so that the ecosystem can seek a new equilibrium.

4.6 INVESTING IN SYSTEM STRENGTH

To strengthen a system actively requires the targeted protection of its *homeostats* (homeostatic mechanisms) and *foundations* (essential features and processes) while it finds a new equilibrium under changed conditions. Both people in social systems and organisms in ecosystems can adapt in this way, provided their homeostats and foundations are secured. These things are clearer and more manageable in small social systems and local ecosystems (the 'local and landscape' levels of human society), and become increasingly hard to understand and organise effectively at larger scales (the 'national, continental and global' levels of human society). The points in Box 4.3 therefore relate primarily to *small* social systems and *local* ecosystems. Again, production systems are not included because of their complexity and remoteness from this level of discussion.

4.7 LEVELS OF SOCIETY

Systems and chaos relate to life and society at different geographic scales, which I divide into:

- *the local and landscape levels*, meaning neighbourhoods and municipal locations, typically bounded by a skyline or watershed and containing

BOX 4.3   **The homeostats and foundations of social and ecological systems**

Homeostats, or Homeostatic Mechanisms

- *Homeostats of social systems.* Forums (councils, associations, cooperatives, unions, etc.) where their members can agree to recognise and relate to each other in the framework of a common system of values, arrangements for conflict resolution and understanding of roles, rights and responsibilities. The forums and appearance of how they work will vary with culture and circumstance, but *what's important is their role in sustaining social integrity.* Hence, forums are the key homeostats for social systems, but their foundations must be preserved as well.

- *Homeostats of ecological systems.* 'Keystone species' and 'top predators' are important, but their roles would be meaningless without producers and recyclers, predators and prey, and all the other species of co-evolved and co-dependent pollinators, symbionts, parasites, commensals, etc. So the equivalent of the 'forum' in an ecosystem is the web of life itself, and this can be preserved only in a working ecosystem. The details vary immensely, but *what's important is the role of all species interacting together to sustain ecological integrity.* Hence, area protections (between total protection and managed harvests) are the key homeostats for ecological systems, but their foundations must be preserved as well.

Foundations, or Essential Features and Processes

- *Foundations of social systems.* Clear tenure over useful territory, and locally accountable management of all the things that are important to livelihoods, security and well-being. These may all mean different things in traditional, rural, modern and urban environments (farmlands, hunting, fishing and gathering areas, fresh water supplies, physical security, money, shops, banks, supply chains, etc., but note that distant dependencies reduce security for small social systems). The key thing is their role in *sustaining livelihoods, security and well-being* (i.e. 'economic' integrity).

- *Foundations of ecological systems.* Physical structure of above-ground biomass (e.g. woody vegetation, climbing plants) and below-ground biomass (e.g. root systems, soil biota), soils, water flow (drainage, aquifers, surface waters), microclimate (stability, light, moisture) and nutrient flow (trophic structure). The details will vary but the foundations of all ecosystems are similar, and the key thing is their role in *sustaining ecological integrity.*

people's homes, gardens and places of work, along with the institutions of local government; and

- *the continental and global levels*, meaning large and multiple countries, major mountain ranges and river systems, whole biomes, and institutions that include national and federal governments or transnational and intergovernmental alliances.

My starting point is that people live in complex ecological and social systems at the local and landscape levels, and experience the effects of climate change where they live. Despite access to global trade goods and services, which give some protection against local events, the challenge for most people is to cope with local conditions where they live. Here there is the problem that climate predictability is least at the local and landscape levels, for two main reasons. The first is that even if the complexity of the climate system was predictively modelled at the continental or global scale, which remains in doubt because of rapid system changes, tipping points and the continuing discovery of major new factors that deeply affect predictions, this would remain extremely hard to achieve reliably at landscape level for the foreseeable future. The second is that the interaction of unpredictable climate change at the landscape level with all the dimensions of natural and human ecology at that level would generate such complexity that detailed predictive modelling is not even foreseeable.

These factors undermine confidence that top-down planning for adaptation to predicted climate change can be effective at local and landscape levels. Yet most of the official planning capacity, power and money for the climate response exists at the national and global levels. This creates a tension between how the problem and potential solutions are seen by those who focus on the big picture compared with more local perceptions, and misunderstandings can arise in the space between them. All that can really be predicted with confidence in most locations is that climate stresses will be unpredictable, so adaptation must mainly be about people coping with chaotic change in detail, rather than with predictable change in general.

If large amounts of money are to be spent on adaptation, with major potential impacts on people's lives and livelihoods, then we need to be confident that investments respond realistically to uncertainty. Above all, investments must be scaled to the local and landscape levels, while also retaining a focus on how large-scale phenomena manifest themselves locally, and what can be done about them locally in ways that are consistent with what is done globally. This requires a kind of dual vision, with local action being significant both locally and globally, and everything needing to be coherent at both levels simultaneously. This need to hold differently scaled aspects of reality in view at the same time has already been highlighted as diagnostic of systems thinking (Section 1.4), anthropology (Section 3.4) and ecology (Section 3.5).

4.8 ALGORITHMS OF ADAPTATION

It is now possible to identify with confidence a constellation of measures that are most likely to safeguard or restore the strength of social and ecological systems, against climate chaos or anything else that challenges them. These include, at their heart, two propositions in the form of algorithms that are expected to work in a similar way in all circumstances, provided that they are applied together:

- that to strengthen social systems requires clear tenure over useful territory for local groups, with forums to sustain their social integrity, environmental education and knowledge sharing to increase understanding of ecological sustainability and options, and locally accountable management of all the things that are important to their livelihoods, security and well-being; and
- that to strengthen ecological systems requires the protective and precautionary management of ecosystems to sustain their ecological integrity, while safeguarding their ability to provide protective services in return, and environmental education and knowledge sharing to increase understanding of why this is all needed among the people involved.

If these things are done, one way or another, then systems will be strengthened, regardless of whatever *else* is done. But if they are *not* done, then systems will continue to weaken, and nothing else that is done will compensate for their lack. Moreover, using this basic package of local empowerment and environmental education, larger structures and programmes can be built, through replication, networking, technical additions from the environmental 'toolkit' and the shielding of local community-ecosystem units by higher authorities. Getting this mixture exactly right for every locality in the world is the essence of adapting to climate chaos, and the need now is for cheap and effective ways to do it.

PART III Practical System Strengthening

5 Community Forest User Groups in Nepal

5.1 HISTORICAL CONTEXT

Geography and People

Nepal is a mountainous, landlocked country located between China to the north and India to the south, east and west (see Caldecott, Hawkes et al., 2012). It lies along the southern flanks of the Himalayas, which were and are being formed by crustal folding so Nepal is prone to earthquakes. The deposition of material eroded from the mountains also gives the country an unconsolidated and fragile geology, making it vulnerable to further erosion and landslides. There are three east–west ecological zones determined mainly by altitude: the cold Himalayan region (2 per cent cultivated); the warm, monsoonal mid-hills region (9 per cent cultivated, and including the Kathmandu Valley); and the southern Terai lowlands, with hot wet summers and mild dry winters (41 per cent cultivated). There are about 28 million people, with 60 languages in use although more than half speak Nepali as their mother tongue. Most are Hindu, although few are strictly orthodox, and there are strong shamanic and Buddhist traditions as well as some adherents to Islam, Christianity, Sikhism and other faiths. The caste Hindu parts of Nepalese society are historically dominant in much of the country, and are associated with pervasive social inequality based on caste, ethnicity and gender. This has contributed to a fatalistic attitude among many Nepalis that has tended to act against the acceptance of modernity and development (Bista 1991).

Political Changes

After 190 years of near-feudal rule, Nepal held its first parliamentary elections in 1959, but King Mahendra soon afterwards removed the elected prime minister from office and imposed a new constitution and a kind of monocultural nationalism upon Nepal's diverse peoples (Shah, 2008). This system ended in 1990/1 when a popular movement forced King Birendra from power, but the parliamentary democracy that followed was unable to control corruption among officials responsible for delivering services, resulting in widespread resentment among the rural poor. Remote, centralised governance, social segregation and official impunity all facilitated the abuse of power. Unequal land distribution, land degradation in the hills and the repeated subdivision of household plots to successive generations further contributed to tensions. Exploiting this, the Communist Party of Nepal-Maoist launched a violent insurgency in February 1996, and the 'People's War' that followed lasted 10 years. During this time there was a wholesale breakdown in social relationships, especially in the western and far-western regions, whereby all males aged 16–40 years were presumed to be absent – in the army, with the Maoist forces or in foreign employment – and traditional social structures and forms of authority changed greatly.

The insurgency took at least 13,000 lives before being resolved through a Comprehensive Peace Accord in November 2006, leading to the abolition of the monarchy and declaration of a federal democratic republic in 2008. The scarcity of young men continued, however, as migration for foreign employment increased, resulting in a sustained feminisation of agriculture and resource governance in rural areas. These changes reduced rural labour capacity during and after the insurgency, but also opened space for social participation and leadership by women at local level. In the 2013–2016 period and beyond there were difficulties of consolidation and sustainability due to fundamental unresolved social issues, and a huge earthquake in 2015 exposed the fragility of Nepal's progress in terms of poverty reduction. Nevertheless, a new constitution was eventually agreed in 2015 and the election of local government bodies occurred in 2017.

Aid in Nepal

Nordic (i.e. at least Danish and Finnish) aid to Nepal dates back at least to the 1960s, when it was often linked to missionary work; British aid goes back even further because of military and colonial connections, and there was also some Cold War interest because of Nepal's location between India and China. Modern development cooperation began to accelerate in the 1980s (1983 for Finland, 1989 for Denmark), with Nepal becoming a favoured recipient of aid from many sources. Total annual official development assistance averaged about US$181 million in the 1970s, US$519 million in the 1980s, US$507 million in the 1990s, US$550 million in the 2000s, US$913 million in 2010–2017 and US$1,173 million in 2015–2017 (all in 2016 dollars; OECD/DAC, 2019). These figures reflect various kinds of interest in the country by different actors at different times, including the quest for political influence, humanitarian assistance during the armed struggle, facilitation of conflict resolution and peace building and willingness to invest in a stabilising and democratising country after the 'third breakthrough' in the late 2000s and into the 2010s (Table 5.1).

5.2   THE DENMARK–NEPAL RELATIONSHIP

Denmark opened its embassy in Kathmandu in 1991, and the resulting development partnership between Nepal and Denmark (represented by 'Danida' – parts of its Ministry for Foreign Affairs that are responsible for aid) was reviewed just before it ended some 25 years later (Caldecott et al., 2017). The aims of this final evaluation were threefold: to describe what Danida had achieved in 1991–2016; to assess the added value of a Danish emphasis on sustainable, rights-based development, including adaptations to changing contexts and the overall effectiveness of the partnership; and to identify lessons learned that might be useful in promoting sustainable, rights-based development elsewhere and in the future (Danida, 2016a).

Table 5.1 *Historical context of development in Nepal*

Milestones	Transformations
1st breakthrough	1951. The Rana regime (in place since 1846) is overthrown and parliamentary democracy and a constitutional monarchy are introduced.
Reversal of 1st breakthrough	1961. Direct royal rule and the semi-feudal *panchayat* system are imposed. India and China fight a border war in 1962 (coinciding with the Cuban Missile Crisis), the outcome (with India supported by the USSR) being to stabilise the frontier and leave Nepal between the two sides.
2nd breakthrough	1991. A people's movement (*Jana Andolan*) leads to the restoration of constitutional monarchy and parliamentary democracy. Cautious reforms of a corrupt and centralised bureaucratic state follow, while externally the Cold War ends. A Maoist insurrection forces the pace of change in 1997–2005, but its partisans are unable to take power due to a lack of decisive external intervention.
Reversal of 2nd breakthrough 3rd breakthrough	2005. Palace coup and direct royal rule. 2006. A new *Jana Andolan* forces the king to restore democracy. Comprehensive Peace Accord signed, ending the Maoist insurrection. Election of a Constituent Assembly, which abolishes the monarchy (2008). Agreement of a new national Constitution (2015). Elections for leadership of the 744 Local Bodies (2017).

Table 5.1 (*cont.*)

Milestones	Transformations
Potential sources of ongoing instability (see note on 'stability')	Parts of the Madheshi Terai are less aligned than the rest of the country with the 2015 constitutional settlement and remain a source of political instability. The whole country is vulnerable to climate change effects (fire, glacier lake outburst floods, rain-fed flooding, etc.).

Note on 'stability': In governance, stability is a condition in which there is peaceful compliance with the prevailing system by most of a country's inhabitants, regardless of its lawfulness, equitability or sustainability. To be stable a system must, for an extended period, meet the needs and satisfy the aspirations of enough of its members to head off organised, large-scale opposition. Since needs and aspirations change over time with internal cultural, socioeconomic and ecological conditions, and can be affected by external context, stability requires adaptation to these changes. Since regimes tend to become more inflexible over time, due to the continuity and 'tiring' of personnel and relationships, a system of governance must be reinvented from time to time, through decentralisation, elections, constitutional revisions or revolutions. Periods of stability and reinvention offer different kinds of opportunity for improvements in lawfulness, equitability or sustainability to be introduced by governments, activists, oppositions, NGOs and donors.
Source: Caldecott (2017b).

A country is a very complex system – in fact, a constellation of many complex ecological and social systems, all affecting each other – and over decades many things happen that can add up to gradual change, but can also occur in transformative outbursts. Gradual progress is seldom clearly directional and can be diverted one way or another by events, often apparently rather randomly, while transformations are seldom entirely irreversible in the face of reactions by those most affected by change or most fearful of it. This gives rise to a sense of 'breakthrough' and 'reversal' when viewed from the perspective offered by values such as a rights- and equity-oriented concept of development, as here with the Danida programme. The key milestones and transformations seen in Nepal are highlighted in Table 5.1.

A sustained partnership such as that between Nepal and Denmark is made up of many parts, including policy dialogue, commercial relationships, educational and research activities, tourism, journalism, friendships and diplomatic relations, which transcend the projects, programmes and other 'interventions' that are usually tracked bureaucratically and evaluated technically. But the interventions do provide a structure for studying the overall partnership, since each is related to some agreed policy priority, formally designed and budgeted, and leaves a paper trail that includes contracts, meeting notes, financial accounts, progress and completion reports and evaluation reports. The Danida interventions in Nepal since 1998 are summarised in Table 5.2. All improved the industrial, urban and agricultural sectors where they worked, and through the peace, rights and governance (PRG) and education interventions the lives of people that they touched, and they often had strong legacy effects. But one series of thematic actions in particular, on enhanced ecosystem management through community forest tenure and local empowerment, has important lessons for strengthening systems against climate chaos.

5.3   NARMSAP (1998–2005)

Catchment Management

Danida's investments in renewable natural resource management in the 1990s focused on soils and forests. Soils were at risk from cultivation and excessive grazing in hilly areas that are vulnerable to soil erosion, often made worse by competition between land users where land ownership was disputed. Solving this, it was thought, would need community groups to be supported by technical advice and laws, and close cooperation between government and local people. Responsibility

Table 5.2 *Danida interventions in Nepal, 1992–2018*

Energy Sector Assistance Programme (1999–2012) and *National Rural and Renewable Energy Programme* (2012–2017). These featured consistent investment in: capacity building at the Alternative Energy Promotion Center; in policy and strategy development to enhance access to renewable energy in rural areas; in systems to promote investment in renewable energy by working with banks and applying subsidies and credit; in promoting renewable energy and relevant capacity at local government levels; in aligning national and external partners to the national rural energy policy and institutional framework; and in the systems needed to sustain and guide the spread of renewable energy technologies and businesses.

Environment Sector Programme Support (1999–2005, cleaner production). This aimed: to promote awareness of cleaner production and occupational health and safety needs and opportunities at businesses in industrial districts by assessing them and inviting them to request a detailed design for cleaner production measures which would then be prepared in collaboration with their own staff; to establish a Cleaner Production Fund offering loans and grants to eligible businesses; and to enable further development of government policy to promote energy efficiency.

Environment Sector Programme Support (1999–2005, urban/industrial). This focused on urban and industrial environments, and is discussed further in Chapter 8.

Natural Resource Management Sector Assistance Programme (NARMSAP, 1998–2005). This focused on community-based forest management, and is discussed in the text.

Support to PRG (1999–2017). This programme contributed: to *peace building*, by supporting the demobilisation of former combatants, and participatory development of the 2015 Constitution; to *democracy*, by supporting voter registration, voter education and inclusive participation by all genders and groups, civil society strengthening and encouraging a free press; to *decentralisation*, by promoting autonomy for local bodies, fiscal decentralisation and performance-based management; to *inclusion*, through new legal protections for Dalits ('low-caste') and indigenous peoples, and measures to ensure their equal access to education and other opportunities; to *human rights*, by helping detainees and poor and

Table 5.2 (*cont.*)

---

marginalised people gain access to justice, abolishing bonded labour and supporting the National Human Rights Commission; and to *tax reform*, notably through the introduction of value-added tax as government's single largest and most reliable source of revenue.

Support to agriculture and production. This comprised two interventions widely separated in time, on dairy development in 1998–2001 and through the Unnati Inclusive Growth Programme, which promoted business innovations in the agriculture sector in 2014–2018.

Basic and Primary Education Programme (1992–2004), *Education for All Programme* (2004–2009), *Secondary Education Support Programme* (2003–2009) and *School Sector Reform Plan* (2009–2012). These contributed: to enhanced inclusion of girls, disabled people and pupils from disadvantaged communities; to development of the education system as a whole, including in planning, decision making, financial management, data collection and assessment of student learning (yielding reduced drop-out rates, better pupil–teacher ratios, increased numbers of trained teachers, reduced repetition rates, increased survival rates and greater commitment among district education officers to monitoring learning outcomes); and to improvements in knowledge management and evidence-based decision making, while building a culture of research and innovation across the education system.

Civil society support. Close partnerships with NGOs were essential to the PRG programmes, important in the renewable energy and urban-industrial environment programmes, and integral to the community forestry programme under NARMSAP and its predecessors.

Source: Caldecott et al. (2017).

from the government side for making this happen lay with the Department of Soil Conservation and Watershed Management of the Ministry of Forest and Soil Conservation (MoFSC). The Nepal-Denmark Watershed Management Project (1996–2001) supported this approach in nine districts. Meanwhile, forests were at risk from overharvesting of timber and fodder, fires and clearance for farming. The underlying challenge here was friction between the ownership of forest lands by government and their use by people to sustain their own livelihoods. Government foresters claimed the right to manage forests, while people needed to use them, often in competition with one another as well as with foresters.

Community Forestry

By the late 1970s the World Bank was predicting widespread desertification in Nepal, and began promoting the idea that forests could be safeguarded and improved if communities had the authority, knowledge and skills to control and use their own forests in their own interests. Thus, community forestry was conceived as a way to reduce deforestation and improve livelihoods, and was endorsed by the government in the late 1980s. Within five years the government had adopted the idea of Community Forest User Groups (CFUGs) and had started handing them patches of the national forest estate to manage. A succession of laws and policies in 1995–2005 provided a firm legal basis for this, making the process simpler and recognising CFUGs as being responsible for the management of community forests according to an approved Operational Plan. This was tested through the Nepal-Australia Community Forestry Project, and then replicated through the Danida-funded Community Forestry Training Project under a World Bank Hill Community Forestry Project (1989–1998), which established five Regional Forestry Training Centres. These efforts were concentrated in the 38 mid-hills districts where most forests were located.

'Improving' Trees

Danida's Tree Improvement Programme (TIP, 1992–1998) was based on the idea that natural forests were in decline, and were increasingly being handed over to local control as community forests, but communities lacked the knowledge and skills to manage them sustainably, and government foresters lacked the skills needed to advise them. At this point, the idea of cultivating trees (i.e. silviculture) joined that of maintaining their genetic diversity by preventing the loss of valuable species and cultivars. Both are necessary to sustainable outcomes but require a long-term strategy since trees grow and breed so slowly. Thus the TIP focused on establishing a seed supply and tree improvement programme for some of the most commonly used species, preparing a seed procurement policy for government and applying techniques of silviculture and selective breeding to chosen species. The TIP involved and interested the communities, since it worked with what were being increasingly seen as their own resources and their own lands, but few benefits were obvious after only five years and it was decided that a longer project was needed.

NARMSAP and Forest Recovery

The Danida projects of the 1990s, on catchment management, community forestry and tree improvement, offered three traditions of fieldwork, all with a strong community orientation but carried out jointly with the responsible ministry. NARMSAP was designed to continue and advance all three traditions. Previous investments by several donors, in alliance with a learning government back to the 1980s, made it possible to consolidate some serious gains, particularly in community forestry, in the 1990s and 2000s, when Nepal became a world leader in the subject (Agrawal and Ostrom, 2001).

Since then Nepal has stabilised its forest area at about 40 per cent nationally (MoFSC, 2014; Government of Nepal, 2016), but this varies among places and forest cover has been increasing mainly in areas subject to net out-migration (Oldekop et al., 2018), or else that are under community forest management. Reports are positive about aggregate effects of the latter, in predicting that 'community forestry is likely to lead to a continued trajectory toward reforestation and regrowth' (Southworth et al., 2012: 62), and in observing that 'Many formerly degraded forests have recovered well after a prolonged period under community management and now represent a significant natural resource with productive potential for yielding a range of forest products and other co-benefits' (MoFSC, 2017: 10).

Forest recovery in Nepal is mainly through natural regeneration, a process known to enhance biodiversity recovery, but the gains for carbon storage are less certain since communities typically continue to harvest the woody growth (Wilson et al., 2017; Luintel et al., 2018). Nevertheless, Nepal seems to be on the verge of joining the tiny number of tropical developing countries that have undergone a 'forest transition' in shifting from net deforestation to net reforestation (Meyfroidt and Lambin, 2011), if it has not done so already.

The 'Palace Coup' and the Integrated Environmental Programme

A Nepalese conservationist, who had spent 14 years in the Forestry Department and six with NARMSAP, estimated the overall distribution of credit for developing community forestry in Nepal to be about 20 per cent to the World Bank and Australia for pioneering it, 50 per cent to Denmark for implementing and replicating it, and 30 per cent to the UK and Switzerland for helping to consolidate it after Denmark's departure from the sector in 2005. This departure in itself requires explanation, since it was unexpected at the time and much regretted by many stakeholders.

By 2004, most parts of NARMSAP were rightly considered to be successful, as were most parts of the Energy Sector Assistance Programme and Environment Sector Programme Support. An Integrated Environmental Programme (IEP) to carry forward all of these plus renewable energy was therefore designed with the intention of implementing it in 2005–2010 (Danida, 2004a, 2004b). This had been agreed with the government, but was suspended by the Danish government as a political response to the 2005 Palace Coup in Nepal. Because people then moved on at the ministry in Copenhagen and embassy in Kathmandu, the importance of the IEP was forgotten and only renewable energy investment restarted when the aid programme was restored in 2006. Thus environmental and forestry cooperation was brought to a premature end.

Aborting the IEP in 2005 had two main consequences. It left unfulfilled the need to prevent the deterioration of environmental and particularly air quality in the Kathmandu Valley, which later became among the worst in the world (see Chapter 8). And it left unconsolidated the positions of the CFUGs and also the Community Development Groups, which had not yet taken off and (with hindsight) would have been very useful in putting the 2015 Constitution into effect. On the other hand, there are two senses in which NARMSAP left very strong and continuing legacies. The first is that the CFUGs had become so well established by 2005 that the momentum of the community forestry programme allowed it to continue without external support in the same form. The second is that the CFUGs had the effect of keeping the idea and practices of participatory democracy alive throughout the suspension of democracy and the civil emergency in 1997–2006. This gave Danida's forestry role a key impact on governance, which would prove vital to the constitutional journey on which Nepal was engaged throughout this period, and which continues today.

5.4   NEPAL'S OWN 'COUNTRY PROGRAMME'

Escape from Fatalism

Nepal in the 1980s was described by the Nepalese anthropologist Dor Bahadur Bista (1991) as a country affected by centuries of feudal rule and influence from its neighbours India and China, and historically from the British Raj and Tibet. In this view, Nepal's peoples were divided into many distinctive and competing groups, and the prevailing attitude was one of fatalism – the feeling that destiny is determined by birth, class, caste, gender, disability or some other accidental, natural, social or supernatural factor, over which no control is possible. This tended to inhibit the sense of agency among the people of Nepal, although they had long been using a number of strategies for making progress despite it, including mutual-support networks (*afno mannche*), sycophancy (*chakari*) and emigration.

The developmental weakness and chronic poverty of Nepal was attributed to divisions among its peoples, the partial and divisive success of mutual-support networks, sycophancy and emigration and the pervasive atmosphere of fatalism. But cracks were already appearing in the traditional patterns of Nepalese society and governance (Whelpton, 2005; Shah, 2008; Jha, 2014). There had been the first phase of parliamentary democracy in 1951–1960, which was suppressed in 1961–1979 but inspired agitation for change in 1979–1988 that culminated in the 'People's Movement' (*Jana Andolan*) of 1989–1991. Although His Majesty's Government remained, Nepal then became a parliamentary democracy and held elections, and began a cautious programme of reforms to defuse opposition to centralised and what was widely perceived as corrupt rule.

Building Democracy

Danida and others began funding projects to promote awareness of democratic practices, and the skills and rights needed for democratic participation. The global context at the time included the end of the Soviet bloc, the Soviet Union and the Cold War, with Western donors being drawn by the opportunity to encourage the emergence of a new and democratic Nepal. Subsequent events in Nepal can be interpreted as milestones on a collective path away from fatalism. These milestones included the 1989–1991 restoration of democracy, which was repeated in 2006 after the 1996–2006 insurrection. They continued thereafter through the 2008 abolition of the monarchy and election of a Constituent Assembly, the 2015 agreement of a new national Constitution and the 2017 elections for leadership of the 744 Local Bodies that are responsible for significant budgets under the supervision of local people.

Local people are keenly interested in how the money is spent, and this attention is facilitated by local NGOs and community-based organisations, including 19,316 CFUGs, a free press, mobile telephones and an internet that exposes local transactions to local view and feeds global experiences into local minds, all of which help deter abuses of power. Accountable local power inevitably undermines fatalism, since it allows people to do meaningful things on their own behalf. Thus, while the Danish programme was going on in 1991–2016, the Nepalese people were undertaking their own agenda that transformed the country through growth and change. All Danida's interventions encouraged this outcome, by supporting elections, decentralisation, transparency, inclusion, equity, literacy and various measures to promote the habits and mechanisms of participation, and introducing new ideas and ways of doing things, often through exposure to global and Danish experience.

Moreover, many of these interventions amplified one another's effects, with decentralisation and local empowerment over schools, forests and renewable energy all contributing to and benefiting from a trajectory towards decentralised, accountable, participatory, lawful and rights-based governance, and all being supported simultaneously by Danida and Nepalese interest groups. Thus, all the Danida interventions helped each other and contributed to the overall process of change by addressing the issues of who owns what, who decides what, how people and groups influence one another and how disputes among them are resolved. This whole sequence joins Danida's efforts together into one consistent process that was in dialogue with changes occurring over the same period within Nepalese society.

Symbiosis and Independence

The essential symbiosis between Danida's actions and Nepal's development was highlighted by a Nepalese interviewee who had spent a dozen years with the World Bank and Danida forestry programmes, right up to the morning of the Palace Coup when he observed the IEP being lost. He described the relationship as Denmark giving Nepal 'milk' to sustain its growing body over time, which he contrasted with the stunting effects of 'alcohol' from quick-fix stimuli. Our study concluded that Danida had helped, and rarely if ever stood in the way of, the Nepalese project of freeing itself from fatalism. The mid-hills, those 38 districts where many of Danida efforts were concentrated (particularly on community forestry), and the Kathmandu Valley, are now strongly aligned with the 2015 constitutional settlemcnt.

Difficulties remain, among them the 'potential sources of ongoing instability' noted in Table 5.1 – including the lesser align-ment of parts of the country with the 2015 constitutional settlement, and the fact that the whole country is vulnerable to climate change effects. To these can be added economic effects of interrupted remit-tance payments from Nepalese workers abroad, which in 2017 were anticipated from the blockade of Qatar by some of its neighbours, and in 2020 from the Covid-19 pandemic. Even so, Nepal was in a much stronger position to cope with *any* challenge in 2017 than it was in 1990, with a viable constitution, widespread progress on both repre-sentative and participatory democracy, regrowing forests and a less fatalistic, more confident and better-educated population that was more aware of and better able to defend its rights than at any time in its history.

5.5 LESSONS FROM NEPAL

It is in the changes highlighted in this chapter that lessons can be found for strengthening social and ecological systems against climate chaos. These can be spelled out in the following terms. First, that CBRM is effective in reducing poverty, increasing equity, maintaining ecosystem goods and services and promoting climate change adapta-tion. The only condition is that it requires close attention both to local cultural and ecological conditions and to the empowerment of locally accountable groups. And second, that renewable natural resources are diverse living systems able to yield multiple goods and services, including cultural, ecosystem and genetic resource services that are subtle, non-obvious and require sustained attention and care to maintain. If it is possible to achieve a unity of purpose between these two truths within one programme, then much of great benefit will likely be achieved.

Human history can be seen as made up of many separate trajectories, each involving one cultural system in one environment as it evolves between major discontinuities – conquests, collapses, forced migrations, etc. In each system there is a sense of direction, even if the process is often slow, intermittent and prone to course changes and temporary reversals. To call Nepal one cultural system is not entirely plausible, but is no less so than for many other countries, and in any case the very idea of a system being a 'country' recognised by other countries is a potent stabiliser of identity over time. Thus, Nepal has existed in a coherent, bounded way since the Rana regime, and it is this system that has evolved directionally, from feudalism and fatalism through various unstable experiments with democracy and warlordism, to arrive at a participatory form of locally accountable governance with a strongly environmental theme. This outcome may be reversed through invasion or foreign-backed uprising, or overwhelmed by natural disasters, or betrayed by corrupt or incompetent leaders, but it is hard to see how it can be *improved* other than by being consolidated through practice and enriched through knowledge, ideas and fair trade.

That being the case, while all the foreign sources of development assistance since at least the 1980s can be assumed to have contributed something, Danida is singled out for targeting the key issue of community and the ecosystems that sustain communities and make them permanent. In other contexts those ecosystems could be farmlands or fishing areas, but in the mid-hills of Nepal they were forests. Through a sequence of projects, Danida built on and then eclipsed other donors' work with the ministry and with forest-dependent communities, taking on a larger and larger share of influence and training, helping the ministry steer changes to policies and laws through the system to enable CFUGs to be established and to obtain forests and a clear mandate to manage them. By these means, to use the vocabulary from Chapter 4, within the social and ecological systems of the mid-hills the social and ecological homeostats were put in place and put down roots, and the foundations for both were preserved.

By Danida staying engaged during the troubles of 1996–2005, momentum was maintained and a degree of participatory democracy sustained, which was later able to influence the emergence of a constitution that recognised the things that were important to social and ecological sustainability in the mid-hills. Perhaps all of this could have been achieved faster if the whole process had been more deliberate, and the abandonment of the IEP in 2005 did unnecessary damage. Regardless of what survived from Danida's previous efforts (and clearly much did), another few years of consolidation investment by Danida would have been extremely helpful. The lack of follow-through may be a sign that Danida failed to recognise that ecosystem tenure security and community empowerment are absolutely central to the strengthening of Nepal's systems against climate chaos. If adaptation had been a key priority in 2005, perhaps this failure would not have happened. But the take-home message for now is that social and ecological strengthening can be done, and done quickly, if it is done just like it was by Danida and Nepal, but without wasting too much time and making too many false moves.

6 Community Land Titling in Bolivia

This chapter focuses on community empowerment, ecosystem tenure and environmental education as the essence of local system strengthening, as already explored in the mid-hills of Nepal but in very different circumstances. And in Bolivia, aside from the proximity of mountains, the context is very different indeed. The social system is rooted in multiple indigenous Amazonian-Andean cultures, Spanish colonial and post-independence governance, foreign interference and the exploitation of natural resources, both non-renewable (minerals and hydrocarbons) and renewable (including coca, *Erythroxylum* spp., the source of cocaine). All the rivalries to be expected in such a diverse system are clearly visible in Bolivia.

As with Nepal, the investigation focused on a 25-year engagement between Danida and Bolivia, over much the same period, thus to some extent controlling for differences that might otherwise have existed between the policies, practices and priorities of different aid institutions. The historical context of the Danish–Bolivian partnership is summarised in Table 6.1, drawing attention to several breakthroughs and reversals from the point of view of Danish rights and governance values. They would presumably look different to a silver mine owner, a narcotics trafficker or a US State Department official. From 1952, Bolivia had experienced revolutions, coups, dictatorships, social and economic instability and repeated foreign interventions, but by 1992 it was seen as stable enough for development cooperation to bear fruit.

Although the Danish–Bolivian aid relationship dated back to the 1970s, the pace increased in 1994 when Bolivia was chosen as a

Table 6.1 *Historical context of Danida's work in Bolivia*

Milestones	Transformations
1st breakthrough	1952. The traditional system (in place since at least 1920) dominated by tin mining and other interests is overthrown by the Revolutionary Nationalist Movement which introduces universal suffrage and carries out sweeping land reform, promotes rural education and nationalises the largest tin mines.
Reversal of 1st breakthrough	1964. A military coup is followed by a succession of US-backed but unstable military dictatorships.
2nd breakthrough	1982. Popular movements lead to the restoration of democracy. Social and human rights reforms follow.
Reversal of 2nd breakthrough	1985. Tin prices collapse (Mallory, 1990), triggering inflation and unemployment. The country becomes committed to neoliberal economic policies. Privatisations and foreign investment programmes guided by the World Bank and International Monetary Fund affect water and other public services. Aggressive US-backed coca eradication programmes alienate indigenous peoples. A decentralisation process is initiated in 1994, including 20 per cent tax retention by the municipalities. Mass protests result in temporary martial law in 2000 but achieve the reversal of some privatisations. National governance is destabilised thereafter through indecisive elections and protests centred on exploitation of gas fields and associated issues of land rights and oppression of indigenous peoples in the historical context of silver and tin mining, and continuing resistance to coca eradication, water issues, corruption and military violence.

Table 6.1 (*cont.*)

Milestones	Transformations
3rd breakthrough	2006. Indigenous leader Evo Morales and the Movement Toward Socialism (MAS) party are elected to power. A process to develop a new constitution begins. In 2008, rightist groups try to establish autonomy in wealthier regions through unilateral referendums (declared illegal), a recall referendum (defeated) and an attempted coup (thwarted). The new Constitution is approved by referendum in January 2009.
Reversal of 3rd breakthrough	2010. The MAS-Morales government continues and the 2009 Constitution is implemented, creating a plurinational state and requiring new structural legislation against political turbulence generated by various interest groups. Under pressure, 'increasingly, the Government of Bolivia appeared to be adopting policies that ran in opposition to its professed commitments to the environment and indigenous values (i.e. caring for the Pacha Mama/Mother Earth). These centred on the strategy to promote extractive industries, in particular hydrocarbons. . . . By 2010, there were clear signs that the land reform policies pursued during Morales's first term were being wound down' (Schwensen et al., 2017: 8–9; also McCormick, 2017). There were also water crises generated by catchment damage, progressive drought and the 2014–2016 El Niño.
Potential sources of ongoing instability	Unresolved issues of locally accountable governance, decentralisation, ethnic competition for resources (including fossil fuels) and the distribution of costs and

Table 6.1 (*cont.*)

Milestones	Transformations
	benefits from resource exploitation. Increasing ecosystem damage linked to fossil fuel exploitation and mining. Vulnerability to drought, glacier melt and other climate change effects. Erratic US relations, economic problems in Venezuela and a compromised model of Bolivarian solidarity.

Sources: Caldecott (2017b); Schwensen et al. (2017); others as cited.

Danida priority country (see Table 6.1). Denmark opened its embassy in La Paz in 1995, and the first cooperation strategy was approved in 1997 (Danida, 2016b). Danish support in 1994–2001 focused on agriculture, environment, indigenous peoples, popular participation and decentralisation, institutional reforms and local grants, and positioned Danida especially well to support indigenous peoples in the context of the government's decentralisation efforts at the time (Schwensen et al., 2017). These themes were further developed thereafter, leading to the breakthroughs discussed in Sections 6.2 and 6.3.

The Danida programme started uneventfully, with Environmental Sector Programme Support that sought rather unsuccessfully to improve industrial and urban pollution and promote cleaner production and renewable energy (Table 6.2). This first phase in 1999–2005 did, however, lay the groundwork for longer-term engagement on these topics after 2006, and also in this later period investment in water catchment planning, public sector reform, agriculture and education. These often had useful effects but struggled to obtain a sustainable impact, much less a transformative one, against deeply polarised governance systems that also tended to exclude indigenous peoples. This last factor was associated with a Danish

Table 6.2 *Danida interventions in Bolivia, 1998–2018*

Environmental Sector Programme Support Phases I and II (1999–2013),
Sustainable Natural Resources Management and Climate Change
(2014–2018). Between them, these programmes focused on the following
three themes:

- *Cleaner production and energy efficiency*: Danida support yielded good
  results on cleaner and more efficient production technologies in certain
  industries, but they proved unsustainable as the businesses concerned
  refused to invest.
- *Urban/industrial environment*: Danida support: (1) helped local
  environmental management, which made a good start but was fading by
  2017; (2) showed that civil society networks were effective in linking the
  central, regional and local levels, raising public awareness and boosting
  the environmental agenda through targeted research and education, but
  this capacity had been lost by 2017 due to a narrowing of the space for
  civil society operations; and (3) in the mining sector led to social and
  environmental improvements, and demonstrated an institutional model
  that linked central-level interventions to regional and local levels, but
  the capacity was evaporating in 2017 due to a lack of interest.
- *Water and climate change adaptation*: (1) The National Catchment
  Plan was initially supported, but Danida withdrew to focus on regional
  and local levels and the plan was being phased out by 2017 due to lack of
  donor funding and limited ownership by government; (2) the second
  phase included support to environmental research on integrated water
  resource management and to civil society through the *Liga de Defensa
  del Medio Ambiente* (LIDEMA) to raise awareness and capacities on
  environmental issues, but LIDEMA fell apart after Danida left.

Environmental Sector Programme Support Phase III (2014–2018). This
focused on renewable energy at a time when it was moving high up the
political agenda, with a commitment by government to ensure access to
energy for all Bolivians by 2015 and that 25 per cent of this energy should
come from renewable sources, and in January 2017 the establishment of a
new Ministry for Electricity and Renewable Energy. Danida provided
technical support for these moves and co-funded (with the *Empresa
Nacionál de Electricidad*) feasibility studies for potential wind, biomass,
solar and geothermal projects that were included in national development
plans for 2016–2020.

Table 6.2 (*cont.*)

Pilot project with the Sub-Secretary of Ethnic Affairs (1995–1997), *Sector Programme for Indigenous Peoples, Decentralisation and Popular Participation* (1998–2004), *bridging support* (2004), *Support for the Rights of Indigenous Peoples* (2005–2010). These programmes are discussed in the text.

Fiscal, judicial and administrative reforms (*Pro-Reforma* 2007–2013, *Pro-Justicia* 2009–2014 and Promotion of Exercise of Rights and Access to Justice 2014–2018). These programmes promoted: (1) the reform of state institutions (e.g. improvements in the operation of the national customs, tax and roads agencies); (2) citizen registration and ID cards; and (3) popular participation and decentralisation reforms (e.g. devolution of budgets, more inclusion of indigenous people in local government).

Agriculture and production (*Programa de Apoyo al Sector Agrícola* Phase I 1999–2005 and Phase II 2005–2011, Sector Support for Agriculture and Production 2010–2013 and Inclusive and Sustainable Economic Growth 2014–2018). These programmes supported: (1) the ministry of agriculture and the Potosí and Chuquisaca regions through technical training, rural enterprises and dairy development; (2) the development of national policies and programmes and improved access to financing small and medium enterprises and rural organisations; and (3) small-scale farming and a Competitive Fund for Innovation.

Education Sector Programme Phases I and II (2005–2015). This contributed to the Education Sector Fund, and provided support to the Ministry of Education.

Civil society. All the Danida programmes took a dual approach with government and NGOs as complementary partners with different strengths. This was particularly useful in the areas of justice reform and indigenous land titling actions, and on social, gender and environmental issues related to mining operations.

Source: Schwensen et al. (2017).

intervention of exceptional significance, since it ensured Danish interest in improving the circumstances of indigenous people right from the start of the country partnership.

6.2   INDIGENOUS PEOPLES' LANDS, RIGHTS AND LANGUAGES

Peoples are 'indigenous' if they identify themselves as such, based on what they perceive as cultural distinctiveness and prior territorial occupancy relative to a population with its own different and often dominant culture. Denmark's experience of governing indigenous peoples is small by the standards of European countries with past empires. Outside the North Atlantic, it involved only trading facilities in what is now Ghana (whence they were expelled by the Akan subgroup of the indigenous Ashanti people), and some plantations in the Caribbean (later sold to the USA) and trading posts in Asia (later sold to Britain, along with part of the area that would become Ghana). The territory of Greenland is a different story, since it is inhabited by indigenous 'Inuit' peoples (an umbrella term covering multiple distinct cultural groups indigenous to northern Canada, Russia and Alaska, as well as Greenland). Here a discontinuous Danish presence started in 1721, later being stabilised with Greenland becoming first a colony and then in 1953 an administrative county of Denmark. The territory was granted home rule in 1979 and self-rule (apart from foreign affairs and defence) in 2009.

Danish interest in indigenous peoples was at least partly shaped by its felt responsibilities for Greenland's 'Inuit' population. Denmark's reputation for being 'on the side' of indigenous peoples was summarised by an indigenous peoples' advocacy group, thus:

> The Kingdom of Denmark has long been a champion of human rights, leading the way for critical policy implementation in the international arena. Its recommendations have yielded some of the most influential conferences and declarations in the realm of human rights. Empowering both domestic and international populations,

Denmark has aided in the improvement of lives across the globe. Especially in the case of Indigenous Peoples, Denmark has galvanized the creation of monumental policy that has allowed Indigenous Peoples to engage actively with mechanisms to protect and to control their land, resources, and culture. (Cultural Survival, 2015: 2)

Pilot Project and the First Sector Programme, 1995–2004

In Bolivia, Denmark's only priority country in South America, these interests and preferences were first expressed through a pilot project with the Sub-Secretary of Ethnic Affairs in 1995–1997, which focused on helping indigenous people engage with discussions and planning around decentralisation, and providing a legal advice service for indigenous people. This led to a Sector Programme for Indigenous Peoples, Decentralisation and Popular Participation in 1998–2004 (Schwensen et al., 2017). In the context of a government policy of national decentralisation, this focused on raising the profile of indigenous peoples' concerns for land tenure, human rights and language freedoms. It helped to build the capacity and confidence of indigenous civil society organisations in protecting and advancing their rights, in establishing bilingual intercultural education in the Bolivian lowlands, in facilitating the 'titling and ordering' of Indigenous Community Lands in the Oriente, Chaco and Amazon regions and in providing legal advice to peasant groups in Potosí.

The Second Sector Programme, 2005–2010

According to the International Work Group for Indigenous Affairs (IWGIA), the emphasis on community land tenure was driven by an appreciation that it is

> a fundamental tool to fight poverty among indigenous peoples and communities since having a territory assures them access to the resources they need for satisfying their material, spiritual and cultural needs, taking into account that the true poverty indicator for indigenous peoples is associated with the deterioration,

reduction, privatization or loss of their territory. The TCO [*Tierras Comunitarias de Origen*, changed in 2009 to *Territorios Indígenas Originario Campesinos*] titling also allows indigenous peoples and communities to exert a greater control on the protection of the environment, through observing and reporting the impacts that may be caused by the exploitation of natural resources and, in the case of the Highlands, allows them to fight against soil and water contamination. (Parellada et al., 2010)

This approach proved so effective, and was so clearly needed and welcomed by the beneficiaries, that Danida 'bridging support' was provided in 2004 while a new programme of Support for the Rights of Indigenous Peoples was designed and agreed with government for 2005–2010 (Parellada et al., 2010). The timing was consistent with the release of a new *Strategy for Danish Support to Indigenous Peoples*, the objective of which was 'to strengthen the right of indigenous peoples to control their own development paths and to determine matters regarding their own economic, social, political and cultural situation' (Danida, 2004c: 8). The new programme was designed with three components to address a number of key challenges faced by indigenous peoples, especially insecurity of land rights arising from uncertainty over the recognition of their land claims by the state, and conflict between those claims and the administrative divisions of the country, as well as a general disconnect between government development strategies and their own needs and values.

The 'Mainstreaming of Indigenous Rights' Component

This aimed to incorporate indigenous rights in the laws and procedures of the state, to build public awareness and support for the rights of indigenous peoples, and to build appropriate capacity within the Ministry of Indigenous Affairs while reaching out to engage the ministries of health, justice, defence, labour, mining and hydrocarbons, education and rural development. It is always hard to point to specific results in this kind of component, but in this case they included:

- wording in the new Bolivian Constitution to incorporate community justice and indigenous autonomies, and more generally an Andean cultural approach called *suma qamaña* in Aymara or *buen vivir* in Spanish, meaning 'living well' in social balance and harmony with nature;
- the approval of 'supreme decrees' that mandated consultation with and supervision by indigenous people over mineral exploitation on their lands, and that prohibited the use of forced labour (which had particularly affected the Guaraní people); and
- a series of new ministerial policies to incorporate indigenous rights in the health, labour, education, rural development and justice sectors.

The 'Regulation and Titling of Indigenous Community Lands' Component

Earlier efforts to conclude TCO claims had run into opposition, including from 'a marked racist tendency that considers indigenous peoples as an inferior population, an object of economic and political exploitation, subordinated to the patrones, who historically have been the owners of the land' (Parellada et al., 2010: 37). This component therefore aimed to conclude the 61 TCO claims already begun and add others, to facilitate the certification of ethnic identity by the Ministry of Indigenous Affairs, to promote the role of organised civil society activism in resolving conflict and to facilitate policy development. By continuing and extending the approach of the previous programme, the component added more than 11 million hectares in 135 TCOs. Of almost 17 million hectares for which TCOs had been completed in 1997–2009 (with more expected to be completed in 2010), '67 percent were supported by Danida' (Parellada et al., 2010: 74). The map of Bolivia as viewed through an 'indigenous lens' was therefore changed dramatically in this period.

The 'Indigenous Territorial Management' Component

This component aimed to use a 'learning by doing' approach to build the capacity of indigenous communities to manage their natural resources productively and sustainably. This involved training to facilitate preparation of a PGTI (*Plan de Gestión Territorial Indígena,* or Indigenous Territorial Management Plan), for each community, based on participatory resource assessment, mapping and agreed management rules. It also aimed to support diverse community projects, while agreements were negotiated on health, education, drinking water, road improvements and electricity, and on networking communities so that they could learn from one another. In the process, existing laws and regulations that obstructed the aims had to be amended through negotiation with administrative institutions in various sectors and different levels of society.

For outreach purposes, territorial management was visualised in an ingenious drawing of an ant, in which the head represents the indigenous organisation, the thorax the territory and culture, the segments of the abdomen the lines of action (education, health, production, resources, territorial control and basic services) and the legs the six active principles of administration, networking, plans, projects, rules and training (Patiño, 2009). In the lowlands a 'school of projects' was set up where people were trained and potential projects explored, and six full PGTIs formulated along with 125 communal and intercommunal plans. In the highlands, less well-organised indigenous groups and more conflicts with mining interests made progress harder and slower, but it was achieved through an emphasis on capacity building among indigenous authorities, supplemented by training and apprenticeships, the use of geographic information systems, community radio and co-venture relationships with the municipalities and private companies before projects could be formulated at the TCOs with a deliberate emphasis on women's participation.

6.3   NET RESULTS AND POLITICAL TRANSITIONS

The Rights of Pacha Mama: 'Mother Earth'

The Danida efforts contributed strongly to outcomes that are summarised by Parellada et al. (2010): indigenous rights were recognised;

regulation and titling of the TCOs changed the map of Bolivia; the Bolivian state was transformed; and the new (or ancient Andean) concept of 'living well' became the basis for a new development model. These things may never have happened if the investments had not occurred in the context of a major political transition in Bolivian society that had been building up for many years. The complexities of this are explained by Parellada et al. (2010), but the result was that newly mobilised indigenous and peasant populations elected to power the MAS government of Evo Morales in 2006. Then, for a few years, and despite multiple attempts by conservative factions to derail or reverse the process, there was enthusiastic government support for a social movement among indigenous and poor people for land reform, community tenure and management of renewable natural resources, and participation in the management of protected areas.

Danida was well placed to encourage and enable this through the environmental sector and indigenous peoples' rights programmes, and the result was a dramatic increase in communal land titling. Suddenly, after centuries of dispossession and marginalisation, indigenous communities found themselves with legal title to a large share of Bolivian national territory. Meanwhile, a new constitution to guarantee recognition of Bolivia as a 'plurinational' state (i.e. with indigenous peoples as partner nations) was drawn up and ratified, and grand policies and laws were devised to safeguard the rights of *Pacha Mama* – 'Mother Earth' and all her ecosystems, species, spirits, peoples, waters and landscapes.

The Influence of 'Bolivarianism'

These events in Bolivia after 2006 were connected with and influenced by Bolivarianism, a broader political movement in Latin America and the Caribbean (Caldecott, van Sluijs et al., 2012). This is a form of social democratic nationalism named for Simón Bolívar (1783–1830), the Venezuelan general who led the struggle for independence from Spain in much of South America. The Bolivarian Hugo Chávez was elected president of Venezuela in 1998 and convened a process to rewrite the national constitution, one outcome of which was that Venezuela was renamed the Bolivarian Republic of Venezuela. The Bolivarian Alliance for the Peoples of Our America (ALBA) was then formed through the Cuba–Venezuela Agreement in 2004, which envisioned the exchange of medical and educational resources and petroleum, and the more general People's Trade Agreement in 2006. Bolivia joined the trade agreement and therefore ALBA in 2006, Nicaragua did so in 2007, Dominica in 2008 and Antigua and Barbuda, Ecuador and Saint Vincent and the Grenadines all followed in 2009.

The existence of ALBA changed the international context for each of the participating countries, some of the governments of which (notably Nicaragua's, since the 'Triumph of the [Sandinista] Revolution' in 1979) had long seen themselves as struggling for socialism alone apart from Cuba in a hemisphere controlled by a hostile USA (Walker and Wade, 2011). But ALBA offered a network of like-minded countries, with at least Venezuela possessing significant wealth and being willing to spend it on practical forms of cooperation. Those years, up to the death of Chávez in March 2013, were marked by a heady combination of what was widely seen as freedom, the righting of deep structural wrongs and a general sense of economic and political optimism. The existence of ALBA had also offered hope and encouragement even to territorial rebellions, such as the Zapatistas in Chiapas and nearby states of Mexico.

In short, the restoration of resource ownership to indigenous peoples, more or less allied to the cause of more equitable distribution of land in favour of the poor, and the quest for freer and better governance that might be expected to go with it, has been the defining issue in much of Central and South America for centuries. Denmark had been encouraging solutions in this direction in many places for many years, and happened in Bolivia to be in an excellent position to respond to a sudden validation of the whole approach by an elected government. This was a government, moreover, whose vision of the state was as a federation of first nations, with each enjoying secure rights over their own land resources. As the minister of autonomies of the plurinational state of Bolivia put it in 2010, the titling of Indigenous Community Lands

> represents the breakup of the power relations imposed by big landowners and local power groups, besides projecting indigenous collective rights as the materialization of the Plurinational State's horizon. In turn, the possibility of setting up the TCOs as territorial entities represents the territorial reorganization of the State on a plurinational criterion. Finally, the territorial management processes form part of the new development model that articulates the state economy, the private initiative and the traditional communal economic system.
>
> *(Bonifaz, 2010: 11)*

What Happened Next

Bolivia is larger than most countries and as complex as any, and transformative social and political changes, such as those that occurred there in 2004–2008, are inevitably opposed and resisted overtly or covertly by reactionary or conservative groups, as well as being undermined eventually by exhaustion, disillusionment or corruption among the ageing regime that brought about change in the first place. Reversals may be anywhere between very partial and almost complete. Assuming that enhanced land tenure and the package of environmental education, mapping and planning that went with it strengthened local social and ecological systems and the links between them, and continued to be supported by the indigenous peoples involved, the question is whether the TCO rights obtained by indigenous peoples throughout the country survived the pushbacks and back-sliding that evidently occurred after 2010. Yet half a decade later the government of Bolivia retained its eloquence on past injustices and the bright future enabled by *buen vivir* (Box 6.1), so who can tell where this particular national insight will lead?

BOX 6.1   **Mother Earth, living well and the rights of peoples**

The structural cause that has triggered the climate crisis is the failed capitalist system. The capitalist system promotes consumerism, warmongering and commercialism, causing the destruction of Mother Earth and humanity. The capitalist system is a system of death. Hence, capitalism is leading humanity towards a horizon of destruction that sentences nature and life itself to death. In this regard, for a lasting solution to the climate crisis we must destroy capitalism.

The capitalist system seeks profit without limits, strengthens the divorce between human beings and nature; establishing a logic of domination of men against nature and among human beings, transforming water, earth, the environment, the human genome, ancestral cultures, biodiversity, justice and ethics into goods. In this regard, the economic system of capitalism privatizes the common good, commodifies life, exploits human beings, plunders natural resources and destroys the material and spiritual wealth of the people.

Thus, Bolivia presents its intended contribution [to the goals of the UNFCCC] consistent with its vision of holistic development, according to the provisions of the State Constitution, Law No. 071 of The Rights of Mother Earth and Law N° 300 of Mother Earth and Integral Development to Live Well, guided by the 2025 Patriotic Bicentennial Agenda and its 13 pillars, as well as national plans for medium and long-term.

Bolivia understands Living Well as the civilizational and cultural horizon alternative to capitalism, linked to a holistic and comprehensive vision that prioritizes the scope of holistic development in harmony with nature and as structural solution to the global climate crisis. Living Well is expressed in the complementarity of the rights of peoples to live free of poverty and the full realization of economic, social and cultural rights and the Rights of Mother Earth, which integrates the indivisible community of all systems life and living, interrelated, interdependent and complementary beings who share a common destiny.

Source: Government of Bolivia (2016: 1–2).

What Happened Later

When Danida ended its direct support for the TCO process in Bolivia, it recruited through a framework agreement Verdens Skove (meaning 'Forests of the World'), a Danish NGO active in Central America since 1992 and in Bolivia since 2008, to continue related work. This agreement lasted until 2015, when Denmark's Civil Society in Development Fund (CISU) took over support for Verdens Skove, whose continuing work on indigenous community land rights in Bolivia in 2019 was summarised in a report to CISU:

> The Movima people have gained their territorial rights to the Chimanes forests [in Amazonian Beni province], which were previously under private forest concessions given by the state. There has been made a proposal for a normative guideline for sustainable forest management in Monte Verde, Santa Monica. It has not yet been implemented due to a debate between the two wings of the National Forestry Authority, and affected by the election process culminating with public protest in Bolivia in October 2019. Indigenous peoples rights have in general been strengthened.
>
> *(Verdens Skove, 2020: 4)*

This is a reminder of how complex land development issues in the Bolivian Amazon continue to be. This is an area where 'Highways are projected to cut through indigenous territories and protected areas, raising concerns about migration, deforestation and contamination. Moreover, large oil and gas deposits beneath the forests where the Andes descend into the Amazon Basin cause general uncertainty about the future of indigenous territories and protected areas there' (Christoffersen, 2018: 1). Nevertheless, recent satellite images and maps reveal that those parts of the Bolivian Amazon where indigenous territories received community land titles with Danish help in 1995–2010, whether through Danida programmes or partnerships with indigenous peoples' associations linked to IWGIA and Verdens Skove and funded by Denmark, have now often become green islands in a sea of new soya plantations (Theilade, 2020).

This and other evidence from Peru and Brazil show that indigenous territories may well be the *only* effective governance mechanism capable of withstanding deforestation pressures under modern conditions in the Amazon Basin (Ida Theilade, personal communication, November 2020). This amplifies earlier understandings that such territories are at least as effective as national parks at protecting biodiversity and natural forests (e.g. Nepstad et al., 2006; Porter-Bolland et al., 2012; Schleicher et al., 2017), and that local communities can mount very effective forest monitoring and protection activities with very modest levels of external support (e.g. Danielsen et al., 2013; Brofeldt et al., 2014; Brofeldt et al., 2018; Theilade et al., 2021; Box 2.5).

According to Thomas Nielsen (personal communication, May 2020), on the basis of his recent involvement with a Wildlife Conservation Society Bolivia indigenous forest management project in Northern La Paz, 'community land titles are still important for the indigenous communities living in these territories in this part of Bolivia'. He also stressed the complex and 'sometimes conflictive relationship between mostly Quechua or Aymara speaking highland indigenous groups who migrated to the lowlands, and the lowland indigenous groups themselves', which presumably offers opportunities for the seeds of political division to be sown between them. He concluded that,

> despite the importance of the indigenous land titling in Bolivia, I don't think it is possible to consider it as an irreversible change in the territorial security of the peoples concerned. The country is more than ever a battleground between different groups or classes based on social, ethnic and/or economic interests. The TIPNIS [*Territorio Indígena y Parque Nacional Isiboro Sécure*, or Isiboro Secure Indigenous Territory and National Park] conflict over the construction of a highway through the indigenous territory is one example, and in the indigenous territory Pilon Lajas where I have

been working the last three years, the main conflict has been between the local T'simane-Mosetene people (owners of the territory) and the central government, promoted by a major hydroelectric project in the territory (later abandoned).

Meanwhile, Carmen Barragan, who was involved with the Danida programme itself, added the observation (personal communication, May 2020) that,

> there is no simple answer [to the question of whether community land titles offer an irreversible change in the territorial security of the peoples concerned], because of the multiple factors involved and the different realities of the indigenous peoples territories in Bolivia. The land titling was a social and political *process* for the organisations and indigenous people concerned. A decade later, there are territories that have achieved their autonomy, but mostly those that lay within one municipality, rather than those which overlapped more than one because of the overlapping institutions and greater complexity involved. There are indigenous territories that were able to continue implementing their Territorial Management Plan, like the Tacanas, but others that faced too much competition from economic and political interests in their territories.

And finally, it must be noted that a complex and contested series of events surrounded Evo Morales' attempts to be re-elected as president over several years, culminating in disputed elections in October 2019, an intervention by the Bolivian military in November and his exile. A caretaker government headed by Jeanine Áñez of the opposition alliance was installed to arrange fresh elections. These were delayed by the Covid-19 pandemic, and while campaigning Áñez was reported to have made racist remarks about indigenous peoples (Weisbrot, 2020), contributing to a spectacular victory by MAS in October 2020 and the election of the former MAS finance minister Luis Arce as president. Morales himself returned from exile in November, but the new government made clear that he would have no role in government (Phillips, 2020). In any case, the reinvigoration of MAS is another surprise from this extraordinary country.

6.4   LESSONS FROM BOLIVIA

As was the case in Nepal (Chapter 5) and Zanzibar (Chapter 7), Bolivian circumstances had matured as a result of other factors to the point where rural or indigenous community land titling was no longer perceived as eccentric and dangerous. Rather it was seen, from the mid-1990s to the mid-2000s, as potentially helpful to the government's decentralisation agenda, and then, after 2006, as central to the main purpose of the government. Why it had taken so long to reach that point is debatable, but there is a temptation to seek an answer in the hostility to social justice that is characteristic of New World power structures, and the source of their founding principles as imports from pre-modern Spain, amplified by post-independence settlements that left so much land in the hands of so few, and political policing in modern times by the United States.

In any case, by the mid-1990s Danida spotted an opportunity and acted on it. As things turned out, by being there doing the right things at the right time, the alliance between Danida and indigenous peoples' fundamental interests allowed good progress to be made. And once the indigenous peoples gained a measure of power, through their own activism and a vacuum in the state's capacity to react effectively after 2002, the result was a dramatic breakthrough. A problem is the lack of follow-through, since Danida withdrew from the TCO process in 2010. Regardless of what survived from Danida's previous efforts (and clearly much did), another few years of consolidation investment by Danida would have been extremely helpful. The lack of follow-through may be a sign that Danida failed to recognise clearly enough that ecosystem tenure security and community empowerment are absolutely central to the strengthening of Bolivian systems against climate chaos. If adaptation had been a key priority then, perhaps this would not have happened. But the take-home message for now is that social and ecological strengthening can be done, and done quickly, if it is done just like it was by Danida and Bolivia.

Coastal Zone and Community Planning in Zanzibar

7.1 TROPICAL ISLANDS, COASTS AND CLIMATE CHANGE

This chapter focuses on the strengthening of ecological and social systems in the Indian Ocean archipelago of Zanzibar. The entry point is the final evaluation of a Finnish-funded project called the National Spatial Data Infrastructure for Integrated Coastal and Marine Spatial Planning in Zanzibar (ZAN-SDI), which ran from January 2016 to April 2019 (Caldecott, Killian et al., 2019). It is included here because many of Zanzibar's most severe challenges are shared with other tropical islands and coastal zones. They include limited land availability, vulnerable and declining fresh water supplies, dependence on fisheries sustained by coastal ecosystems that are themselves under pressure, conflicting land uses and excessive or poorly planned tourism development, all aggravated or jeopardised by the effects of climate change. The solutions being developed in Zanzibar are therefore useful in many other contexts, being based on networks of collaborating governmental and non-governmental institutions, local participatory assessments of resources, risks and opportunities and detailed spatial planning supported by drone and digital technologies.

Tanzania = Tanganyika + Zanzibar

Tanzania has an area of about 945,000 km². The much larger mainland part, once known and sometimes still referred to as Tanganyika, consists of a low-lying eastern coastal area, a high central plateau and scattered mountainous zones especially in the north-east, where Africa's highest mountain (Kilimanjaro) is situated. To the north and west are Africa's largest and deepest lakes (Victoria and Tanganyika respectively). Central Tanzania comprises a large plateau, with plains and arable land. The eastern shore is hot and humid, with the islands of Zanzibar offshore. The number of Tanzanians was 45 million at the 2012 census, and an estimated 56 million in 2018, of whom about 1.5 million lived in Zanzibar. In Tanzania as a whole, the most fertile lands are now densely populated, there are conflicts over land between pastoralists and cultivators, land degradation accompanies rapid deforestation, soil erosion and nutrient depletion, and the country is increasingly water stressed (World Bank, 2017, 2019). Signs of ecosystem degradation and biodiversity loss are becoming obvious in parts of Tanzania's coastal and marine environments.

The United Republic of Tanzania was formed in 1964 through the union of two states that had recently gained independence from Britain, Tanganyika and Zanzibar (Killian, 2008a). The Zanzibar independence process was unusual, in that it had been a British protectorate rather than a colony since 1890, and this status was ended by an act of the UK Parliament in December 1963 which made Zanzibar an independent state with its sultan as a constitutional monarch. In January 1964, however, the sultan was violently deposed and a socialist regime established under the Revolutionary Government of Zanzibar (RGoZ, an official title still in use today). Federation with Tanganyika, whose president Julius Nyerere was also concerned with promoting social justice, followed in April, under which the Union government and the Zanzibar government each has its own executive, judiciary and legislature. The Zanzibar government exercises sovereignty over all domestic and non-union matters, while the Union government has jurisdiction in areas that include defence and security, foreign affairs, police, emergency powers, citizenship, external borrowing and trade, mineral oil resources, higher education, the court of appeal and the registration of political parties. In common with other former British territories that had been independent but were united with others when the British left (e.g. Sarawak within Malaysia), there are groups that desire greater autonomy, or even a return to independence.

In Zanzibar this restlessness mixes with ethnic rivalries and other tensions to generate political turbulence. Thus, there was significant social discord in the 1990s and following elections in 2000, 2005 and to a lesser extent in 2010 (Killian, 2008b, 2014). The 2015 elections were annulled (Throup, 2016), leading to renewed questions over the political future of the archipelago. Fresh elections were scheduled for October 2020, and since election problems have led to aid suspensions in the past, their conduct was expected to affect the willingness of aid agencies to continue investing in Zanzibar. This is important, since development financing remains significant for Zanzibar and made up 42 per cent of the total budget for 2015–2016 (ZAN-SDI, 2015). As is turned out, the election in Zanzibar was bitterly contested and marred by some deadly violence (e.g. Burke, 2020; EU Councils, 2020; Scotland, 2020), but whether this will affect donor attitudes is currently unclear.

Many younger Zanzibaris are less concerned than their elders to perpetuate the ethnic and religious divisions of the past, however, and the independence debate is also mitigated by an increasing awareness that without the Union it would be hard for Zanzibaris displaced by water shortages and rising seas to relocate to the mainland. There is also a sense that Zanzibar has experienced generational and mindset change, and has an atmosphere of youthful energy and initiative that was absent in former times. So there is the feeling that whatever the future holds it will be very different to the past. The quality of that future will depend on ecological as well as social factors, however, with climate change effects being particularly influential.

Zanzibar's Coastal and Marine Ecosystems and Vulnerabilities

Zanzibar is a small, low-lying, coralline archipelago comprising the two large islands of Unguja (where Stonetown is located) and Pemba and many small ones, lying in the Indian Ocean 25–50 km off the eastern shore of mainland Tanzania. Coastal and marine ecosystems include coral reefs, sea grasses, coastal forests, mangroves and sandy beaches, and these form the basic foundation of the local culture and economy of coastal Zanzibaris. The design of the project (ZAN-SDI, 2015) identified some of the key challenges to the well-being of local people:

- Demand for land is driven mainly by the expansion of settlements and the growth of tourism, and is increasingly competitive.
- Urban growth is creating a need for additional infrastructure, utilities and services, while also causing land use conflicts with agricultural and environmental interests as coastal towns expand into forest and reserved areas, as well as rural areas, while population growth drives higher demand for food and firewood.
- An expanding tourism industry, and the informal settlements and in-migration associated with it, are increasing pressures on land, water and forest ecosystems in the coastal zone, while adding problems of solid waste management, biodiversity loss and deteriorating air and water quality, and challenging the traditional cultures of coastal communities.
- Overfishing, destructive fishing practices and removal of sand and vegetation all degrade the coastal zone and the vulnerability of its ecosystems.
- Coastal erosion is facilitated by mangrove cutting, the building of sea walls to protect tourism facilities and coral and sand mining for construction.
- Conflicts caused by competing and overlapping uses of land and coastal/ marine resources occur in some areas, and planning processes are inadequate to prevent them.
- As salt water infiltrates water tables and arable lands, the islands are running out of fresh water.

Zanzibar's economy largely depends on climate-sensitive activities such as farming, fishing and tourism, but the ecosystems supporting these activities are increasingly threatened by changes in rainfall and temperature patterns as well as sea level rise, and many residential, productive, water, energy and transport facilities are also vulnerable (Watkiss et al., 2012; UNDP, 2018; MANRZ, 2019; World Bank, 2019). Although the issues are perhaps more acute in an island context than in a continental edge, and Zanzibar's climate vulnerability profile has been described as 'more similar to that of the Small Island Developing States' (World Bank, 2015: 23), these same issues arise across the entire coastal zone of Tanzania and nearby countries such as Mozambique.

7.2 THE ZAN-SDI PROJECT

As an institutional cooperation project funded by Finland, ZAN-SDI required a Finnish partner and this was the Finnish Environment Institute. The project's aim was to build the capacity of partner agencies within Zanzibar to manage and share spatial information among themselves and with the public, and to use it in developing wise and popular spatial plans, by providing the guidelines, tools, training and experience needed for effective integrated coastal and marine spatial planning. Although a small place, Zanzibar has a flourishing system of institutions with powers and responsibilities relevant to development planning, so the lead partner of the project – the Zanzibar Commission for Lands (COLA) – could only make progress collaboratively. So the first aim of the project (*spatial data infrastructure (SDI) development*) was to build capacity at COLA and a number of other key actors, including the departments of forestry and non-renewable natural resources, fisheries development (DFD) and environment, and the Zanzibar Environmental Management Authority (ZEMA), to develop and maintain a widely used, interoperable, cross-thematic and fully functional SDI system.

The second, parallel, aim (*spatial mapping*) was to build capacity at the Department of Urban and Rural Planning (DoURP) for ecosystem-based planning and management of coastal zones, maritime activities and the marine environment, especially by populating priority spatial data layers and drawing up local area plans with inclusive local participation. This, likewise, could not be done by DoURP alone, so required collaboration with the same institutions that were already working with COLA, as well as with other government departments (e.g. of surveys and mapping) and the State

University of Zanzibar (SUZA). And the final aim (*geospatial capacity*) was to improve geospatial information management capacities at all the partner organisations, mainly through expert networking and training in areas of weakness.

Since capacity building was to be advanced largely through 'learning by doing', it involved DoURP working with the local authorities in parts of the north-east coastal zone of Unguja island to develop detailed local area plans for two specific areas – Kiwengwa (DoURP, 2018a) and Pongwe (DoURP, 2018b) – as well as a Special Area Plan for the whole region (DoURP, 2019). Findings on the design quality of the ZAN-SDI project are given in Table 7.1, and on its performance summarised in Table 7.2. The ways in which design quality and performance are presented and scored here are explained in Chapter 12. These details offer a way into the larger and more diverse institutional system concerned with planning and building resilience to climate chaos in Zanzibar, which is discussed further in Section 7.3.

7.3   THE INSTITUTIONAL SYSTEM
         FOR SPATIAL PLANNING

Multisectoral and Multilevel Cooperation

The ZAN-SDI project offered a way to bring many interests together around a clearly defined purpose. The main impact lay in enhancing the capacity of the RGoZ system to appreciate and plan for emerging challenges in adapting to climate change effects in the coastal zone, supported by the SUZA research and knowledge management system, the community dialogue and planning system built up by the Mwambao Coastal Community Network (MCCN) and interagency knowledge sharing. The plans and associated ideas for managing conflicts among user groups, and for meeting needs among coastal communities, represented potentially irreversible improvements, but as the *shehia* head of the Kiwengwa community (one of those covered by spatial plans) observed: 'The plan is a good thing, and we provided inputs to it, but we'll see how it is put into effect.' He went on to

Table 7.1 *Design quality of the ZAN-SDI project*

Theory of change. 'The Project will improve access to spatial information and thus possibilities for inter-agency cooperation and public participation in environmental and other spatial planning and management processes. Easy access to spatial information is a precondition for private sector involvement in economic development, participatory democracy, good governance (transparency), reduction of inequalities and fair sharing of benefits from natural resources, which all contribute to poverty reduction. More efficient data sharing will create savings by reducing duplication of work by various agencies. In particular, accurate coastal and marine spatial data also support the planning of adaptation to climate change, which is imperative for small islands communities like Zanzibar. ... It is assumed (but it is outside of the control of the ICI [Institutional Cooperation Instrument] Project) that the data holders are willing to share their data with others, [that] competent authorities and other decision-makers take the improved spatial data and spatial plans into consideration, and [that] spatial plans are implemented and enforced in Zanzibar' (ZAN-SDI, 2015: 13).

Assumptions underlying the theory of change	Judgements on the validity of assumptions
Assumption 1: More efficient data sharing will create savings by reducing duplication of work by various agencies.	*Plausible.* But stakeholders must be persuaded to trust each others' data, and that using data from other institutions will relieve demands on their own resources.
Assumption 2: Accurate coastal and marine spatial data will support the planning of adaptation to climate change.	*Strongly plausible.* But planning standards must ensure that the use of spatial data is required, and the idea of adaptation must be clarified and related in specific terms to planning objectives.
Assumption 3: Data holders are willing to share their data with others.	*Plausible.* Zanzibar is a small society with socialist traditions and although the state sector is relatively large many officials

Table 7.1 (*cont.*)

	know each other personally, and are familiar with each others' duties. Barriers to data sharing within government are thus more likely to arise from personality issues or institutional capacity constraints than from structural-institutional rivalries.
Assumption 4: Improved spatial data and spatial plans will be used by decision makers.	*Plausible*. But stakeholders must be persuaded that the new planning tools are helpful in practice, and that their use will be recognised and rewarded.
Assumption 5: Spatial plans are implemented and enforced.	*Weakly plausible*. From policies and enthusiasm among collaborators, the intent to implement seemed favourable. Implementation itself would depend on sustained commitment by supporters relative to the intransigence of competing interests, which were both unknown.

Overall conclusion. Accepting the validity of assumptions 1–5, it was reasonable to expect that improved and open access to spatial data through partner institutions, and integration of those data into spatial plans by inclusive and participatory means, would increase understanding and compliance with planning aims while also tending to reduce planning failures and harmful impacts by private investments, and to promote transparent and accountable governance and fairer and more sustainable development outcomes. *Score for design quality*: high (5/6).

Source: format and content modified from Caldecott, Killian et al. (2019).

Table 7.2 *Performance of the ZAN-SDI project*

Relevance to Zanzibar. Zanzibar has effectively run out of land and competition for renewable natural resources is intensifying (e.g. with large foreign fishing boats offshore and local fishers displaced by tourism development from their traditional fishing areas and forced to raid those of other communities), while climate change and sea level rise are reducing fresh water availability and eroding coastlines. This puts an increasing premium on detailed spatial planning and the building of consensus and compliance around constraints on competing development initiatives, making the ZAN-SDI project highly relevant to Zanzibar's needs. The capacity-building needs of the partner institutions reflected their convergent roles within the RGoZ, where they coalesced around a common understanding that intense competition for fragile and degrading natural resources required detailed spatial planning and the building of consensus and compliance around constraints on competing development initiatives. The project responded precisely to this situation (*Score 7*).

Relevance to Finland. There was a feeling among the Finnish Environment Institute informants that the availability of cheap, cloud-based, open-sourced and other digital technologies now makes spatial planning far more doable than before, and that the ZAN-SDI project would offer Finland a proving ground for how to induce transformative change through modern and 'light-touch' technologies and networking for practical and locally valued ends, combined with the use of Finnish digital and environmental expertise for capacity building. The project was also in line with Impact Area 1 of the *Country Strategy*, by promoting improved governance, leadership and civil society participation in addressing urgent and important social and environmental challenges, while also building on Finland's excellence in digital technologies and the involvement of related programmes by Finnish institutions and others (*Score 6*).

Efficiency. In *SDI development,* progress was inhibited by a shortage of skilled personnel which led to individuals being 'poached' from the project, and some hardware and data management issues (*Score 3*). In *spatial mapping,* cooperation among stakeholders was excellent in practice at a technical level, despite some director-level lack of participation (*Score 6*). In *geospatial capacity,* the main users of the SDI were internal to the COLA system, and although other departments and local governments also

Table 7.2 (*cont.*)

contributed and used data, the aim of 'single data hub' for use by all parts of the RGoZ was not entirely achieved (*Score 4*).

Effectiveness. Major achievements in capacity building were seen in *spatial mapping* (*Score 6*), but were offset to some extent by networking issues in *geospatial capacity* (*Score 4*) and more seriously by staffing and associated hardware and data management issues in *SDI development* (*Score 3*).

Impact. World Bank investments in sustainable tourism in Zanzibar are now restricted to areas covered by ZAN-SDI spatial plans, which therefore had a strong leveraging and guiding role. Because of its strategic and holistic nature, similar effects can be expected of the spatial planning effort across all sectors that engage with the mapped area, with a significant long-term influence. The project had a major impact in promoting interagency collaboration and joint visioning around an ambitious but largely successful spatial knowledge management approach, and the participants learned many new things in the process and are now used to thinking together in new ways about how to use and develop this approach in the future, for the likely benefit of all stakeholders (*Score 6*).

Sustainability. Communities and government agencies involved with ZAN-SDI seem convinced that spatial planning is a vital tool that should have been used long ago, that it is particularly necessary now in light of past experience (especially by communities, of the impacts of planning failures and tourism) and that Zanzibar needs to concentrate on 'smarter' and more climate-proof use of all its resources. The RGoZ appears very willing to extend the ZAN-SDI approach, using its own new capacities, to the southern and western parts of Unguja island, and also to Pemba. Even with offsetting RGoZ human resource capacity issues, sustainability was judged to be high (*Score 5*).

Source: format and content modified from Caldecott, Killian et al. (2019).

comment enthusiastically about the economic and ecological value of fishing and octopus closures, but also to draw attention to the failure and salinisation of fresh water supplies as the key problem in his area. This draws attention to the fact that social and ecological systems must be strengthened in multiple ways because they are threatened in multiple ways.

The strategic value of multisectoral cooperation in planning for coastal resilience and ecological integrity lies in the process of bringing together many points of view, sources of knowledge and interpretation, and life experiences, to create a meaningful and useful tool. The more inclusive the process, the more valuable the output, as long as it rests firmly on environmental and livelihood security for people at the base of the pyramid. With the right foundations the superstructure can reach the sky (drones) or beyond (satellites). This conclusion tends to validate the wishes of all the institutions involved to extend a version of the ZAN-SDI project approach to cover other parts of Unguja island and all of Pemba.

A weakness of the project was the lack of an institutional mechanism, such as a standing committee, with a specific mandate and resources to provide leadership and coordination across the RGoZ partners and hence a pathway to extension and replication. But the ambition and enthusiasm to find such a pathway existed and active dialogue was underway, both within RGoZ and involving other development partners, both Finnish (e.g. the University of Turku (UTU) and the geospatial and information and communication technology (Geo-ICT) programme) and international (e.g. the World Bank and the Southwest Indian Ocean Fisheries (SWIOFish) programme). Moreover, the practical challenges arising from climate change impacts are now very much appreciated (particularly at the Department of Environment, ZEMA, SUZA and the community level) and there is commitment to applying the new planning techniques to help resolve them, although doubts remain at the community level and elsewhere over what will be the effect of plan implementation.

The State University of Zanzibar

The government institutions responsible for terrestrial planning (i.e. COLA, ZEMA and the departments of forestry and environment among others) and marine planning (i.e. DFD) are important but not the only actors in seeking to strengthen Zanzibar's systems against climate chaos. Thus SUZA teaches on climate change issues, for example with 15 students in 2019 on a two-year course with dissertation subjects including 'tourism' and 'floods' in relation to climate change, and 'the role of NGOs in supporting local adaptation in Pemba'. The ICT training delivered at SUZA also has a gender equity effect in attracting women into a quickly growing sector with significant employment potential. There are few other such opportunities for young women to train in Zanzibar, and they are often discouraged from studying elsewhere by family pressure.

More technically, SUZA was also involved in developing digital mapping systems through the Zanzibar Social Environmental Atlas for Coastal and Marine Areas (ZanSea) project. This was funded by the Norwegian company Statoil, and developed an SDI-like platform to support 'sensitivity mapping', a coastal and marine resources atlas and oil-spill contingency plans (ZanSea, 2020). It had also collaborated with UTU in Finland on the World Bank-funded Geo-ICT programme. The involves a network that includes three other Tanzanian universities: Sokoine University of Agriculture in Morogoro, Ardhi University in Dar es Salaam and the University of Dar es Salaam itself (Geo-ICT, 2020).

An extension of Geo-ICT is the Resilience Academy Tanzania, which promotes training for community mapping of buildings, sewage systems, bridges, areas and resources at risk of floods and other environmental hazards (Resilience Academy, 2020) – an extremely useful approach that brings together the technologies of risk assessment and digital mapping with the key strengths of community action in city neighbourhoods (see Chapter 8). A World Bank 20 cm drone imaging survey and mapping project was also underway, in which SUZA has been involved in a training role (World Bank, 2020; ZMI, 2020). These various initiatives made SUZA a key hub in a network of institutions that were collectively able to offer support to government partners in the ZAN-SDI project and related initiatives. This swarm of initiatives was being coordinated through informal dialogue among the various partners, rather than through any formal RGoZ arrangement, and observers commented that these arrangements work better in practice than on paper – again reflecting the small scale of Zanzibar's technical community.

The Mwambao Coastal Community Network

All this activity, however, is at the more technical, digital and official level of Zanzibar society. Connections with local people living in real-life environments occur through the 'participatory' dimension of resource assessment, mapping and planning. This often makes use of *shehia* structures which are the basis of community organisation in coastal and rural Zanzibar. Although the *shehia* are not necessarily 'traditional', since their heads are appointed and are often former police or military personnel, their familiarity with local conditions can supplement the opinions of other local informants and any 'scientific' knowledge that may exist on the distribution and value of marine and coastal ecosystems. The MCCN is an NGO that promotes environmental education and community action throughout Tanzania's coastal zone, but is based in Stonetown and has several projects on coral reef recovery, monitoring blast-fishing and octopus co-management in Zanzibar (MCCN, 2017, 2020).

Working with volunteers and employees in the villages, MCCN planted ideas in several places that resulted in community fishery management zones being established, with fishing sometimes prohibited in favour of snorkelling tourism, and 'octopus reserves' having been established where harvesting is limited to one month in four, resulting in more valuable catches of larger octopuses. These initiatives were largely fitted into the area plans at Pongwe and Kiwengwa, and the trials continued after the ZAN-SDI project with MCCN encouragement and spontaneous replication among coastal communities attracted by income and fishery benefits. The involvement of MCCN thus gave access to ideas and practices for CBRM, and offered a valuable reality check for the remote images and digital mapping process. And MCCN also works with DFD on two projects – to prepare management plans for octopus, small pelagic and reef fish fisheries, and to deliver training for reef octopus-harvesting closures – through its participation in the SWIOFish programme, a regional network concerned with promoting sustainable and community-based fishery management (MCCN, 2019; SWIOFish, 2018, 2019). These connections add complexity and depth to the adaptation community of interest focused on Zanzibar.

The Institutional Cooperation Instrument

In considering the larger process and the ZAN-SDI project's contributions to it, it is worth being aware of how this project was funded. This was done through the ICI, which is one of the aid mechanisms used by the Ministry for Foreign Affairs (MFA) of Finland. The ICI supports mutually beneficial capacity-building relationships between Finnish institutions and their partner institutions in developing countries, including the exchange of training and research visits, in-country meetings and the purchase of essential equipment. In ICI projects, guided by an agreement between the institutions and the rules of the ICI, the local partner is responsible for adapting and using ICI resources to meet its own needs. A study of the ICI (Bäck et al., 2014) concluded that the best way to use it was to support the goals of a country strategy, and since then the cooperating institutions have been guided to areas of work in line with the priorities of the partnership agreement between Finland and the country concerned.

From MFA's point of view, valuable roles of the ICI include as a 'bridge' to provide connectivity and continuity between bilateral projects and phases, as a way to build the capacity of 'sibling' institutions to participate more effectively in the projects themselves, and as a mechanism that does not require a lengthy process of consideration by either government. It is thus seen as a fast, cheap and effective way to deliver Finnish added value, but certain constraints on its use are also recognised. Importantly, partner institutions must be responsible for their own salary costs, so an ICI project must have enough support at the non-Finnish partner to guarantee local salaries and other counterpart costs, making demand, ownership and commitment key factors. All in all, though, the ICI has a specific role that can be extremely valuable because of its flexibility and responsiveness to local needs and opportunities. In an increasingly chaotic world, with diverse actions required everywhere, often on short notice or in response to new experiences and new thinking, support for flexible but fast-acting and inherently equal partnerships can only be useful.

7.4 LESSONS FROM ZANZIBAR

The main strengths of the ZAN-SDI project lay in its success in building enthusiasm among diverse government stakeholders for participating in a technical process that involved research to support planning, and digital mapping to visualise its findings. This was helped by the obvious value of the product in using organised knowledge to make it easier to avoid and reconcile conflicting demands on resources. The technical process was not wholly new, however, since ecological and participatory research had long been underway in the Zanzibar coastal zone through the efforts of MCCN and others, and digital mapping had been pioneered in Zanzibar by the Finnish-supported Sustainable Management of Land and Environment (SMOLE) project up to 2013 (Caldecott, Valjas et al. 2012), and later the ZanSea project. It must also have been 'in the air', as the SWIOFish, Geo-ICT and drone-mapping projects were being developed by some of the same consultants and university partners. It was also helpful that the various RGoZ agencies were inclined to cooperate and share information, and that community leaders were already concerned and knowledgeable about the impacts of competitive resource exploitation and climate change on local livelihoods.

This sense that all the social systems of Zanzibar were 'ready' for this particular approach contains perhaps the most important lesson here. An ideal package of adaptation measures may contain just the right mix of social, ecological and technical measures, all finely balanced and mutually reinforcing yet robust and flexible. But it cannot work while stakeholders are naïve regarding what is at stake, and officials are more concerned to promote their own institutions and sectoral interests than to cooperate meaningfully. Those barriers must have been weakened before a project can make real progress quickly. In Zanzibar this had already happened through SMOLE, but the same pattern was also seen in Bolivia and Nepal, where other projects in the same fields had preceded the 'main acts' described in Chapters 5 and 6. But a similar effect can happen through exposure to the global discourse on modernity, environment, sustainability and climate change, and the influence of overseas visitors and visits. As we run out of time to strengthen systems against climate chaos, learning to appreciate and recognise 'adaptation-readiness' will help to prioritise adaptation efforts in locations that can respond to them best, while also accelerating and focusing preparatory work in other places.

8 Liveable and Sustainable Cities

8.1 AIR QUALITY IN THE KATHMANDU VALLEY

Environment Sector Programme Support (1999–2005)

The entry point here is another Danida intervention in Nepal, this one focused on safeguarding air quality in the Kathmandu Valley. The story may be useful to the administrations of any number of quickly growing cities in developing countries. It starts with ideas that were being developed in the early 1990s, in discussions surrounding the drafting of the Rio Treaties and Agenda 21: the 'polluter pays' principle and its variant of 'user charges', which were later developed into PES and REDD+ (reducing (GHG) emissions from deforestation and (forest) degradation, with internationally-agreed forestry, bio-diversity and social safeguards). These helped to shape Nepal's Eighth Five-Year Plan for 1992–1998 and the Nepal Environmental Policy and Action Plan (NEPAP) of 1993, both of which stressed environmental management and awareness in the context of sustainable economic growth.

These ideas were much needed in Nepal, where serious pollution and low occupational health and safety standards were the norm in industrial enterprise zones, and businesses were uninterested in environmental issues and energy efficiency (Danida, 2005c, 2005d). Policies like NEPAP called for action, but little was done until Danida designed an Environment Sector Programme Support (ESPS) package in the late 1990s (Danida, 1999). This had components on institution building, promoting cleaner and safer industry and building a waste-water treatment plant, but another on air quality was added later (Danida, 2000), followed a year or so after that by an energy efficiency theme.

The Air Quality Component

Air pollution had long been a serious environmental problem in the Kathmandu Valley, posing a health risk to residents especially during the dry winter months. In the 1990s, high concentrations of pollutants in the lower atmosphere were causing respiratory disorders, eye, throat and skin problems and cardiovascular diseases in Kathmandu (Pradhan et al., 2012). The main sources were obsoletely designed brick kilns and old, poorly maintained, dirtily fuelled petrol, liquified petroleum gas (LPG) and diesel vehicles. Policy, technology, public opinion and pilot programmes during the 1990s had converged to make it feasible to begin phasing out many sources of air pollution, and replacing some of them with electric vehicles (EVs).

The air quality component of the Danida intervention was designed in light of government priorities that would shortly lead to the 2002 National Transport Policy and the 2003 Sustainable Development Agenda for Nepal. It drew on preceding activities to promote EVs in the area (Moulton and Cohen, 1998), and aimed to head off some of the worst consequences of unregulated growth in pollution sources. The theory of change and assessments of the design of the intervention are given in Table 8.1, and how these are presented and scored is explained in Chapter 12.

Thus, the purpose of the component was to reduce air pollution from vehicles in the Kathmandu Valley, which would be achieved by: promoting the use of EVs; improving fuel quality; establishing and enabling the enforcement of vehicle emission standards; establishing an ambient air quality monitoring system (plus awareness creation); and setting up a vehicle engine maintenance training centre to support the other activities. Danida paid for several advisors to support strategy, training, advocacy, monitoring and testing, and also equipment, operating costs of the monitoring system and training centre, and other facilities and running costs. The component orchestrated and resourced a number of partnerships between governmental institutions, local bodies, NGOs, the private sector and academic

Table 8.1 *Design quality for cleaner air in the Kathmandu Valley*

Theory of change. Air pollution in the Kathmandu Valley was largely driven by increasing numbers of vehicles. The component was established to head off some of the worst consequences of unregulated growth in pollution sources, and to facilitate the emergence of regulatory and monitoring systems that would continue to work after project's end.

Assumptions underlying the theory of change	Judgements on the validity of assumptions
Assumption 1: Air pollution in the Kathmandu Valley was serious, increasing and unpopular, and the government was committed to reducing it.	*Strongly plausible.* Evidence in Danida (2000).
Assumption 2: The main sources of air pollution in the Kathmandu Valley were obsoletely designed brick kilns and old, poorly maintained, dirtily fuelled petrol, LPG and diesel vehicles.	*Strongly plausible.* Evidence in Danida (2000).
Assumption 3: Policy, technology, public opinion and pilot programmes in the 1990s had converged to make it feasible to begin phasing out many sources of air pollution, and replacing some of them with EVs.	*Strongly plausible.* Evidence in Danida (2000).

Overall conclusion. Accepting the validity of assumptions 1–3, it was reasonable to expect that enough sectoral actors could be recruited into a cooperative partnership, and public demand for change increased, for a systemic process to be advanced of replacing the older and fossil fuel vehicles with EVs and/or by newer, better-maintained and more cleanly fuelled units, and that this would result in improved air quality. *Score for design quality*: very high (6/7).

Source: format and content modified from Caldecott et al. (2017).

institutions, and these were encouraged and enabled to deliver complementary parts of a complex but coherent programme that contributed to and demonstrated improving air quality (Table 8.2). This is told in the form of a 'contribution story' in Box 8.1, which also covers the other parts of the ESPS.

Replicability and Lessons Learned

In the context of Kathmandu in the late 1990s, air quality was being degraded by a number of factors but dirty brick kilns and dirty vehicle were identified as the key problems. Kilns could be regulated much more easily than vehicles, since they were licenced, large and immobile, and could be inspected easily and if necessary closed down. On the other hand, cleaning up the kilns while allowing continued growth of dirty vehicles would have been unsatisfactory, and Danida attempted the complex task of cleaning up the vehicle fleet by facilitating the substitution of cleaner technologies, monitoring outcomes and supporting the necessary enforcement, training and technical upgrades. It then became feasible for government to embark on a more ambitious regulatory programme.

As can be seen from Table 8.2 ('effectiveness' and 'impact'), multiple changes were needed at the same time if any one of them was to work well. Meanwhile, the component and its predecessors demonstrated that although multiple processes have to happen at once (involving battery and power unit technology, battery charging, exchange, reuse and recycling, electricity supply and distribution, training, targeted investment incentives, institutional cooperation, etc.) EVs can take off within an urban economic system, and with other things going on as well (legislation, monitoring, enforcement, awareness raising, etc.) they can have an important role alongside an integrated AQMS that actually demonstrates improving air quality. The principle that it *can* be done, and the combination of things that are needed to make it work, provide an extremely replicable model for use in other urban situations.

Moreover, the whole process offers a conceptual model for mitigation efforts in the global climate response, with GHG concentration

Table 8.2 *Performance of the intervention for cleaner air in the Kathmandu Valley*

Relevance. The aims of the component were fully in line with the 2003 Sustainable Development Agenda for Nepal, which called for setting strictly enforced ambient air quality standards, encouraging the shift towards zero emission vehicles, especially in dense urban areas, and the shift towards clean sources of industrial energy, while also promoting domestic monitoring and research capacity for domestic and cross-border pollution, to provide necessary data for effective international negotiations and to contribute solutions to the global environmental issue of climate change (Danida, 2005e). Thus, relevance was by definition high (*Score 7*).

Efficiency. Danida (2005e) found that all planned activities were achieved at acceptable cost, and that the timely establishment of the Vehicle Anti Pollution Programme (VAPP) and its training centre had allowed emissions testing and monitoring activities to intensify over time (*Score 4*).

Effectiveness:
- For *institutional support to EVs and other clean vehicles*: (1) an EV lobbying group was established and helped to influence government policy, regulations, electricity tariffs, etc. in favour of the EV sector; (2) training activities were delivered to develop the skills of mechanics working in charging stations to enhance battery life, to creating job opportunities for women as EV drivers (a total of 289 women received EV driver training, of which around 100 were later employed as EV drivers) and to enhance the skills of mechanics and technicians working in relevant manufacturing industries; and (3) a Clean Vehicle Promotion Fund was established to make grants to promote EV use (e.g. EV bus routes, public information materials).
- For *improving fuel quality,* research found that emissions were due more to the condition of the vehicles than to the quality of the fuel, so this effort was discontinued.
- For *vehicle emission standards*, the component encouraged amendment of existing vehicular emission standards and continued to develop proposals and recommendations to the Ministry of Population and Environment, and a system for continuous review and modifications on vehicle emission standards was established.

Table 8.2 (*cont.*)

- For *vehicle emission control and enforcement*, equipment and training were provided to the Kathmandu Valley Traffic Police and the Department of Transport Management for testing vehicles under the VAPP, which then began testing about 60,000 vehicles per year.
- For *ambient air quality monitoring*: (1) an air quality monitoring system (AQMS) was established in the Kathmandu Valley, with six permanent and one mobile stations used to monitor particulates, benzene, and nitrogen and sulphur oxides; (2) operational capacity and ways to inform the public on air quality (e.g. electronic notice/warning boards for monitoring results) were built and maintained; (3) status reports on air quality were prepared; (4) institutional arrangements for the AQMS were made by establishing a permanent working group with the participation of all five municipalities of Kathmandu Valley (Kathmandu, Lalitpur, Bhaktapur, Madhyapur Thimi and Kirtipur), the Ministry of Population and Environment, other knowledge holders and contracted laboratories; (5) reports on the health implications of air quality were prepared; and (6) the AQMS was used as a management tool, in which information was used to guide policy and regulatory change (see impact).
- For *public awareness*, major activities included seminars, broadcasts, posters, electronic hoardings and website content.
- For *vehicle maintenance training*, the VAPP was established, including a training facility that delivered a series of popular training courses on vehicle engine maintenance and emission control; by the end of 2004 a total of 1,653 people had received 550 training days.
  Cooperation with the five municipalities of Kathmandu Valley was a priority, both in the AQMS and through the placement of an ESPS advisor with Lalitpur municipality and establishment of a permanent working group in which all five municipalities participated. The dissemination of air quality data to the public was very effective and led to many policy decisions aimed at reducing air pollution in the valley. Air pollution levels declined over time despite a rapid increase in the number of vehicles numbers in the valley (Danida, 2005e) (*Score 7*).

Impact. In response to declining air quality, the government made a number of policy decisions, including: (1) introducing the Nepal Vehicle Mass Emission Standard, 2056 B.S. (only vehicles complying with these standards can be imported to the country); (2) banning the import

Table 8.2 (*cont.*)

of second-hand and reconditioned vehicles, banning the import
of two-stroke engine vehicles and phasing out three-wheeler diesel
'tempos' from the valley; (3) phasing out three-wheeler two-stroke engine
vehicles from the valley; (4) phasing out taxis 20+ years old from the
valley (later extended to all vehicles 20+ years old); and (5) banning the
new registration of Bull's Trench Kiln brick manufacturing industries in
the valley and requiring all those already in operation to be changed to
cleaner technology. Meanwhile, public awareness activities built concern
over air pollution in the Kathmandu Valley, leading to media coverage
and policy influence. A 10 per cent annual reduction in the air
concentration of dangerous PM10 particulates in the residential area of
Kathmandu Valley was confirmed (Danida, 2005e: 7) (*Score 6*).

Sustainability. The component was based on a strong trend in
government policy and public opinion in favour of reducing air pollution
in the Kathmandu Valley, of which an important feature was the use of
EVs and the introduction of regulations to favour them and other cleaner
vehicles. By continuing, amplifying and resourcing this trend through
training and equipment, the component was likely to be associated with
multiple sustainable outcomes (*Score 7*).

Source: format and content modified from Caldecott et al. (2017).

standing in for air quality. Here there is the same daunting array of
interlocking, interdependent actions that all need to be done at once –
saving high-carbon-density ecosystems, changing land use priorities,
decarbonising whole economies through technological innovation
and new investment in multiple sectors driven by profit, regulation,
public demand, carbon taxes, lifestyle choices, consumer preferences,
etc. – all implemented by competing and cooperating governments in
line with the Paris Agreement. The only thing missing is to measure
GHGs in the air, rather than mean global surface temperature, as the
basis for a practicable and useful monitoring system. That aside, the
same conclusion can be reached as for improving air quality in the
Kathmandu Valley: that it *can* be done.

BOX 8.1   The 'improving urban environments' contribution story from Nepal

Summary

In 1999–2005, Danida supported the government in a spectacular demonstration of how to build urban environmental awareness, how to regulate environmental standards and encourage both cleaner production and energy efficiency in partnership with business and how to improve air quality in a growing metropolis. The programme was not perfect and ended prematurely, but some of the ideas, approaches and skills lived on within government, among the public and in the form of programmes supported by other donors.

Why It Was Needed

In the late 1990s, there was a serious lack of capacity for environmental management in the industrial sector, and industrial zones had become extremely polluted because businesses had been protected from regulation to encourage their growth, with the result that they used resources excessively, with low productivity and generating unnecessary waste, and seemed unlikely to be able to comply with any environmental standards set under the 1993 NEPAP without considerable technical and other support. Meanwhile air quality was becoming a problem in the Kathmandu Valley with the rapid growth of poorly maintained fossil fuel vehicles, and generators to compensate for irregular electricity supplies. To correct these trends a high-impact programme to encourage and enable compliance with environmental standards was urgently needed, and would have to involve setting standards, formulating regulations to mandate them, training to enable them to be met and various incentives and disincentives to encourage them to be complied with.

What Was Done

Danida's ESPS targeted the following main areas:

- establishing a new Institute of Environmental Management (IEM) to take the lead on awareness raising, training, research and advising on standards;

BOX 8.1 **(cont.)**

- assessing industrial businesses for cleaner production needs and opportunities, and establishing a credit mechanism to finance investments in cleaner production;
- helping businesses identify self-financing ways in which they could increase their energy efficiency;
- building a wastewater treatment plant (WWTP) in one of the industrial zones, along with a district-level sewer system for industrial wastewater and an extended and rehabilitated drainage system;
- institutional strengthening of the three ministries responsible for environmental management; and
- improving air quality in the Kathmandu Valley, by promoting use of EVs and establishing an ambient AQMS.

What Was Achieved

Impact

The WWTP was rather unsuccessful due to design issues, but the other components of Danida's ESPS achieved the following.

- *Institute of Environmental Management.* By 2005, the IEM had created high levels of public awareness, including through training, posters, brochures, manuals, schools curricula and 'baseline study reports' on various industrial subsectors (leather tanning, vegetable oil and ghee, soap, wool dyeing, sugar, fermentation, textiles, dairy, jute mills, brick-making, pulp and paper, paint, plastic bags and cement).
- *Cleaner Production.* Almost all industrial businesses were assessed for cleaner production needs and opportunities, and many cleaner production options were identified and responses designed. A Cleaner Production Fund was established with Himalayan Bank Limited as manager, and loan approvals increased steadily (the fund was later transferred to another component to compensate for a shortfall there). Monitoring and reporting occurred at numerous businesses, and confirmed annual reductions of 9,600 tonnes of solid waste and 43,600 tonnes of GHGs in total, and cost savings of over 805,000 Nepali rupees per business per year.
- *Energy efficiency.* Some 65 training events had about 1,700 participants, energy efficiency audits were done of 360 business units, over 2,100 options for energy and emission savings were identified, significant energy and

BOX 8.1   **(cont.)**

emission savings at 117 business units were demonstrated, an industrial energy efficiency policy was developed and public awareness was raised.

- *Institutional strengthening of ministries.* Improved institutional capacity was achieved through or in terms of: staff training; formulation and enforcement of environmental standards; preparation, evaluation and monitoring of compliance plans for industries; monitoring of ambient pollution levels; managing data on cleaner technologies among industries; industry and public awareness raising; demonstration and integration of environmental management systems and cleaner production and energy efficiency policies; and review and revision of legislation.
- *Air quality management in the Kathmandu Valley.* There were intense efforts to promote use of EVs (lobbying groups, training of mechanics and drivers, and a Clean Vehicle Promotion Fund), and better maintenance and emission control for other vehicles. Vehicle emission standards were established, and enforcement and testing systems introduced to cover about 60,000 vehicles per year. An ambient AQMS was established in the Kathmandu Valley and used to monitor particulates, benzene and nitrogen and sulphur oxides, advise the public, and inform policy, and improved air quality was achieved (e.g. a 10 per cent reduction in PM10 – particulate matter with a diameter of 10 μm or less). A huge impact on public awareness was achieved.

Lessons Learned

- *Cleaner production and energy efficiency really work.* The concept of a training institution like the IEM, and the principle that businesses should be required, and expected to be willing, to pay for compliance with environmental standards and to invest in cost savings associated with cleaner production and particularly energy efficiency (which has an immediate effect on profitability) were in the process of being established when the programme ended in 2005. That the approach is valid is shown by the global system of National Cleaner Production Centres sponsored and monitored by the United Nations Industrial Development Organization and UNEP, which embraces several Asian countries not including Nepal.
- *Urban air quality can be improved,* but multiple processes have to happen at once, including legislation, monitoring, enforcement, awareness raising, targeted investment incentives, institutional cooperation and for EVs, battery

BOX 8.1 **(cont.)**

and power unit technology, battery charging, exchange, reuse and recycling, electricity supply and distribution and training.

- *Historic opportunities do occur and can be recognised*, to allow new ideas, regulations and enforcement and compliance systems to head off a predictable deterioration in environmental quality, but if the timing is missed irreversible damage may occur and problems could become impossibly severe later on, as happened with air quality in the Kathmandu Valley.

Sustainability

Sustainability is likely for legacies of the programme where environmental management regulations exist, including:

- *compliance* based on ISO14001 – the core set of standards used by organisations for designing and implementing an effective Environmental Management System – which was taken up by the private sector through the Federation of Nepali Chambers of Commerce and Industry and the Confederation of Nepalese Industry;
- *energy efficiency activities* taken over by the Nepal Energy Efficiency Programme funded by Germany; and
- an *AQMS* in the Kathmandu Valley, which was later funded by the USA and others.

Some aspects of institutional strengthening also suggest sustainability, including the agreement of roles and responsibilities between the ministries, the establishment of new regulations and standards and the creation of the Nepal Occupational Safety and Health Association by the ministries.

Conclusion

All in all, the ESPS was an excellent demonstration of how to create mass awareness of urban and industrial environmental issues, how to introduce the idea that businesses must pay to comply with regulated environmental standards and adopt clean technology and energy efficiency measures, and how to implement and monitor environmental quality standards.

Source: adapted from Caldecott et al. (2017), annex G.10, 'Improving Urban Environments'.

What Happened Next

As described in Chapter 5, one effect of the 'palace coup' in 2005 was to abort an IEP that was to have carried forward the gains of the ESPS and other dimensions of Danida's partnership with Nepal. This put a stop to many of the things that the ESPS had been doing, and in the case of the air quality component had the effect of removing what had been set to become a key safeguard for the urban environment of the Kathmandu Valley. Competition from EVs declined and registration of fossil fuel vehicles (many of them second-hand imports) took off from 2006. During the 2000s the number of vehicles in the Kathmandu Valley tripled, industrialisation doubled and urbanisation increased at 4–5 per cent annually, all accompanied by growing use of diesel generators to compensate for an increasingly irregular power supply. Deterioration in air quality since the ESPS was ended means that Kathmandu now has some of the most unhealthy air in the world (Lodge, 2014; Saud and Paudel, 2018), contributing to Nepal's worldwide environmental performance rank of 145th of 180 countries (or 178th by air quality alone; EPI, 2020).

A US Department of State and Environmental Protection Agency joint initiative to restart ambient air quality monitoring began in Kathmandu in 2017 (US Embassy in Nepal, 2020). There was also a boom in EV sales in Nepal linked to the political interruption of fossil fuel imports from India in 2015, and the neighbouring country of Bhutan meanwhile made a commitment that 70 per cent of all its vehicles will be electric by 2025 (Basnet, 2016). But the issue is not whether air quality can be monitored after the event, or if EVs have a future as an alternative to fossil fuel vehicles, but whether an emerging environmental threat can be identified well in advance and headed off through a series of careful interlocking steps before it acquires overwhelming momentum. The story of Danida, the government of Nepal and the Kathmandu Valley municipal authorities working together on air quality shows that it is indeed possible, but that mere politics can derail even the most promising and necessary of initiatives. In this sense the story can also be seen as a microcosm of the whole global climate change response since 1992.

8.2 SELF-ORGANISED URBAN COMMUNITIES

An Urban Species

More than half of all people now live in cities and the proportion may be two-thirds by 2050, with most growth having occurred in urban areas in developing countries that now house fewer than 500,000 people (Jarvie et al., 2015). Cities are potent aggregations of infrastructure, industry, human and financial capital and voting power, so they are influential in many ways, including by emitting 70 per cent or more of the world's GHG emissions (Dhakal, 2010; Pichler et al., 2017; UNEP, 2019c, 2020; Ghanbari and Daneshvar, 2020), but they also have special needs in the context of adaptation to climate change. As the C40 Cities organisation observes, on behalf of the mayors of 96 large cities: 'Cities are as vulnerable as they are powerful. 70% of cities are already dealing with the effects of climate change, and nearly all are at risk. Over 90% of all urban areas are coastal, putting most cities on Earth at risk of flooding from rising sea levels and powerful storms' (C40 Cities, 2012; see also C40 Cities, 2020; UCLG, 2019).

The built urban environment means that system strengthening must involve design and engineering as well as human cooperation. This makes neighbourhood activism important as an essential part of the adaptation process, yet city governance mechanisms are often at odds with the kinds of adaptive, learning-oriented processes that are at the heart of climate resilience (Friend et al., 2014). This can be corrected through the mass involvement of urban residents in identifying, mapping, considering and reinforcing vulnerable points in their own environments, an approach that the Resilience Academy and SUZA are using in Zanzibar (Chapter 7). This is appropriate because in cities many of the most significant contributions to resilience are related to knowledge, networks, information and greater engagement of citizens (Reed et al., 2014).

A Rural Bias

Like almost all 'aid literature', this book gives special attention to natural, traditional, rural and coastal systems, rather than to people living in modern city environments. The astonishing urbanisation of our species, as we aggregate in cities to benefit from their density of people, jobs and opportunities to interact with others, is relatively neglected, both here and in general. This is because of a number of concerns that combine to create a rural bias, including special interests and feelings of loss, fear or outrage due to:

- the decline and endangerment of ecosystems, biodiversity and ecological services;
- the decline and endangerment of cultural and linguistic diversity – cosmologies, mythologies, languages and ways of thinking about and running societies;
- the tendency for people in rural areas to be 'left behind' in what may be seen as squalor, poverty and ignorance by the advance of cosmopolitan urban modernity;
- the tendency for those 'left behind' to be marginalised, exploited and oppressed by urban elites bolstered by various entitlement myths; and
- the fact that cities depend upon the renewable and non-renewable goods and services provided by the ecological and social systems and physical fabric of their hinterlands (which have grown to regional and global scale in recent decades), often obtained through unfair terms of trade and creating an unsustainable and offensive regional or global 'footprint'.

Just one example of how such feelings originate is provided by scuba diving in Indonesia, the experience of which since the early 1990s has been degraded in area after area by the advance of plastic waste and other detritus into the sea from urban areas. Coral reefs around islands near Java, for instance in the Pulau Seribu ('thousand islands') marine park north of Jakarta, which were clean and valuable ecotourism resorts in the early 1990s, steadily became waste fields over the following decade. This kind of effect is a global phenomenon, downstream, downwind or down-current of any major, rapidly growing and poorly administered city. The net effect is to reduce affordable and satisfactory recreational opportunities for the city dwellers them-selves, while inducing feelings of loss among those who remember a cleaner world.

To correct these things city dwellers and urban elites must somehow be persuaded to pay more money and attention to preserve the complex hinterland systems that sustain them. But this can be hard to do when people are struggling to make a living in a largely asocial and often brutally competitive urban environment, hampered by weak educational services that barely teach literacy, let alone an appreciation of biological and cultural diversity or where food and water come from, and where wastes go to and with what consequences. There have been occasional efforts by rural people to impose their priorities directly on city dwellers (e.g. in Bolivia, where indigenous peoples' marches on La Paz in 2002 led to the rise of the MAS, see Chapter 6), but some did not end well (e.g. in Cambodia, where the rural-based Khmer Rouge regime forced the total evacu-ation of Phnom Penh in 1975). More general and stable solutions must lie in education and leadership within the cities themselves.

8.3   SUSTAINABLE CITIES?

Meanwhile, though, if it is argued that a city needs its hinterland systems, and should treat them better, the question also arises of whether, in the longer term, it is possible for cities to be self-sufficient. Part of the answer is that cities can go a long way towards true self-sufficiency by putting all of the following thoroughly in place:

- local food supply from intensive organic horticulture on roof-tops and in private gardens, public parks and community allotments, and from neighbourhood and household hydroponic, aquaculture, microlivestock and invertebrate protein farms;
- local water conservation through household and neighbourhood rain harvesting, water purification and water recycling;
- local sourcing of biogas and fertiliser from household and neighbourhood waste and sewage fermentation;

- local sourcing and distribution of energy through household and neighbourhood biomass, solar, wind and geothermal energy systems;
- more efficient use of personal energies and social behaviour, for example through use of local currencies based on the trading of time spent in social care, labour commitments and local produce (TBUK, 2020; Letslink UK, 2020), or through residents' associations and other self-organised local networks (see next subsection, 'Self-organised Local Networks in Bath and Edinburgh');
- trading of distinctive city products and features with other cities (honey, textiles, designs, educational and recreational opportunities from specialist universities to casinos, etc.);
- the deliberate redesign of cities to be much more liveable, efficient and self-sustaining, though public transport, architecture, pedestrianisation, etc.; and
- powerful and locally accountable self-government.

All these things are highly desirable, and many cities are doing some or all of them (e.g. C40 Cities, 2020; C40 Cities et al., 2019). In these ways they can minimise their environmental and harmful social footprints, and maximise the well-being of their inhabitants. A wise city administration would give much thought to – and reach out to their own hinterlands and other cities to discuss – the issue of sustainable access to those goods and services that the city cannot now, or cannot ever, provide for itself. These include the educational and recreational opportunities offered by landscapes, nature reserves and communities in the hinterlands. They also include the issues of how to pay a fair price for the water supplies and environmental security services offered by large and fully functional ecosystems, as well as protections against many of the larger-scale manifestations of a chaotic climate. But because it will take decades to organise all cities in all these ways, and meanwhile natural and political disasters, both slow and fast acting, will occur, it would make sense for all cities:

- to recognise that neighbourhoods can and should have effective residents' associations to build local solidarity, to protect their own interests, to guide the city administration to help meet their special needs and to prepare for emergencies of any and all kinds;

- to accept that these residents' associations are more useful to city administrations the more self-organised and independent-minded they are; and
- to take action to encourage, enable and partly fund residents' associations without strings attached, as accepted components of a partnership between organised neighbourhoods and accountable local government.

Self-Organised Local Networks in Bath and Edinburgh

Two cases are described here, in the cities of Bath in England and Edinburgh in Scotland. In both the initiative arose from individuals reaching out to neighbours, who saw the potential advantages and joined in. The membership of both stabilised at about 50 household contacts representing 100–150 people in total. The Bath group was founded in 2010 and grew up in a suburban area of detached and semi-detached houses. It started off with social gatherings before starting to lobby the city council to address issues such as overhanging branches, parking permits and gritting roads in icy weather. It later became more formally organised, registering with the city council, appointing officers and joining the Federation of Bath Residents' Associations. In this form it was better able to lobby for improved services and comment on local planning applications and plans for city-wide traffic and air quality management.

Among the other topics that the network addressed were the setting up a memorial nature reserve (for one of the founders, Don Grimes), rehabilitating a local park so that it earned 'Green Flag' recognition from the Department for Communities and Local Government (which has been renewed in every year up to and including 2020) and the issue of people replacing their front gardens on a local hill with parking spaces, because of its implications for water run-off and flash flooding. It was agreed to hold an open meeting every few weeks and some kind of larger event every six months, whether a street party for a royal wedding or birthday, or a 'social' in a local pub or garden, which gave the residents an opportunity to help plan the event and contribute food, equipment or services.

Meanwhile, email addresses and phone numbers were exchanged so that a network of people arose who had an interest in spreading news, for example warnings about a burst water pipe and criminal activity, or notices of firework parties, funerals, planning-permission appeals or litter collections in local parks. Vulnerable individuals were identified and the group made aware of who would need to be checked on in the event of power cuts or extreme weather, and some funds were pooled to allow road salt to be stored in the sheds of able-bodied members against severe winter conditions on the local hill.

The Edinburgh group was set up in early 2020 in response to the Covid-19 lockdown, and covers an inner-city crescent of Victorian apartment buildings. It started with paper notes through a few letter boxes, but email contacts were then shared and smartphones later became the main means of networking. Vulnerable and 'shielded' people were identified and volunteers came forward to provide them with special help. Separate groups were set up and used for chats (e.g. exchanging information about services and sources of good food during the lockdown) and emergencies (e.g. an alert that a man was behaving threateningly in the local park). The weekly 'clap for carers' event in from late March to late May 2020 provided an opportunity for a collective display of solidarity.

For both groups an inventory was done to identify the skills available within the network, which would be available for any purpose that might arise, from entertainment to survival. The lists of these skills for the two groups are given in Box 8.2.

Thus, with almost no public or private financial investment a neighbourhood turned from a scattering of strangers into a group that was capable of discussing risks and opportunities, acting collectively to protect and advance its own interests and sharing information about its own capabilities. All it would take for either network to become a local climate change adaptation and disaster preparedness group would be to allocate responsibilities and do some training, for example by adapting disaster preparedness guidelines described in

BOX 8.2 **Skills available within two self-organised local networks**

Bath, England (46 contacts, as of June 2012): painting-decorating, gardening, DIY, personal fitness training, languages (Italian, Indonesian, Mandarin, Cantonese, French, German, Polish, Russian, Welsh), general building (all trades), curtain/blind-making, doula (birth companion, post-natal care), musical performance and teaching, driving for the disabled, marine engineering, business and charity start-ups, pet care, environmental consultancy, cranial osteopathy (people and horses), first aid, marketing consultancy, chainsaw operation, electrical engineering, fixing computers, cooking, organising (events, activities, institutions), advice on home security, boules coaching, helping the elderly (e.g. filling in forms), craft teaching, kitchen design/installation, Braille transcription, public speaking, English teaching, model railways, office skills, yacht crewing, babysitting, driving, engineering, insurance brokerage, IT consulting, sports team management, architecture, children's physiotherapy, and chartered surveying (e.g. for building projects).

Edinburgh, Scotland (53 contacts, as of June 2020): household DIY, bicycle and motor-vehicle repairs, errands, eye care advice, emergency electrical repairs, social contact/chatting/reassurance, maths tuition, ecological gardening, cooking, shopping, science/physics teaching, book/film/music lending, home care, dog care, DJing, innovation and leadership, database design, osteopathy, counselling and anxiety management, creativity, soup-making, teaching English (to native speakers and as a second language), journalism (writing, subediting, proofreading, teaching), languages (Spanish and French interpreting/translating, some Italian and Indonesian), food bank management.

Source: author, based on declarations by network members.

handbooks from UNEP's Awareness and Preparedness for Emergencies at Local Level programme (UNEP-DTIE, 2015) or the EC's equivalent (ECHO, 2003; ECHO and ADPC, 2004). Since these measures greatly increase community connectivity and recorded knowledge on who lives where, who does what, who needs what and who can contribute what, communities that have prepared in this way are likely to cope better than others during and immediately after disasters, and to need little in the way of 'registration' assistance in the first stage of disaster relief, thus potentially speeding up the delivery of external assistance.

PART IV **Global Perspectives**

9 Changing Ideas of Adaptation

9.1 INCONCLUSIVE DEBATE, 1998–2008

Both mitigation and adaptation are essential parts of the climate response, yet they have very different underlying mechanisms and implications for designing actions and judging their success. Mitigation is about reducing absolute GHG emission rates, and/or recapturing free GHGs, in order to stabilise and then reduce their concentration in the air. This is technically, financially and politically demanding but in concept straightforward. Not so with adaptation, where solutions must be found that involve supremely complex systems of climate, ecology and human society all interacting under increasingly chaotic conditions at all levels of global society, including especially the local level where people and ecosystems coexist in any number of different configurations. Moreover, our understanding of mitigation is far deeper than it is for adaptation, since efforts to understand it started earlier and with greater urgency. The timeline and key outputs of the process of trying to work out what adaptation means, and how it might be designed for and measured, are described briefly in this chapter.

Rio Climate Markers

The UNFCCC created a need for countries to show their peers that they are doing their best to comply with it. Among other things, this required official donors to record their expenditure in line with the aims of the treaty. A process of research and dialogue resulted, to define and agree what was meant by a 'relevant' investment. The resulting checklists represent the kinds of investment that qualify to be recorded, and are a reasonable way to classify the flow of aid funding into the broad areas concerned. In the case of mitigation and adaptation, the checklists are known as the Rio Climate Markers, and those for tracking aid invested partly or wholly in mitigating climate change were agreed in 1998 (Box 9.1).

Debate continued around how best to reduce net GHG emissions by technological means (e.g. renewable energy and energy efficiency) and ecological means (e.g. REDD+ and organic farming), and who would pay for these mitigation actions. Progress was agonisingly slow, and the issue inevitably arose regarding whether it was ethical or wise to invest in adaptation while the chief drivers of climate change remained unmitigated. As the need for adaptation came to be accepted, however, discussions began over what it meant and how to make it fair and effective (e.g. Pielke, 1998; Adger et al., 2006; OECD, 2006; Roberts and Parks, 2007; UNDP, 2007b). As frustration grew with the complexity and indecision of the mitigation discussions, initiatives began with the aim of clarifying how to programme adaptation investments as well or instead.

9.2   ADAPTATION MARKERS, 2009–2010

Agreement was eventually reached in 2010 on Rio Climate Markers for adaptation (Box 9.2), although they only provide guidance on actions to be encouraged, such as 'promotion of climate resilient agriculture, food security and basic services', 'road construction to account for climate change impacts and variability' and 'infrastructure and hubs that would support improved business continuity during and after extreme weather events' (OECD/DAC, undated). These make sense, but do not make it easy to judge progress, or to decide on relative contributions or priorities for investment, and in any case are considered unreliable due to a lack of quality control which would require much better criteria and indicators to correct (Weikmans et al., 2017).

BOX 9.1 Actions that promote mitigation

Practical Actions for Climate Change Mitigation

- *Technology*. Reducing or stabilising GHG emissions in the waste and sewage management, transport, energy, agricultural, construction, industrial and other sectors through application of new and renewable forms of energy, measures to improve the energy efficiency of existing generators, machines and equipment, or demand-side management.
- *Ecology*. Protecting or enhancing GHG sinks and reservoirs through forest protection, avoided deforestation, sustainable forest management, reforestation, restoration of disturbed ecosystems (including soils through organic farming), rehabilitation of areas affected by drought and desertification and sustainable management and conservation of oceans and other marine and coastal ecosystems, wetlands, wilderness areas and other ecosystems.
- *Capacity*. Developing, transferring and promoting emission-reducing technologies and know-how, including building capacity to control, reduce, prevent or reverse emissions of GHGs in waste and sewage management, transport, energy, agriculture, forestry, construction, industry and other sectors.

Enabling Frameworks for Climate Change Mitigation

- *Mainstreaming*. Integrating mitigation concerns and priorities within development processes, through preparation of national inventories of GHGs (emissions by sources and removals by sinks), mitigation-related policies, legislation and economic analysis and instruments, low-carbon development strategies and plans, needs assessments on mitigation technology and building relevant institutional capacity.
- *Regulations and incentives*. Strengthening regulatory frameworks to put in place fiscal, economic, legal and other incentives that discourage GHG emissions and encourage investment in reducing them.
- *Education and training*. Promoting mitigation-related education, training and public awareness.
- *Research and monitoring*. Promoting research and monitoring efforts focused on mitigation and the understanding of oceanographic and atmospheric systems and processes.

Source: OECD (2011).

BOX 9.2   **Actions that promote adaptation**

Practical Actions for Climate Change Adaptation

- *Resilience.* Making landscapes, farming systems and communities more resilient to environmental change, including (as appropriate to changes anticipated in each location) through measures to safeguard or restore the ecosystem services of water catchments, floodplains, wetlands, mangroves, coral reefs, beach dunes and aquifer recharge areas, conserving water and introducing water-saving irrigation methods, introducing crops that are resistant to heat, drought, submergence and salinity, prophylaxis against vector-born and other diseases, amending fishery management practices in response to new ecological conditions and changing fish populations, promoting diverse forest management practices and species, developing emergency prevention and disaster preparedness measures (including insurance and engineering works to relieve known threats, e.g. from glacial lake outburst floods and sea-borne storms).
- *Knowledge.* Promoting stakeholder environmental monitoring and networking to enhance sharing of knowledge on environmental change, threats, solutions and adaptation best practices (as appropriate to changes anticipated in each location), including the building of social capital, cooperation and adaptation/disaster preparedness, and the production and dissemination of public information materials on the principles and practices of adaptation.

Enabling Frameworks for Climate Change Adaptation

- *Mainstreaming.* Supporting the integration of adaptation into national and international policies, plans and programmes, and strengthening the capacity of key national institutions (including finance and planning ministries) to coordinate and plan for adaptation activities, and integrate adaptation into planning and budgeting.
- *Disaster risk reduction.* Building capacity for disaster risk reduction, preparation and management at local, national and regional level, by making disaster-relevant information and tools more accessible to all, by promoting disaster consciousness in adaptation policies, strategies and programmes, and encouraging systematic dialogue, information exchange and joint working between climate change and disaster reduction institutions and experts, in collaboration with policy makers and development practitioners.

BOX 9.2 **(cont.)**

- *Education and training.* Promoting adaptation-related education, training and public awareness raising.
- *Research and monitoring.* Promoting research focused on environmental change, and weather, climate and water monitoring and information systems, including observation and forecasting, impact and vulnerability assessments and early warning systems, and on how to make landscapes, farming systems and communities more resilient to detected or anticipated changes.

Source: OECD (2011)

9.3 DIVERSE FRAMEWORKS, 2010–2014

There followed a period that saw the development and trial of diverse criteria, indicators, tools and frameworks for guiding adaptation investment and assessing results, including:

- Making Adaptation Count (Spearman and McGray, 2011);
- Learning to ADAPT (Villanueva, 2011);
- the Adaptation Fund Manual (Adaptation Fund, 2011);
- the UK Climate Impacts Programme/AdaptME Toolkit (Pringle, 2011);
- the Adaptation Monitoring and Assessment Tool (GEF 2010, 2012);
- Adaptation Made to Measure (Olivier et al., 2013);
- Tracking Adaptation and Measuring Development (Brooks et al. 2011, 2013; IIED 2012);
- the TANGO Resilience Assessment Framework (Frankenberger et al., 2012);
- the IISD Climate Resilience and Food Security Framework (Tyler et al., 2013);
- the Global Programme of Research on Climate Change Vulnerability, Impacts and Adaptation (Hinkel et al., 2013);
- the PPCR (Strategic Climate Fund) Monitoring and Reporting Toolkit (CIF, 2012);
- the Community-Based Resilience Assessment tool (UNDP, 2014a, 2014b, 2014c); and

- Participatory Monitoring, Evaluation, Reflection and Learning for Community-Based Adaptation, for CARE International (Rossing et al., 2012).

In addition, a series of high-level analyses and indices of risk and adaptation compiled from national data and oriented to the global level began to be prepared, including the *World Risk Report* that has been published annually by insurance companies since 2011 (e.g. BEH, 2017) and was later adopted by the World Economic Forum (WEF, 2020), and the preliminary *Adaptation Gap Report* by UNEP (2014) and its successors (UNEP, 2017, 2018). In this period too, Moser and Boykoff (2013a) edited the work of 39 contributors and 24 peer reviewers in order to explore the issue of how to recognise successful adaptation. The editors concluded that the systems involved were so complex and dynamic, and defining adaptation success therefore so hard, that an particularly pragmatic approach was required, based on principles such as the need to assess and communicate all risks, to make full use of all sources of knowledge, to involve all stakeholders and to consider all potential side effects (Moser and Boykoff, 2013b). The contributors also produced a checklist of topics in the 6–10 key realms where solutions were needed, often simultaneously (see Box 9.3 for an extract). This further illustrated the challenge of making sense of complex systems, even without considering that the realms are all 'deeply value-laden and perpetually contested' (Moser and Boykoff 2013c: 24). The overall conclusion was that the science and practices of adaptation were still in their infancy.

Meanwhile, the vocabulary of adaptation was proving to be just as slippery as the topic itself. Moser and Boykoff (2013c: 11), for example, distinguished between: a short-term and sectoral *adaptation* approach, to patch things up or harden them against known threats at least cost; a *vulnerability* approach to protect and empower the vulnerable and excluded in each society; and a future-oriented *resilience* approach, which 'focuses on large-scale coupled social-ecological systems with the goal of enhancing overall system capacity to persist,

BOX 9.3 **Dimensions of potential adaptation success**

The contributions collected here address many of the dimensions that shape common notions of success:

- *Economic dimensions* – how to minimize or avoid losses, damages, and adaptation costs, while maintaining, creating, or banking on possible benefits and opportunities;
- *Institutional and policy dimensions* – how to formally account for obligations to each other and to non-human beings when establishing and promoting particular actions and behaviors;
- *Ecological and environmental dimensions* – how to value and foster resilience, cultivate diversity and health in the biosphere, and continue to provide vital ecosystem services;
- *Social dimensions* – how to reduce inequities, vulnerabilities, and, in turn, how to strengthen communities, livelihoods, and justice;
- *Political and procedural dimensions* – how to support transparency, inclusiveness, and collective learning via democratic and legally defensible responses to climate change; and
- *Cultural and psychological dimensions* – how to create and retain the highest quality of life, meaning, and happiness, sense of community, and connection to place.

Each of these dimensions, and the interactions among them, must be the focus of further research in years to come ... Of course, each of these dimensions is deeply value-laden and perpetually contested on the never-level playing field of social relations in any on-the-ground adaptation process. Our goal in this volume is not to resolve the tensions among these dimensions, but to name them, and to compile encouraging evidence that resolutions can be found.

Source: Moser and Boykoff (2013c: 23–24).

recover, and renew after disturbance, and thus minimising the probability of rapid, undesirable, and irreversible system changes [with an emphasis] on the continuation of ecological functioning and the provision of ecological services'. They noted that these approaches had elsewhere been termed 'resilience', 'transition' and 'transformation' respectively, and that smaller interventions had previously been called 'adjustments' while deeper and lasting ones had been called 'adaptations'. In the same volume, Preston et al. (2013: 160–1) described levels of adaptation ranging 'from incremental adaptation (adjustments to existing systems) at one end of the spectrum to transformational adaptation (creation of new systems) at the other'.

9.4   KNOWLEDGE CONCERNS, 2014–2015

While the 2015 Paris Agreement was being negotiated, progress on adaptation was being reviewed and conclusions drawn (e.g. by Bours et al., 2014a, 2014b, 2014c, 2014d, 2014e; Smith et al., 2014; Leagnavar et al., 2015; Silvestrini et al., 2015). It was found that a range of mixed methods, participatory procedures, qualitative narratives and theory of change-based approaches were in use, of necessity since other methods of design and evaluation were compromised by unresolved issues that included the following:

- *Attribution and complexity of determinants.* The difficulty of attributing change is characteristic of policies and programmes that address complex issues.
- *Accounting for maladaptation.* Adaptation interventions can result in negative outcomes for people or the environment, for example by increasing GHG emissions, adding burdens to the most vulnerable, having high opportunity costs, reducing incentives and capacity to adapt and/or by setting paths that limit future choices and increase vulnerabilities.
- *Counterfactuals.* Establishing adaptation success often requires comparison against hypothetical scenarios, which are hard to formulate robustly.
- *Shifting baselines.* Comparing effects to baselines is hard because underlying conditions are themselves changing in uncertain and emergent ways.
- *Variable time horizons.* Impacts take time to work and before they can be measured, while changes in adaptive capacity, vulnerability and socioeconomic conditions are also important.
- *Adaptation as a moving target.* Exposure to hazards can change during a project, so the target at the beginning might not be the same as at its end.

- *Cascade of uncertainty*. Uncertainty gets worse as adaptation is considered at each level of a path from GHG emission projections to climatic models and types of adaptation response.
- *Lack of a conceptual agreement on definitions and aims*. There is no uniform definition for adaptation, or for what successful adaptation should look like. As 'success' implies achieving an outcome, while adaptation is a process that lacks an end point, an emphasis on success may not be quite to the point.

About this time, too, a 'realist' synthesis review of UN agency adaptation project evaluations by Miyaguchi and Uitto (2015) laid the foundations for the approach taken here (Chapters 5–8 and 12). They described theory-based evaluation as an approach that examines the assumptions that underlie the expectation of outcomes, and which are therefore folded into the theory of change. Thus, the aim is to explain not just what the programme was expected to achieve, but also how and why it was likely to achieve it (or not). The importance of this lies in the clarity which it brings to the question of how desired changes are activated and delivered, often in complex systems and circumstances and therefore particularly suitable for evaluating adaptation projects.

Realism takes the theory-based evaluation approach further, by identifying underlying causal mechanisms and exploring how they work and under what conditions, thus offering a way to make sense of projects in 'high causal density' environments, where there is much 'uncertainty coming from the differing types and interests of stakeholders and openness to exogenous influences' (Miyaguchi and Uitto, 2015: 4). Because interventions only work to the extent that they offer ideas and opportunities to people in appropriate social and cultural conditions, the realist approach is based on a search for mechanism, context and outcome relationships, where:

- *mechanisms* are the possible lines of causation, not the interventions themselves but the ways that they create change;
- *contexts* are the external and internal circumstances in each case that render the mechanisms active or that prevent them from working; and

- *outcomes* are the intended or unintended consequences, which can take many forms so the aim is to work out why and how certain outcomes emerge depending on the context.

Applying this to adaptation project evaluations in nine countries, the authors found enough (577) remarks in the reports that could be used to shed light on relevance, efficiency, effectiveness and sustainability, and used them to reach conclusions similar to previous portfolio synthesis studies using rather similar methods (e.g. Caldecott et al., 2010):

- that *relevance* to policy is a high priority of all concerned so is typically high, but this does not necessarily lead to strong design or performance;
- that *efficiency* is typically undermined by high staff turnover and poor financial delivery, and improved by investing effort in stakeholder involvement and partnership building, and through competent and effective leadership;
- that *effectiveness* is often poorly defined (e.g. as output delivery), but it can be high if based on careful capacity building, systematic mainstreaming and strong public awareness and support for the aims of the intervention; and
- that *sustainability* is best guaranteed by stakeholder engagement and demand for the continued delivery of whatever services have been introduced.

9.5   ADAPTATION RESEARCH AFTER PARIS

Building on the work that helped to shape the Paris Agreement (e.g. Aerts et al., 2012; Boulter et al., 2013; Bob and Bronkhorst, 2014; Field et al., 2014), as it entered into force there began an explosion of research to explore the economic, institutional and policy, ecological and environmental, social, political and procedural, and cultural and psychological dimensions that had already been identified by Moser and Boykoff (2013c), as well as the issues of attribution, complexity, maladaptation, counterfactuals, baselines, variable horizons, moving targets, uncertainty and key concepts raised by Bours et al. (2014a) and others. This yielded a large literature, including many book-length syntheses and compilations of articles and case studies organised around such themes as:

- *the utility of traditional knowledge* (e.g. Bryant-Tokalau, 2018; Nakashima et al., 2012, 2018);
- *issues special to, or studied with a focus on, specific geographies*, including Latin America (e.g. Leal Filho and de Freitas, 2018), North America (e.g. Leal Filho and Keenan, 2017), the Asia-Pacific (e.g. Leal Filho, 2017), Africa (e.g. Yaro and Hesselberg, 2016; Leal Filho et al., 2017, Leal Filho et al., 2021), small islands and coastlines (e.g. Petzold, 2017; Bush, 2018; Leal Filho, 2018) and the limits of adaptation in Asia, Africa, the Pacific and elsewhere (e.g. Leal Filho and Nalau, 2018);
- *education, communication and knowledge management* (e.g. Azeiteiro et al., 2018; Leal Filho et al., 2018a, 2018b);
- *disaster risk reduction* (e.g. Kelman et al., 2017; Yokomatsu and Hochrainer-Stigler, 2020); and
- *many other development issues*, from health, farming, cities and science to conflict, gender, human rights and governance (e.g. Leal Filho, Azeiteiro et al., 2016; Leal Filho, Adamson et al., 2016; Kabisch et al., 2017; Lipper et al., 2017; Tangney, 2017; Alverson and Zommers, 2018; Alves et al., 2018; Berck et al., 2018; Carstensen, 2017; Klepp and Chavez-Rodriguez, 2018; Aboulnaga et al., 2019; Castro et al., 2019; Iizumi et al., 2019; Marselle et al., 2019; Sarkar et al., 2019; Siegel, 2019; Indrawan et al., 2020).

In short, the entire body of global knowledge about development, created over decades of aid investment, was being re-evaluated from the point of view of what climate change threatens to disrupt, and what can be done about it. This was necessary, and remains so, since climate change clearly has the potential to undermine, halt or reverse all previous development gains. But from the point of view of aid professionals wishing to invest in adaptation, the process has evidently not yet reached a useful end point.

The research that began in the early 2010s with a bewildering array of adaptation toolkits and frameworks went on to raise serious epistemological questions in the mid-2010s, and has since yielded a mass of detail by thousands of authors. But there is the feeling that it has now become immobilised by efforts to reconcile contextual detail with narrow technobureaucratic conceptions, without the emergence of a useful synthesis, and that therefore climate change research is at an impasse (Nightingale et al., 2020).

The continuing lack of useful synthesis is consistent with the call in 2018 by Sweden's Expert Group for Aid Studies (EBA), for proposals to evaluate the 2009–2012 adaptation-focused Climate Change Initiative of the Swedish government. It was specified that:

> Tenderers are given an open mandate regarding the design of the analytical framework, methodological approach, delimitations and evaluation model to fulfil the objective and overall aim with the study. The EBA encourages tenderers to let their expertise guide the choice of approach in answering the evaluation questions. We hope that this open task will be attractive and encourage innovation in submitted proposals.
>
> *(EBA, 2018: 4)*

This implies that the challenges of concepts, indicators, baselines and certainties surrounding adaptation had not yet been resolved to the satisfaction of those professionals who manage the public financing, implementation and reporting of the global adaptation effort.

But these people do not themselves need large numbers of case studies or theoretical discussions. Rather they need confidence that they understand the meaning of those studies and can tell 'what successful adaptation should look like' (Leagnavar et al., 2015: 11), and why, alongside simple, defensible ways to make judgements about how to design and evaluate investments that are likely to perform well in an extremely wide variety of contexts. This calls for an approach that is responsive but robust to the details of particular cases, that tends to promote 'no-regrets' actions, and that also complies with the principles of sustainable development. The difficulty of deriving such simple, reliable ('algorithmic') prescriptions, and the abundance of pathways that lead only into thickets of conditional and contextual uncertainty, was demonstrated by the Moser and Boykoff (2013a) study.

9.6   IDEAS OF ADAPTATION AID

Faced with political pressure to spend money fast on adaptation while not quite knowing how best to do it, the aid community has sought to focus on:

> a broad-based approach, which not only addresses the immediate impacts of particular climate change hazards and risks, e.g. acute response to disasters such as floods and droughts, but also helps build the socioeconomic and institutional foundations for a transformation towards long-term resilience. This requires support at different levels and across sectoral boundaries, and attention to how climate change adaptation is mainstreamed into development to address an unacceptable large and growing adaptation deficit. The notion is that the adaptation deficit can be more effectively addressed by combining the work of the Climate Convention with the development aid process and mainstreaming climate risks. By developing a more coherent and operational adaptation regime collective success in climate change adaptation is believed to be more likely.
>
> *(Danida, 2019: 2)*

Thus, adaptation was to be added to the cross-cutting themes and mainstreamed, by requiring it to be considered in all investment programmes – that is, written into needs analyses, project designs, operational procedures and evaluation criteria – while also, where possible, being the subject of targeted investment. Citing a draft of Funder et al. (2020a), the Danida document just mentioned proposes that such investment might be organised around three themes:

- *poverty and vulnerable groups*, where initiatives target people who have been marginalised by poverty, location, gender or faith, or who are vulnerable because they live in flood-prone, drought-prone or storm-prone areas;
- *resilient livelihoods*, where initiatives promote management of renewable natural resources and ecosystem services, climate-smart farming and

forestry-based livelihoods, and the extension services, infrastructure and training that support them (and where most recorded adaptation aid is concentrated in the agriculture, forestry and fishing sector, and the water supply and sanitation sector; OECD, 2019a); and

- *transformative responses*, where initiatives promote a general strengthening of societies by building institutional capacity (e.g. for monitoring and responding to climate variability and resisting its impacts), disaster preparedness, integration of climate change thinking into development planning, new forms of knowledge exchange and social organisation (e.g. farmers' or women's groups), processes for devising new adaptation policies and economy-wide shifts towards low-carbon, climate-resilient and biodiversity-friendly forms of development.

In all three themes it can be hard to distinguish 'adaptation' investment from any other kind, since there is so much overlap with 'ordinary' aid, which routinely aims to improve the fortunes of vulnerable groups, to promote sustainable livelihoods and to encourage and enable capacity improvements, policy development and economy-wide transitions in line with the SDGs. This is not always or necessarily a problem, but the adaptation elements might be in competition with other efforts and this can take two forms. First, when what is done for short-term economic gain is vulnerable to climate risks that were underestimated when the investment was designed. This is probably a frequent occurrence at a time when the perceived severity of climate risks has been increasing so fast, thereby overtaking and undermining the rationales of earlier projects. An example is a Danida-funded water treatment plant in the Hetauda Industrial District of Nepal, the design of which was inadequate to cope with exceptional rainfall (Caldecott et al., 2017).

Of more significance, perhaps, is that traditional aid projects, especially in the 'resilient livelihoods' theme, may simply be reclassified as adaptation projects without really changing their design, thus undermining incrementality for adaptation financing (see Section 9.7).

Moreover, adaptation projects may now be being designed by a new generation of adaptation professionals, including old consultants rebranding themselves, and new staff recruited from university adaptation courses who lack a firm grounding in the lessons learned from the experience of 'ordinary' development projects over previous decades.

In these circumstances, driven by a rush to demonstrate increased investment in adaptation, projects may be designed that are weak in participation or economic appraisal, may involve significant but unanticipated realignments of people, their resources, economies and ecologies, or the entanglement of new and pre-existing relationships, and may therefore fail to avoid pitfalls of design and performance. This issue is discussed with examples from Nepal (such as an unmaintained greenhouse, a cider press without electricity and a defunct drinking water post) by Nightingale (2020).

The authors of Danida (2019) also point out that because adaptation is embedded in local as opposed to national circumstances, it is hard to address without also addressing the issue of decentralisation. This is consistent with the view that with climate chaos most threatening to local social and ecological systems, adaptation must involve empowerment and environmental action at the local level. The Danida study on adaptation was underway at the time of writing, but its draft report confirmed the importance of community empowerment, ecosystem-based approaches and engagement with community-based organisations and local government, while also noting that adaptation needs are 'highly situation specific, complex and subject to uncertainty, low levels of capacity among partners and wavering levels of political support at country level' (PEMconsult and ODI, 2020: 13). Meanwhile, recommendations by Funder et al. (2020b) highlight the scale of changes needed to upgrade the strategy of Danida (and by extension of other progressive aid donors) in response to climate change (Box 9.4).

BOX 9.4   **Steps to integrate adaptation and development**

Priority needs include:

- *for a clear overall strategy to support the climate response,* with equal attention to mitigation and adaptation since 'the world will most likely move beyond both the 1.5- and 2-degree goals' and 'adaptation efforts must be expanded if the SDGs are to be achieved';
- *for the climate response to be treated as a key aim* of development cooperation, 'alongside poverty alleviation and other core issues';
- *for an ambitious, consistent, integrated and mainstreaming approach* to adaptation across all aid portfolios (and extending it to health 'in the wake of the Covid 19 crisis', and other areas that include 'private-sector support, employment, peace, migration and governance');
- *for more systemic thinking,* meaning that support to agriculture, water and infrastructure, for example, should 'be complemented with a stronger focus on integrating adaptation into ecosystem and resource management, including links to the parallel biodiversity crisis';
- *for adaptation in programming from the outset* through 'early and clear analysis of how support will address climate risks and associated vulnerability';
- *for inclusive decision-making* on adaptation as it becomes integrated into public planning across society, 'including an emphasis on the principle of subsidiarity';
- *for an increased focus on partner countries' own NDCs* and their associated national climate and development policies;
- *for learning from others while promoting adaptation as a priority* among international and intergovernmental organisations.

Source: Funder et al. (2020b).

9.7   ADAPTATION AID BEFORE AND AFTER PARIS

The scale of public funding for the climate response has meanwhile been increasing along with the share of it going to adaptation (OECD, 2019b). The Paris Agreement at the end of 2015 focused and

Table 9.1 *Public climate finance commitments, 2013–2017*

	2013	2014	2015	2016	2017
Total climate response (US$ billions)					
Multilateral	22.5	23.1	25.9	28.0	27.0
Bilateral	15.5	20.4	16.2	18.9	27.5
Total	**37.9**	**43.5**	**42.1**	**46.9**	**54.5**
Mean total	Before Paris Agreement: 41.2 After Paris Agreement: 50.7				
Adaptation, dual purpose and mitigation (US$ billions)					
Adaptation	7.8	8.1	8.1	9.7	12.9
Dual purpose	3.5	4.9	4.7	5.9	4.8
Mitigation	26.6	30.5	29.3	31.3	36.8
Mean adaptation	Before Paris Agreement: 6.0 After Paris Agreement: 11.3				
Adaptation, dual purpose and mitigation (per cent)					
Adaptation	20.6	18.6	19.2	20.7	23.7
Dual purpose	9.2	11.3	11.2	12.6	8.8
Mitigation	70.2	70.1	69.9	66.7	67.5
Mean adaptation	Before Paris Agreement: 19.4 After Paris Agreement: 22.3				

Source: OECD (2019b).

re-energised the global climate response, and reminded donors to seek parity between mitigation and adaptation objectives. In response, climate financing commitments increased from an average annual total of US$41.2 billion in 2013–2015 to US$50.07 billion in 2016–2017, and the share of these funds allocated to adaptation increased from 19.4 per cent to 22.3 per cent (Table 9.1). Later data suggest that the trend is continuing; for example, of US$40.3 billion recorded as allocated in 2018 to either adaptation or mitigation by Development Assistance Committee members (i.e. without separating 'dual purpose'), 40.6 per cent was assigned to adaptation (OECD, 2018).

The sums involved in public climate aid amount to about a third of total aid flows from all sources to all purposes, but are only part of total climate response financing. The latter, including non-aid (e.g. private decarbonisation) investments, rose from a mean of US$365 billion annually in 2013–2014 to US$463 billion in 2015–2016 and US$579 billion in 2017–2018, of which about 43 per cent (± c. 3 per cent) were public (including non-aid) investments (Buchner et al., 2019). This diversity of sources, terms and purposes for financing, all lumped together as 'climate financing', makes it hard for those with a specific interest to reach firm conclusions, for example on exactly what is being done to help developing countries cope with climate chaos.

Meanwhile, the climate response commitment at CoP 15/2009 of US$100 billion per year by 2020 to support developing countries turned out to comprise a mixture of funding from a wide variety of sources, public and private, through various channels, and dominated by loans and mitigation investments (Oxfam, 2020). From CoP 16/2010 to CoP 20/2014, growing commitments seemed on track to meet the goal (OECD, 2015), but CoP 21/2015 postponed it from 2020 to 2025, with a later CoP required to establish a new goal of at least as much.

Especially among developing countries that are suffering climate impacts, and their advocates, the whole system is a source of numerous misgivings. Problems include the mixing of public and private finance flows, unclear disaggregation of aid flows, dominance of mitigation financing (rather than at least parity with adaptation), lack of incremental commitments to adaptation (rather than double accounting between 'ordinary' aid and 'climate' aid), excessive fragmentation of institutional vehicles for disbursing 'climate money' and the absence of a comprehensive 'loss and damage' provision to compensate for climate change impacts for which developing countries do not consider themselves responsible (Khan et al., 2019; Oxfam, 2020). On the last point, it is widely felt that 'those least responsible for the problem of global climate change, namely poor countries and communities, are most vulnerable to its impacts [and those] who have a greater responsibility for cumulative emissions that have driven up

GHG concentrations in the atmosphere should, as a matter of fairness, assist those less responsible' (Government of South Africa, 2016: 10). Such issues are unlikely to be resolved through negotiations without there also being a significant shift in the *Zeitgeist* (Chapter 2) to transform the context for them.

9.8   DEVELOPMENTAL DELAY IN THE ADAPTATION AGENDA

The Entanglement of Mitigation and Adaptation

In embryology and paediatrics, 'developmental delay' describes a foetus or baby that lags behind the normal pathway of physical and behavioural growth. The shortfall in progress on adaptation, relative to what might have been, can be described similarly. If it had been accepted from the beginning that climates change naturally, from one prevailing state to another through unsettled transitions, then the UNFCCC might have been based on the premise that natural climate change was underway. The job of humanity would then have been to cooperate in adapting to it, without much reference to what caused it beyond a degree of scientific curiosity.

In this scenario, informed by the concerns of the parties, successive IPCC assessment reports and CoPs would have considered the changes known to have occurred and to be underway, and their impacts on development, and the findings might simply have been accepted by everyone. Attention would have focused on ways and means for safeguarding and building upon the development gains of the past decades, with the same progression of focus from poverty and sustainable development to the Millennium Development Goals and then the SDGs. Aid costs would have risen, as climate change began hybridising with ecological, biodiversity, desertification, pollution and population issues of various sorts. But adaptation would have been mainstreamed from the start, and the only real debates would have concerned willingness to pay for adaptive measures among the aid agencies, as part of their sustainable development mandate, and they would have continued to search for the most cost-effective ways to adapt as part of their eternal interest in obtaining value for taxpayer money.

Meanwhile, the techniques of adaptation would have evolved quickly, perhaps arriving by about 2000 at the point where we are now, with the idea that strengthening ecological and social systems (rather cheaply) at the local and landscape level is the key to a major part of the puzzle, especially during a chaotic climatic transition. A similarly fast progression might have unfolded if two separate treaties had been signed in 1992, one on 'mitigating anthropogenic climate impacts' and one on 'adapting to climate change'. The two agendas could then have proceeded on separate tracks, perhaps with assessment reports and CoPs on the two subjects in alternate years. That way, all the debate about who caused what and who should pay for it, and the associated delay, would have bedevilled only one of the treaties.

Controversy and 'Wickedness'

But none of this happened, so the two agendas remain entangled as they always have been, due to the hunch, hypothesis, theory and later the certainty that humanity is largely responsible. But now, nearly 30 years later, we know that 'natural things' are going on in climate change as well as 'human things'. We also know that of everything people do and have done, the exact timings and shares of responsibility, by industry, sector, country and company, have never been clear, and they become less so the more we know. And we also know that the ways in which the biosphere buffers, tips, flexes and otherwise responds to GHG emissions, and everything else that humanity does to it, are also unclear, and they too become less so the more we know.

Although what we should do in practice has long been obvious in general and precautionary terms, the existence of uncertainty inevitably fed controversy, and controversy has favoured the rich and powerful who often have least interest in promoting change. The whole thing has been described as a 'wicked' problem, being deeply complex, intractable and resistant to solution (Rittel and Webber, 1973; Bours et al., 2014a; but see Chapter 13). An independent observer might say that we have added to this 'wickedness' through unclear intentions, political naïveté and a specious faith on our own knowledge and ability to understand and predict the behaviour of extremely complex systems.

The Separation of Mitigation and Adaptation

All that said, adaptation has developed a life of its own since the Paris Agreement, being now placed (almost) on equal terms with mitigation. Thus the way is clear to focus on how best to adapt, regardless of whether the mitigation track can make significant progress in reversing the accumulation of GHGs in the atmosphere. A preparatory move was for all the parties to the UNFCCC to be invited (in 2013), encouraged (in 2014) and where necessary enabled to submit INDCs before the Paris CoP if possible, and for these INDCs to contain, if the parties so wished, information about the countries' adaptation plans, needs and desires (Chapter 1).

This was the first major effort to gather adaptation perspectives at the same time from all parties. In the Paris Agreement itself, the parties then agreed to submit adaptation communications in various formats from time to time, including the adaptation parts of National Communications that had always been required under the UNFCCC, and the INDCs themselves (sometimes iNDCs, but all officially now renamed NDCs). NAPs were also being prepared by many countries, so the period since 2015 has featured a very rapid growth in adaptation reporting.

Chapter 10 explores the patterns emerging from the adaptation communications, starting with the first NDCs as a baseline, and then looking for signs of actions to strengthen ecological and social systems as a way to reduce the vulnerability of countries to climate chaos. This is expected to be a growing trend post-Paris, as people think through what the manifestations of chaotic climates mean for rural and urban populations. If so, it may also shed light on the implied question arising from the case study chapters: why, if aid projects had made such good progress on sorting out community tenure and community-based forest and land resource management systems (and solving rather similar issues in cities), were they not appreciated enough to be continued to a point where it was safe to leave them?

10 Learning from the Adaptation Communications

10.1 ANALYSIS OF THE EARLIEST NDCS

This section summarises the findings of a review by the UNFCCC Secretariat of the adaptation components of 137 NDCs submitted up to April 2016 (UNFCCC, 2016b). These give a flavour of how government officials (often with advice from national NGOs and consultants) saw the issue at that historical moment. The material showed that interest in adaptation varied greatly in intensity, according to how vulnerable and threatened the countries saw themselves. Sections 10.2–10.4 extend the analysis to documents prepared up to 2020, where there are signs of explosive growth in understanding of climate trends and impacts, rapid deepening of concern for adaptation and a corresponding increase in the ambition and detail with which it was addressed.

Goals and Visions

The Secretariat's first conclusion was that adaptation aims were often related to the constitution of the country or governing political party, or to national laws, strategies and plans. Extensive intertwining was noted with development objectives such as poverty eradication, economic development or the improvement of living standards, environmental sustainability and security and human rights. The stakeholders considered most vulnerable and most in need of support were sometimes highlighted, such as women, children, the elderly, people with disabilities and environmental refugees, and in Bolivia's case Mother Earth. In other cases the vision was expressed in broader and non-climate or adaptation-specific terms, such as a commitment to safeguarding security, territory and population, human rights and/or nature and biodiversity, as well as advancing development goals in the light of projected climate impacts.

Vulnerabilities and Impacts

The geographical characteristics that increase vulnerability were identified as including low elevation, small land area, insularity and being landlocked or with discontinuous territory. Specific environmental changes over recent decades were noted, including the disappearance of major water bodies, high deforestation rates and rapid desertification. Some parties reflected on how their increasing populations create conditions of additional vulnerability, and others noted added dangers arising from concentrations of people or infrastructure in vulnerable areas such as low-lying coasts. Special constraints on adaptation included political crises, the need to stabilise ongoing conflicts and to provide humanitarian assistance to areas affected by conflict or refugees, and stresses caused by disease outbreaks and major hurricanes. The main reported features of the vulnerabilities, impacts and hazards associated with climate change are given in Table 10.1.

Some national vulnerabilities have regional and even global effects. For instance,

> one Party explained that it is the home of four major rivers of West Africa, which are threatened by the impacts of climate change, and that its geographical situation could make it a shelter for neighbouring countries, in particular nomadic pastoralists, increasing the pressure on river basins already affected by drought and changing rainfall patterns. Two major food exporters reported on their contribution to global food security and the global risk induced by the vulnerability of their agriculture and livestock sectors.
>
> *(UNFCCC, 2016b: §281)*

The costs, losses and damage attributed in whole or part to climate change were estimated in various ways, either over a given period or due to a specific extreme event, mostly droughts, floods or storms,

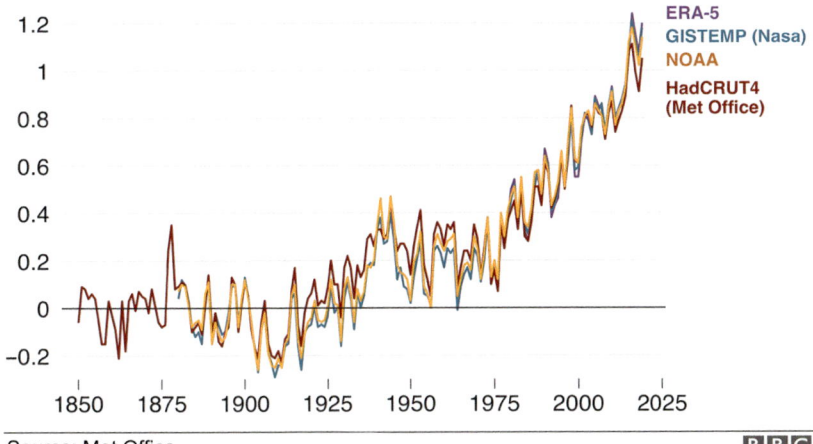

Temperature rise since 1850

Global mean temperature change from pre-industrial levels, °C

Source: Met Office

BBC

FIGURE I.I **The heating biosphere**

Notes: Data processed by the Met Office. Graphic from BBC News at www.bbc.co.uk/news. Reproduced with the permission of BBC News and the Met Office. (A black and white version of this figure will appear in some formats.)

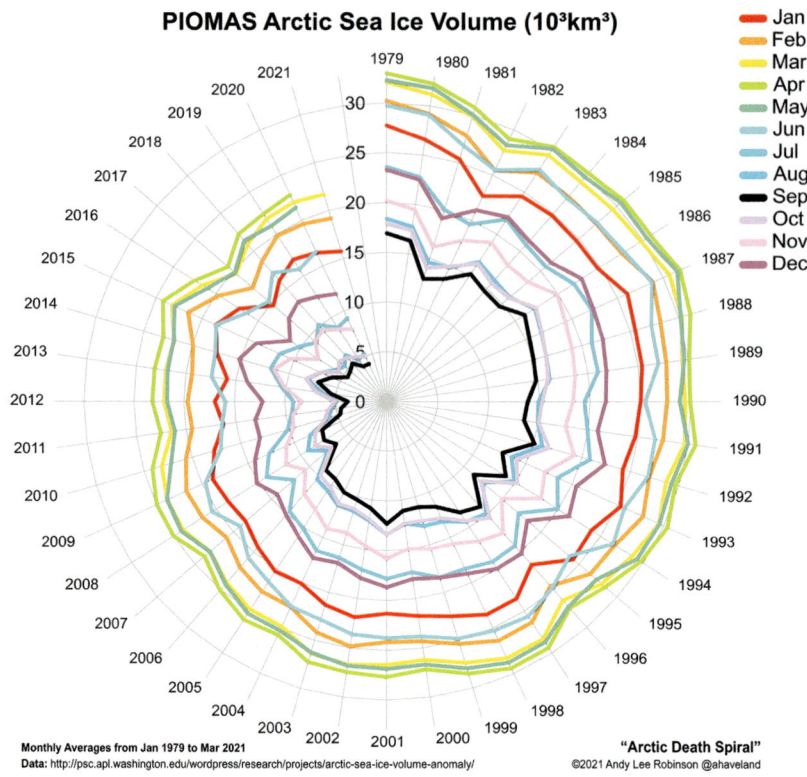

FIGURE 2.1 **The melting Arctic, 1979–2021**

Notes: (a) PIOMAS is the Pan-Arctic Ice Ocean Modelling and
Assimilation System developed by the Polar Science Center at the
University of Washington (PSC, 2020a, 2020b); (b) the Arctic Sea Ice
visualisation is by Andy Lee Robinson, and is used with permission.
(A black and white version of this figure will appear in some formats.)

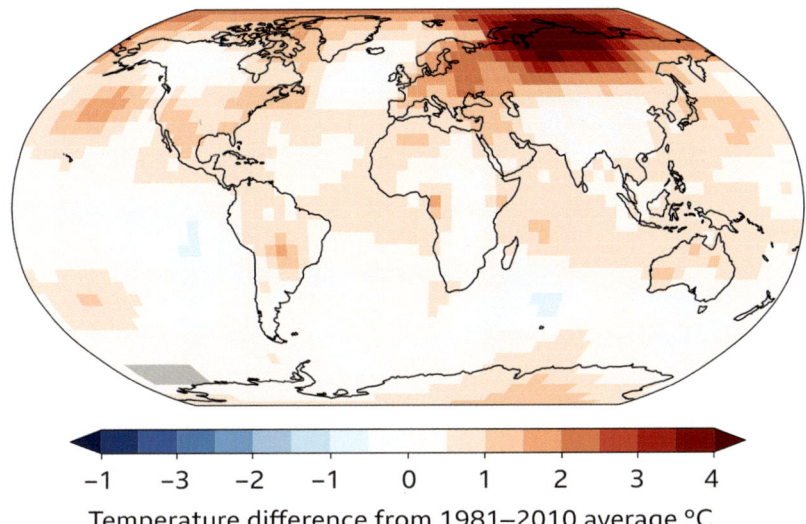

2020 temperature anomalies
Based on HadCRUT5 (Met Office) data set

Temperature difference from 1981–2010 average °C

FIGURE 2.2 **The heating Arctic, 2020**

Notes: Data processed by the Met Office. Graphic from BBC News at bbc.co.uk/news, reproduced with permission. Contains public sector information licensed under the Open Government Licence v3.0, © Crown copyright, Met Office. (A black and white version of this figure will appear in some formats.)

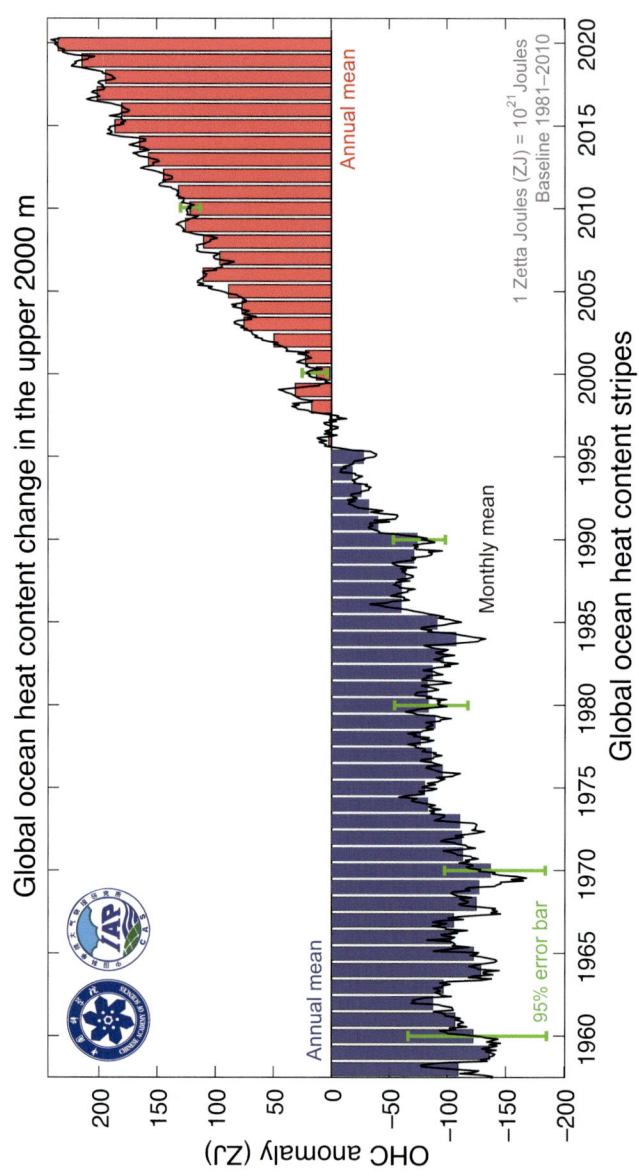

FIGURE 2.3 **The heating ocean, 1956–2020**

Notes: (a) Ocean heat is measured in zettajoules (ZJ), with 1.0 ZJ being equal to approximately double humanity's annual world energy use in 2020; (b) the graphic is from Cheng et al. (2021). Upper ocean temperatures hit record high in 2020. *Advances in Atmospheric Sciences*, reprinted by permission from Springer Nature. (A black and white version of this figure will appear in some formats.)

Table 10.1 *Vulnerabilities, impacts and hazards linked to climate change*

Main feature	Examples/summaries from the NDCs
Non-climatic vulnerabilities (examples)	Post-conflict fragility of the state. Poverty and low-skilled human resources. High prevalence of HIV/AIDS in adult population. Host country to displaced persons.
Observed changes (examples)	Average temperature increase of 0.26°C per decade in 1951–2012. Annual sea level rise of 1.43 mm. Loss of 50 per cent of glacier surface.
Projected changes (examples)	National average temperature increase of 2.4–3.2°C by 2080. Sea level rise of 0.81 m by 2100. Rainfall reduction of 30–50 per cent by 2090.
Key climatic hazards	*Number of 137 countries that highlighted*: floods (83); droughts (76); higher temperatures (74); sea level rise (57); storms (48); reduced precipitation (46); changes in precipitation timing (43); vector or waterborne diseases (34); increased precipitation intensity (33); desertification/ land degradation (32); ocean acidification (28); coastal erosion (28); and salt water intrusion (20). *Specific mention of*: extreme weather, including stronger wind and rain in general and hurricanes (cyclones, typhoons) in particular, along with their associated storm surges; sand-storms and heat waves; and glacial lake outburst floods in the Himalayan region. *Appreciation of slow-onset impacts*: ocean acidification, coral bleaching, salt water intrusion and changes in ocean circulation patterns.

Table 10.1 (*cont.*)

Main feature	Examples/summaries from the NDCs
Types of impact (examples)	52 per cent increase in flood occurrence. Sea level rise of 1 m threatens 67 per cent of sea ports, 50 per cent of airports, 10 per cent of tourism properties and 1 per cent of major roads. Drought projected to affect 27 per cent of country by 2030. 12 out of 75 districts at risk of glacial lake outburst floods.
Vulnerable sectors	*Sectors most often seen as vulnerable*: water (e.g. groundwater levels sinking by 2–7 m per year and possibly depleted by 2035–2050); *agriculture* (e.g. production could fall by 15–19 per cent by 2050), *biodiversity* and *health* (e.g. risk of outbreaks of locally known, or of globally or nationally novel diseases). *Other sectors seen by some countries as vulnerable*: forestry, energy, tourism, wildlife, infrastructure and human settlements.
Vulnerable zones	Arid and semi-arid zones. Low-lying coastal areas and small islands. Mountains. River deltas (e.g. 12 per cent loss anticipated in one country, 70 per cent in another).
Vulnerable populations	Rural populations (e.g. 72 of 75 municipalities highly vulnerable in one country, 115 of 272 vulnerable in another). Poorest segments of society. Women, youth, the elderly and the disabled.
Socioeconomic consequences (summary)	Collapse of the productive apparatus of the country (e.g. through food price volatility, declining productivity of fishing or farming

Table 10.1 (*cont.*)

Main feature	Examples/summaries from the NDCs
	systems, declining water security due to scarcity or contamination and/or the loss of grazing land and the disruption of agricultural calendars). Specific threats to infrastructure and property, including mention of the social justice implications of the poorest and most marginalised people living in high-risk areas. Conflicts between pastoralist groups due to water scarcity. Cultural heritage at risk.
Costs of impacts (examples)	Estimated annual cost of extreme events in 2000–2012: US$1.4 billion. Estimated loss of gross domestic product (GDP) due to drought and floods: 3 per cent. Estimated loss of 20 years' investment in road and water infrastructure at cost of US$3.8 billion (equivalent of 70 per cent of GDP per year) due to one extreme event.

Source: UNFCCC (2016b).

and either in monetary or GDP terms. Some reported losses and damage in non-financial terms, for example as lives lost, people or areas affected.

Strategies and Plans

Although the same documents could have been written with much the same content 15 years earlier, all of this gives an indication of why the adaptation issue is now considered so crucial by so many governments and peoples. The UNFCCC parties are engaged with the climate challenge through national adaptation planning, implementation processes and coordination mechanisms, all generating a diverse stream of documents under various titles and demonstrating the intention 'to enhance the enabling environment for addressing adaptation, [and] the efforts of Parties to address adaptation in a coherent and programmatic manner' (UNFCCC, 2016b: §287). Key goals and examples of quantitative targets for adaptation that had been developed by the parties up to April 2016, along with a summary of the priority needs, are given in Table 10.2.

Many parties reported that they had integrated adaptation into their national plans and policies, or into some of their sectoral plans; at least one had mandated planning at the subnational level, and others saw the opportunity to align national adaptation strategies with regional adaptation strategies and action plans. Finally, several parties reported that they were preparing NAPs, which would receive their main attention for adaptation planning purposes. The combination of refreshed or new NDCs with the NAPs and other adaptation communications prepared since 2016 offers a resource for looking in more detail at current adaptation thinking among the parties as they prepare for the global stocktake in 2023.

10.2   ECOSYSTEMS AND COMMUNITIES IN THE FIRST NDCS

Scoping Exercise on National Preoccupations

Almost all the 158 first NDCs were submitted in 2015–2016, but 23 arrived in 2017–2018 and two in 2019–2020 (the dates being based on the documents themselves and their cover letters rather than the date of loading to the UNFCCC database[1]). To gauge the extent to

---

[1] States that have signed but not ratified the Paris Agreement are: Angola, Eritrea, Iran, Iraq, Libya, South Sudan, Turkey and Yemen. Turkey has an issue over its status as a 'developed' or 'developing' country, but has submitted an INDC (www4.unfccc.int/ sites/submissions/INDC/Published%20Documents/) and several National Communications (unfccc.int/documents?search2=&search3=&page=%2C%2C3). Eritrea submitted its first NDC in 2018, and South Sudan its first National Communication in 2018, while Angola attended AOSIS meetings as an observer in 2020.

Table 10.2 *Goals, targets and support needs for adapting to climate change*

Sector/area	Examples/summaries from the NDCs
Priority sectors	*Number of 137 countries that highlighted the following sectors for action*: water (119); agriculture (106); health (86); ecosystems (65); infrastructure (65); forestry (65); energy (47); DRR (46); food security (45); coastal protection (44); fisheries (41).
Water	*Overall priorities*: e.g. mainstreaming climate change adaptation in the water sector, implementing a national water master plan, putting in place integrated water management systems or building a water-saving society. *Specific measures*: e.g. water-saving irrigation systems, desalination, water conservation facilities for farmlands, reservoirs for glacier meltwater harvesting, catchment management, rainwater harvesting, substituting water withdrawal from aquifers with surface water. *Quantitative targets*: e.g. ensure full access to drinking water by 2025; increase water storage capacity from 596 million m^3 in 2010 to 3,779 million m^3 in 2030; increase desalination capacity by 50 per cent from 2015 by 2025.
Agriculture	*Overall priorities*: e.g. promoting sustainable agriculture and land and resource management, implementing integrated adaptation programmes for agriculture, developing climate criteria for agricultural programmes and adapting agricultural calendars. *Specific measures*: e.g. integrated pest management; heat-, drought- and disease-resistant crops, fodder types and livestock breeds; enhanced irrigation systems, drought management and methods to reduce erosion; affordable insurance; climate research; use of traditional knowledge;

Table 10.2 (*cont.*)

Sector/area	Examples/summaries from the NDCs
	and early warning systems. *Quantitative targets*: e.g. convert one million ha grain fields to fruit plantations to protect against erosion; increase the amount of irrigated land to 3.14 million ha; reduce post-harvest crop losses to 1 per cent through treatment and storage.
Health	*Overall priorities*: e.g. an overall integration of climate impacts and/or the identification of priority actions in the health sector; an enhanced understanding of climate–health connections and changing disease patterns; and enhanced management systems or contingency plans for public health to improve the adaptive capacity of public medical services. *Specific measures*: e.g. protecting pregnant women and children under five against vector-borne diseases, suppressing mosquito populations and distributing test kits for vector-borne diseases; early warning systems with epidemiological information; health surveillance programmes and contingency plans in the event of heat waves.
Ecosystems and biodiversity	*Overall priorities*: e.g. rehabilitating and protecting forest, coastal and marine (especially mangroves and corals) and other ecosystems for increased resilience; establishing biodiversity corridors; protecting species, populations and biodiversity and monitoring efforts and recoveries. *Specific measures*: e.g. prepare a biodiversity index and atlas of biodiversity centres; protect wildlife species; establish watering points for wildlife; stop coastal mining. *Quantitative targets*: e.g. protect 20 per cent of marine environments by 2020; regenerate 40 per cent of degraded forests and rangelands; establish 150,000 ha of marine protected areas.

Table 10.2 (*cont.*)

Sector/area	Examples/summaries from the NDCs
Forestry	*Quantitative targets*: e.g. increase forest coverage to 20 per cent by 2025; maintain 27 per cent forest coverage; achieve 0 per cent deforestation rate by 2030.
Disaster risk reduction (DRR)	*Overall priorities*: e.g. develop DRR strategies, policies, plans, platforms, frameworks and systems against storms, floods, sea level rise and glacial lake outburst floods. *Specific measures*: e.g. early warning systems; risk management institutions; hazard maps; building codes and other standards; infrastructure protection measures and contingency plans; insurance schemes to protect the most vulnerable communities and to incentivise climate-proof construction; resettlements to safer areas; preparations for national evacuation. *Quantitative targets*: e.g. ensure that all buildings are prepared for extreme events by 2030; reduce the number of the most vulnerable municipalities by at least 50 per cent; relocate 30,000 households.
Energy	*Quantitative targets*: e.g. ensure that hydropower generation remains at the same level regardless of climate change impacts; increase the proportion of renewable energy to 79–81 per cent by 2030.
Other	*Quantitative targets*: e.g. ensure that 100 per cent of national territory is covered by climate change adaptation plans by 2030; reduce moderate poverty to 13.4 per cent by 2030 and eradicate extreme poverty by 2025.
Priority needs	*Favourable enabling environments, institutional arrangements and legislation*, e.g. for mainstreaming climate change in development, gender mainstreaming and strengthening the

Table 10.2 (*cont.*)

Sector/area	Examples/summaries from the NDCs
	engagement of subnational communities and the private sector.
	Sufficient financial resources, with which to assess, plan, implement, monitor and evaluate adaptation actions.
	Technologies for adaptation, e.g. for climate observation and monitoring, early warning systems, water resource management, irrigation and wastewater management, coastal zones, resilient transportation systems, sustainable or climate-smart agriculture, forestry and forest fires and land management.
	Training and building of institutional and human capacities and technical expertise, e.g. for vulnerability and adaptation assessments, cost–benefit analysis and the development of sectoral finance plans.
	Research, data and information, e.g. for climate forecasting and modelling, satellite data, regionally downscaled climate data and research into international energy markets.
	Education, raising awareness and outreach, on climate change impacts and adaptation.
	South–South cooperation, on the basis of solidarity and common sustainable development and adaptation priorities.

Source: UNFCCC (2016b); water storage capacity targets from Government of Bolivia (2016: 11).

which community- and ecosystem-oriented thinking had penetrated the world of adaptation communications, the first NDCs were searched for 'community' and 'ecosystem' (and their plurals and variants in English, French and Spanish). Each case was examined for substantive references, to exclude uses like 'international, scientific or business community', or 'community of interest', or in the title of an institution or alliance, or where it was used to mean 'general public', or in phrases in a list with no details provided. But countries were included if there was substantive text on 'community involvement', 'community planning', 'community-based resource management' (or similar) under 'adaptation measures', or where there were multiple references to 'community' in relation to water, ecosystems, insurance, farming, planning, etc., or where 'ecosystem-based adaptation' (or similar) was given as a priority, or where there were strategic and/or or multiple references to 'ecosystem' in adaptation priority or fragility contexts.

On this basis, fewer than a third (30 per cent) of the first NDCs contained 'community', nearly half (49 per cent) contained 'ecosystem' and nearly a quarter (23 per cent) contained both. Reading through the NDC material, it was clear (and to be expected) that the authors had covered the expected content in a general way, but gave particular attention to matters of special concern to their own country. For example, while natural and modern cultural heritage was of concern to many countries, only Egypt, Cuba and Morocco mentioned the protection of ancient historical sites, and only Egypt, with its unique (and exceptionally valuable) heritage of ancient sites, arts and artefacts, gave special prominence to 'the harm inflicted on national heritage as result of temperature rise, sandy winds and ground water' (Government of Egypt, 2015: 6).

An understanding that the content of the NDCs might reflect national preoccupations led to the hypothesis that prominent reference to 'community' might be expected in small-scale intimate societies, or those with socialist traditions of governance, or those with serious tensions between rival ethnicities, or those in recovery from war, dictatorship and genocide. All of these might be expected to give extra priority to social and hence community agendas such as participation, equity, conflict resolution and peace building.

Similarly, it was thought that prominent reference to ecosystems might be expected in countries where ecotourism and/or payments for ecosystem services have had important economic roles, or where there has been a sustained policy interest in biodiversity, or which have a history of slow ecological disasters (e.g. desertification), or experience of ecosystem protection against quick disasters (e.g. storm surges and hurricanes). In other words, countries with a good reason to value ecosystems might be expected to mention their role in adaptation.

About half of all the countries in the 'community' and 'ecosystem' categories were found to comply with these expectations, which are based on general knowledge, so the finding merely suggests that a more detailed study might be rewarding. In any case, many other factors are likely to be at work in shaping countries' preoccupations and interests, including influences from neighbouring countries, or the effects of having one consulting firm or UN agency supporting multiple governments as they each prepared their own NDCs.

10.3   ECOSYSTEM-BASED ADAPTATION IN 2015–2020

Signs of System Thinking

That some countries have been thinking about the systems involved in vulnerability and strength is shown by Mali's recognition, 'at the national level that combating the threat of climate change means increasing the resilience of ecological systems, production systems and social systems' through the integration of priority measures focused on forest, farming, grazing and water management systems, and also Senegal's recognition of much the same while adding biodiversity and coastal and fishing systems as well as culture and health (Quevedo et al., 2019: 29).

There is other evidence that systems or systemic thinking is being adopted, pioneered by the EC, as might be expected considering the system-oriented approach of the Water Framework Directive 2000 (which has also influenced thinking in Norway and Turkey, to judge from sections of text in Government of Norway, 2018: 192, and Government of Turkey, 2018: 163). Examples include the European Research and Innovation for Food and Nutrition Security programme was begun in 2016 and 'is calling for a systemic approach to future-proofing our nutrition and food systems towards becoming sustainable, resilient, diverse, responsible, inclusive and competitive in the longer term' (EC, 2017: 101), and the Regional Knowledge Network on Systemic Approaches to Water Resources Management project (2011–2015). A synthesis review of EU development cooperation also noted that systems thinking was being promoted through the use of the 'energy, food and water nexus' concept to promote a fully integrated approach to managing water, food and energy sources, supplies, demands and the effects of uses on each other and on the systems as a whole, and also links between delivery of more reliable energy and the productive, income-generating uses of it (Caldecott, Clark et al., 2019).

Signs of Ecosystem-Based Adaptation

Systemic approaches become more explicit in terms of ecosystem-based adaptation, some references to which in the adaptation communications are given in Box 10.1. The aim here is to provide a flavour of the diverse contexts in which the conceptual approach is used. As the government of Fiji explained in its National Adaptation Plan (2018: 37):

> Operationalising an ecosystems-based approach to adaptation planning can have multiple interpretations and the NAP makes no attempt to control its usage. However, this NAP notes that ecosystem-based approaches to adaptation could mean: (1) Adopting an 'ecosystem approach' which uses as a strategy the integrated management of land, water and living resources to support delivery of ecosystem services in an equitable way. (2) Adopting a 'socioecological system' for development planning which supports understanding of the interactions between human activities and their impacts on the whole ecosystem and vice versa. (3) Conserving and restoring flora, fauna, and habitats as a basis to enhance adaptive capacity or reduce the adverse socioeconomic

BOX 10.1   'Ecosystem-based adaptation' as used in some adaptation communications

- *Afghanistan.* 'Ecosystem-based adaptation, which integrates the use of biodiversity and ecosystem services into climate change adaptation, can provide a cost-effective approach that both maintains biodiversity and reduces negative impacts from climate change. Examples of ecosystem-based adaptation applicable in Afghanistan include: reduction of habitat loss and fragmentation, as well as habitat conservation through establishment of protected areas; afforestation to stabilize slopes, enhance soil integrity and regulate water flow; the promotion of agroforestry systems using diverse crops and plant species; and the sustainable management and restoration of watersheds linking upstream and downstream areas' (Government of Afghanistan, 2017: 11).
- *Armenia.* 'This document builds on the principle of "green economy" and an ecosystem-based approach towards mitigation and adaptation actions' (Government of Armenia, 2020: 50).
- *Botswana.* 'The reduced rangeland productivity has been identified as one of the impacts of climate change on rangeland ecosystems in Botswana. It is therefore critical to enhance the ecosystems' ecological structures and functions that are essential for ecosystem functioning to enable people to adapt to multiple stressors, including land degradation and climate change. This process is referred to as rangeland ecosystem-based adaptation (EbA) and its key components are conservation of biodiversity that results in resilient ecosystem services leading to improved livelihood' (Government of Botswana, 2019: 108).
- *Costa Rica.* 'Ongoing efforts to develop adaptation measures for the water and biodiversity sectors, including the National Conservation Areas System (SINAC), have resulted in the launch of a National Ecosystem-based Adaptation Strategy. Increase focus will be given to building resilience from a sustainable development, food security and rural productivity perspective' (Government of Costa Rica, 2015: 15).
- *Dominica.* 'Output 2: Governments implement concrete adaptation measures using ecosystem-based approaches where appropriate. This will demonstrate enhanced direct access in the public sector through an on-granting mechanism that aligns GCF-financed concrete local area adaptation projects to climate-proof ongoing investments and co-financing from the Government' (Government of Dominica, 2020: 161).

BOX 10.1   **(cont.)**

- *Dominican Republic*. 'The elements of the strategic planning approach to adaptation are: Ecosystem-Based Adaptation/Resilience of Ecosystems; Increase of Adaptive Capacity and Decrease of Territorial/Sectoral Vulnerability; Integrated Water Management; Health; Food Security; Infrastructure; Floods and Droughts; Coastal and marine areas; Risk Management and Early Warning Systems' (Government of the Dominican Republic, 2015: 2).
- *EC*: 'The European Commission is continuing to explore all the potential ways at its hand to enhance the adaptation capacity of European infrastructures, from mainstreaming to standardisation, to ecosystem-based approaches or providing further guidance to project developers' (EC, 2017:118).
- *Fiji*. 'Ecosystem-based Adaptation (EbA) is a holistic approach to adaptation planning that seeks to harness the potential of healthy ecosystems and biodiversity to strengthen social and ecological resilience. EbA is a nature-based solution and is increasingly seen as a pragmatic and sustainable option for securing resilience in social and ecological systems impacted by climate change' (Government of Fiji, 2020: 102).
- *Germany*. 'The priority areas of the German Ministry for Economic Cooperation and Development's support are ecosystem-based adaptation (EbA) and adaptation of agricultural production and food security, water management and adaptation, risk management instruments in connection with climate change impacts, for example through innovative insurance solutions, and the development and implementation of national adaptation strategies in the context of countries' National Adaptation Plans and Nationally Determined Contributions' (Government of Germany, 2017: 149).
- *Ghana*. 'The [NAP] process seeks to promote community-based and ecosystem-based approaches and works to ensure that it delivers multiple co-benefits to sustainable development, poverty reduction and climate change adaptation' (Government of Ghana, 2020: 229).
- *Grenada*. 'To expand and broaden the objectives of the Marine Protected Area network, Grenada has also initiated a programme of coastal ecosystem-based adaptation (EbA), in partnership with the [UNEP]. This activity, funded by the European Union, has developed a national strategy for EbA and has initiated two pilot projects for the restoration of coral reefs in Grand Anse and Carriacou' (Government of Grenada, 2017b: 188).

BOX 10.1   **(cont.)**

- *Jamaica.* 'The country has also prepared, in collaboration with other regional governments and the UNEP, a proposal to support the implementation of an Urban Ecosystem-Based Adaptation project in the capital city of Kingston. This Project will increase the resilience of Kingston using ecosystem based approaches' (Government of Jamaica, 2016: 8).
- *Kenya.* 'Adaptation Indictors: Number of national and county level programmes/projects incorporating ecosystem-based adaptation and community-based adaptation approaches' (Government of Kenya, 2016: 49).
- *Lao PDR.* 'Managing Watersheds and Wetlands for Climate Change Resilience. (a) Strengthen the protection of watersheds to safeguards and moderate downstream flow during periods of high and low flow. (b) Study and promote the conservation of wetlands as part of a climate resilient ecosystem-based approach' (Government of Lao PDR, 2015: 17).
- *Madagascar.* 'Actions to be undertaken between 2020 and 2030: … (h) Implementation of ecosystem-based adaptation to cope with sand-hill progression (multiple causes but phenomena aggravated by climate change) by leveraging research findings and best practices' (Government of Madagascar, 2016: 8).
- *Mexico.* 'Ecosystem-Based Adaptation. In Mexico there is a large diversity of ecosystems that provide society with a vast amount of environmental services such as carbon sequestration, provision and maintenance of water, habitat conservation for the permanence of species, reduction of impacts caused by meteorological disasters, and the formation and maintenance of soils. These environmental services are seriously threatened by human activities and by the effects of climate change. Ecosystem-based adaptation consists of the conservation of biodiversity and ecosystem services as part of an integral adaptation strategy to assist human communities to adapt to the adverse effects of climate change' (Government of Mexico, 2016: 7).
- *Morocco.* 'The protection of natural heritage, biodiversity, forestry and fishery resources, through an ecosystem-based adaptation approach. Morocco commits to restoring ecosystems and strengthening their resilience, to combat soil erosion and prevent flooding' (Government of Morocco, 2016: 20).
- *Myanmar.* (Government of Myanmar, 2015):
  o 'Ecosystem-based planning. Among other issues which are being addressed include a focus on townships planning for adaptation. Ecosystem-based approaches to adaptation at township level will be tested during the course

BOX 10.1 **(cont.)**

of 2016–18, including a vulnerability analysis of the following elements: urban planning, infrastructure development, environmental risk and livelihood patterns. This will be followed by implementation of solutions to identified issues' (page 6).

o 'Approaches on ecosystem-based adaptation are being explored, for instance by the UNDP [United Nations Development Programme] and by the [Myanmar Climate Change Alliance] programme with the [Ministry of Environmental Conservation and Forestry]' (page 11).

- *Namibia.* (Government of Namibia, 2020):

o 'The use of the ecosystem-based approach in fisheries management with its modules on environment, social, governance, and ecosystem health will improve the State of the Marine Environment reporting' (page 10).

o 'Sector: Terrestrial ecosystems – Adaptation priorities: Ecosystem-based adaptation' (page 148).

o ' The [Benguela Current Commission] promotes an ecosystem-based approach to the management of the large marine ecosystem as an adaptive and resilience capacity-building approach' (page 166).

- *Nepal.* 'At present, Nepal climate change support programme (NCCSP), community-based flood risk and [glacial lake outburst flood] risk reduction programme, ecosystem-based adaptation programme, including enhancing capacity, knowledge and technology support to build climate resilience of vulnerable communities, Hariyo Ban Project (climate adaptation component), and Multi-stakeholder Forestry Programme (adaptation co-benefits) are under various stages of implementation. Localising climate adaptation actions has been deeply rooted in planning and implementation of NCCSP target areas' (Government of Nepal, 2016: 6).

- *Norway.* (Government of Norway, 2018):

o 'Norwegian Environment Agency coordinates the work of establishing a cohesive, ecosystem-based water management in Norway' (page 191).

o 'An ecosystem-based monitoring program for land ecosystems in the Norwegian (Arctic Climate-ecological Observatory for Arctic Tundra – COAT) has been developed during the last years. COAT is particularly designed to be able to detect impacts on climate change' (page 206).

- *Palestine.* 'In addition, it was suggested that some options might be ecosystem-based, i.e. helping people adapt to the impacts of climate change through the conservation, sustainable management, and restoration of ecosystems' (Government of Palestine, 2016: 17).

BOX 10.1   **(cont.)**

- *Saint Lucia.* 'Strategic objective[s]: Minimise agriculture-related climate change risks by adopting Ecosystem-based Adaptation solutions. . . . Minimise water-related climate change risks by adopting ecosystem-based adaptation solutions. . . . Improve the national legal and regulatory framework to facilitate natural resource management and ecosystem-based adaptation under a changing climate' (Government of Saint Lucia, 2018: 3–5).
- *Seychelles.* (Government of the Seychelles, 2015):
  - o 'Projects address issues such as sustainable tourism, watershed management, sustainable agriculture and fisheries, disaster planning, research and a shift toward ecosystem-based adaptation approaches to biodiversity conservation' (page 5).
  - o 'Seychelles is currently implementing three ecosystem-based adaptation projects funded by the GEF Climate Change Adaptation Fund, UNEP and the Government of China. The projects focus on management of coastal ecosystems, protection of mangroves, and sustainable watershed management' (page 7).
  - o 'Seychelles is in the process of implementing an ecosystem-based approach to watershed management and its implications for food supply as well as water security. . . . The Ministry anticipates additional resources being committed to enhance human capacity development at the Seychelles Agricultural Agency, revitalising the extension services and also providing opportunities for young Seychellois to study climate-smart and ecosystem-based approaches to agriculture, put in place programmes for sustainable industrial and artisanal fisheries, sustainable mariculture, promote home gardening, improve port infrastructure for artisanal and industrial fisheries, reduce illegal, unreported and unregulated activities; and continue to support the insurance scheme for farmers and fishers' (page 12).
  - o 'The ecosystem-based watershed project mentioned previously is currently being implemented to address water supply from an ecosystem perspective. Another demonstration project is being implemented on La Digue focused on integrated water resource management. Both of these projects represent an integrated approach to water security that address issues such as ecosystem health, waste management, water treatment and supply, sewage, agriculture, etc. It is advocated that this approach is mainstreamed throughout island water resource management' (page 13).

BOX 10.1 **(cont.)**

- *South Sudan.* 'Forested landscapes and important wetlands in South Sudan
  have been degraded through clearing and resource extraction, exposing this
  area to impacts of climate change. Landscape restoration and ecosystem-based
  adaptation through the scaling up of current afforestation and reforestation
  efforts should be promoted. Wetland wise-use principles should be applied in
  wetland resource management to enable people to adapt to the adverse effects
  of climate change and to promote the sustainable development of wetland
  resources. Projects that place strong emphasis on ecological and nature
  solutions and the maintenance of genetic diversity should be promoted as part
  of the solution to impacts of climate change' (Government of South Sudan,
  2018: 192).
- *United Arab Emirates (UAE).* 'The UAE is also undergoing significant
  restoration and plantation efforts of both mangroves and sea-grass, supporting
  ecosystem-based adaptation as well. In 2013, the UAE initiated the Blue
  Carbon Demonstration Project, which provided decision-makers with a
  stronger understanding of the carbon sequestration potential in the Emirate of
  Abu Dhabi. In 2014, the project's scope was expanded to cover the entire
  country, and is known as the UAE's National Blue Carbon Project'
  (Government of the UAE, 2015: 4).
- *Uruguay.* 'Among those pilot actions it is worth noting the capacity-building
  initiatives in terms of ecosystem-based adaptation, both at national and local
  levels, to strengthen and restore coastal ecosystems that provide buffering and
  wave energy dissipation services during extreme climate events and help to
  reduce vulnerability. As of 2016 there have been ecosystem-based restoration
  and adaptation works along the waterfront of the six departmental
  governments that sit on the River Plate and the Atlantic Ocean. These works
  have produced positive results in the face of extreme storm surge events'
  (Government of Uruguay, 2017: 15).
- *Vietnam.* 'Climate change adaptation in the period 2021–2030 – Ensure social
  security: Implement ecosystem-based adaptation through the development of
  ecosystem services and biodiversity conservation, with a focus on the
  preservation of genetic resources, species at risk of extinction, and important
  ecosystems' (Government of Vietnam, 2016: 10).

Sources: adaptation communications to UNFCCC Secretariat, as indicated.

impacts associated with environmental and climate events. However, it also requires associated values to be explicitly considered and incorporated into decision-making system processes, as well as programme and project design.

Additional values of ecosystem-based adaptation are revealed in Box 10.1 and include: cost-effectiveness (Afghanistan, Dominica), infrastructure protection (EC), risk management (Germany, Nepal, Uruguay), poverty reduction (Ghana), urban resilience (Jamaica, Myanmar) and monitoring and research (Norway, Seychelles). In its own NAP, the government of Grenada (2017a: 60) explained that 'ecosystem-based solutions that will always have a positive impact on livelihoods regardless of how exactly the climate changes, are a priority for Grenada, Carriacou and Petite Martinique'. This is exactly the point of this book: that solutions of this sort will strengthen systems against *anything*, so are particularly valuable when chaotic and near-random stresses are expected, as they are now during a climatic transition.

10.4   COMMUNITY-BASED ADAPTATION IN 2015–2020

A similar approach is taken to illustrate the diversity of uses of 'community-based adaptation' in the stakeholder literature (Box 10.2). In its NAP, the government of Suriname (2020: 16) explained the growing significance of this set of approaches:

> Increasingly, government policies and business strategies at the national and local levels are influenced by environmentally aware citizens, with a trend toward local self-reliance and stronger communities. International institutions decline in importance, with a shift toward local and regional decision-making structures and institutions. Human welfare, equality, and environmental protection all have high priority, and they are addressed through community-based social solutions in addition to technical solutions, although implementation rates vary across regions.

BOX 10.2   **'Community-based adaptation'**
as used in some adaptation communications

- *Australia.* '*Disaster risk management*. The Emergency Management Climate Change Program (commenced in 2017) will help communities, businesses and local governments integrate climate change considerations into emergency management. It includes: community-based emergency management planning; incorporating climate change projections into risk data; climate change and emergency management forums' (Government of Australia, 2017: 135).
- *Belize.* '*Adaptation Measures*: Implement community-based participatory approaches to empower local communities to manage disease vectors in an integrated manner and thus increase their capacity to protect their health and climate resilience' (Government of Belize, 2014: 114).
- *Costa Rica.* 'Community-based adaptation seeks to empower the population to face climate change impacts, by increasing the resilience [of] agriculture producers, developing safeguards for securing water supply and sustainable coastal zone development. Costa Rica is committed to promote Green and Inclusive Development, which favours the implementation of sustainable productive systems, in rural areas with lower human development indexes and vulnerable to climate change in priority productive territories over a 10 year period between 2016 and 2026. Since 2014, Fundecooperación has been implementing a program financed by the Adaptation Fund which will provide resources and technical assistance to over 30 community-based adaptation projects. Learnings from these pilot projects will enable feedback into Costa Rica's National Adaptation Policy' (Government of Costa Rica, 2015: 16).
- *Dominica.* 'Community climate change vulnerability, risk and capacity assessment were undertaken in 2011 as a collaborative initiative between the SPACC [Special Program on Adaptation to Climate Change] program and the GEF-funded Sustainable Land Management (SLM) project – under this initiative Dominica pioneered: (a) the vulnerability mapping and 'climate proofing' of National Parks Management Plans; and (b) community-based vulnerability mapping and the development, through community engagement, of community adaptation plans' (Government of Dominica, 2020: 17).
- *Ethiopia.* 'Participatory forest management and community-based rehabilitation of degraded forests will be implemented' (Government of Ethiopia, 2019: 27).

BOX 10.2 **(cont.)**

- *Fiji.* (Government of Fiji, 2018):
  - o 'It should be noted that a gender and human rights-based approach has cross cutting benefits. For instance, this approach improves inclusivity and the quality of participation. It also matches community-based approaches to adaptation, which itself has many linkages to ecosystem-based approaches to adaptation' (page 39).
  - o 'Vertical integration is an important opportunity to integrate ecosystem-based approaches to adaptation. ... It is also a vital opportunity to enhance participation and inclusivity of decision-making and development planning processes to ensure they meet the needs of all social groups. In this regard, supporting and linking community-based adaptation efforts to development planning can provide for a more effective, empowering, and holistic strategy for tackling environmental and climate risk' (page 50).
  - o 'Traditional knowledge and associated customary practices are often about using resources in a sustainable way and as such are the basis of many community-based resource management activities. Consequently, traditional knowledge has strong linkages to community-based and ecosystem-based approaches to adaptation' (page 54).
- *Ghana.* 'The [NAP] process seeks to promote community-based and ecosystem-based approaches and works to ensure that it delivers multiple co-benefits to sustainable development, poverty reduction and climate change adaptation' (Government of Ghana, 2020: 229).
- *Guinea-Bissau.* 'The main objective of the flood and drought risk mapping will be to contribute to the vulnerability assessment by identifying flood and drought high-risk areas and use the information to support effective community-based flood and drought risk reduction planning. By reducing flood and drought risk, the communities in the districts will build greater resilience to the negative impacts of climate change' (Government of Guinea-Bissau, 2018: 131).
- *Ireland.* 'In recent years, the [National Waste Prevention Programme] has seen an increased focus on areas such as social enterprises and community-based activities. This is due to strategic nudges from national and EU policy (such as the Circular Economy package)' (Government of Ireland, 2018: 278).
- *Kenya.* 'Adaptation Indictors: Number of national and county level programmes/projects incorporating ecosystem-based adaptation and community-based adaptation approaches' (Government of Kenya, 2016: 49).

BOX 10.2 **(cont.)**

- *Kiribati. 'Key national adaptation priority – environmental sustainability and resilience. Priorities*: Identify required enhancements to the frameworks and institutional capacity to implement community-based protected areas and other natural resource management and licensing and enforcement measures. Develop outer island community-based protected areas' (Government of Kiribati, 2019: 121).

- *Lesotho.* 'Intended policy based action[s]: Establish a national integrated water resource management framework that incorporates district and community-based catchment management; Strengthen the implementation of the national Community-Based Forest Resources Management Programme' (Government of Lesotho, 2017: 11–12).

- Namibia. (Government of Namibia, 2020):
  o 'The Vulnerability and Adaptation assessment have two primary focus areas, the sector and constituency levels. The first was to review and prioritize the most significant climate change risks, vulnerabilities, and adaptation for the sectors agriculture, water resources, tourism, health, coastal zones, human settlements, ecosystems and biodiversity. The second was an in-depth analysis of the vulnerability and adaptation of human settlements (constituencies) in Namibia to climate change impacts' (page 4).
  o 'Most of the current climate change adaptation projects, and programs in Namibia are directed in the areas of agriculture, fisheries, sustainable land management, government, climate information and research, ecosystems and biodiversity, forestry and energy. The projects tend to focus on capacity building, knowledge communication, field implementation, and policy formation and integration, with all nationally implemented projects supporting community-based adaptation' (page 9).

- *Nepal.* 'At present, Nepal climate change support programme (NCCSP), community-based flood risk and [glacial lake outburst flood] risk reduction programme, ecosystem-based adaptation programme, including enhancing capacity, knowledge and technology support to build climate resilience of vulnerable communities, Hariyo Ban Project (climate adaptation component), and Multi-stakeholder Forestry Programme (adaptation co-benefits) are under various stages of implementation. Localising climate adaptation actions has been deeply rooted in planning and implementation of NCCSP target areas'. (Government of Nepal, 2016: 6).

BOX 10.2 **(cont.)**

- *Nigeria.* (Government of Nigeria, 2020):
  - o 'Besides the efforts made by governments and corporate organizations, the contributions of NGOs, educational institutions as well as communal efforts in advancing community-based adaptation in areas ravaged by drought, notably the North East and the North West, deserve mention. A typical example is the water harvesting project embarked upon by the University of Maiduguri in Tosha community in Borno State' (page 184).
  - o 'Strengthen the implementation of the community-based natural forest resources management programme. The department of Forestry of the Federal Ministry of Agriculture and Rural Development is responsible for the full implementation of this option' (page 197).
- *Palestine.* 'Major issues limiting adaptive capacity are: increasing poverty and unemployment rates; lack of alternative plans for emergency situations, including financial shortages; and insufficient resources to develop the water and sanitation infrastructure, and to expand community-based behaviour-centered programmes that promote improved hygiene practices at the community and household level' (Government of Palestine, 2016: 31–32).
- *Rwanda.* 'Land use actions: Employ community-based disaster risk reduction (DRR) programmes designed around local environmental and economic conditions, to mobilise local capacity in emergency response, and to reduce locally-specific hazards' (Government of Rwanda, 2015: 12–13).
- *Suriname.* 'For the Interior it is recommended to use a community-based adaptation approach with the following elements: Enhancement of the adaptive capacity of the Interior communities under extreme climatic conditions and incorporate climate change considerations into tribal and community decision making; Conduction of appropriate research programs into community and tribal land tenure systems and the effect that recognition of formal land rights would have on adaptive capacity; Incorporation of gender issues and political decision making in communities into adaptation measures to avoid elite agenda capture' (Government of Suriname, 2015: 152).
- *Solomon Islands.* 'It is the intention of the Solomon Islands Government that a community-based vulnerability mapping, adaptation planning and management approach (tied to direct access to financing for community-based resilience-building projects) be employed on a whole of island basis that will build capacity in vulnerable villages for localised adaptation actions which represents a critical contribution to the implementation of adaptation.

BOX 10.2   **(cont.)**

The Solomon Islands Government will establish the institutional structures and strengthen capacities at the community level in order to support the country-wide implementation of community-based vulnerability mapping and adaptation planning, and the community-based design and implementation of priority resilience measures through direct access to financing for such measures' (Government of the Solomon Islands, 2016: 13).

- *South Sudan.* (Government of South Sudan, 2018):
  - o 'Policy calls for strengthening of the local community's role in environmental management. Community responsibilities include establishing community-based organizations (CBOs), which are expected to play a pivotal role in advocating for the sustainable management of biodiversity and ecosystems through mobilizing and sensitizing local people, supporting local group participation in biodiversity management, and ensuring that the concerns of the underprivileged are integrated into national development plans' (page 83).
  - o 'At the community level, traditional authorities play an important role in natural resource management. In South Sudan, traditional regulations are often used to regulate the use of forest fires and protection of certain wildlife and tree species. These community-based approaches should be an integral part of natural resource management in South Sudan' (page 192).
- *Tonga.* (Government of Tonga, 2019):
  - o '*Other needs*: Re-establish some marine tenure to reinforce the concept of community-based management of the coastal resources' (page 183).
  - o 'The purpose of the Climate change Trust Fund is to finance small community-based climate adaptation and mitigation projects and fund the climate component of non-community-based projects. It will also provide supplementary financial support to small scale community-based, climate-related projects proposed by other organizations such as church groups, charities and non-government organizations' (page 198).
- *United Kingdom.* '*Enhancing Community Resilience Programme*: To achieve sustainable disaster-resilient communities through community-based best practices, public awareness and policy change' (Government of the United Kingdom, 2017: 361).
- *Vietnam.* '*Climate change adaptation in the period 2021–2030*: Allocate and mobilise resources for community-based climate change adaptation and disaster management; raise awareness and build capacities for climate change

BOX 10.2   **(cont.)**

adaptation and disaster risk management. . . . Implement community-based
adaptation, including using indigenous knowledge, prioritizing the most
vulnerable communities' (Government of Vietnam, 2016: 10).

Sources: adaptation communications to UNFCCC Secretariat, as indicated.

Meanwhile, the government of Chile (2020: 18) recognised that:

> Local communities and municipalities are also key, as they will
> suffer the direct impacts of climate change, and their ability to
> respond to such impacts is essential to reducing the damage and
> losses caused by extreme events. Therefore, strengthening the
> responsiveness of the population and institutions facing the
> challenges of an uncertain and changing future is a crucial element
> of the process of adapting to climate change.

Additional or specific values of community-based adaptation are identified in Box 10.2, and include: the management of disaster risks
(Australia, Rwanda, Solomon Islands, UK, Vietnam), health risks
(Belize, Palestine), protected areas and community lands and harvesting areas (Dominica, Fiji, Ghana, Kiribati, Suriname, South Sudan,
Tonga), forests and degraded lands (Ethiopia), wastes (Ireland), flood
and drought risks (Guinea-Bissau, Nepal, Nigeria) and water catchments (Lesotho).

The utility of community-oriented small grants schemes as a
way to scale down major financing to an effective level is recognised
(e.g. by Tonga among others). Synergies between community-based,
ecosystem-based, traditional knowledge-based and human rights/
gender-based approaches are emphasised, for example by Fiji,
Suriname and South Sudan. Finally, Namibia stands out as a leading
practitioner and exponent of settlement/community-based or
constituency-scale participatory vulnerability assessment, planning

and systems thinking, observing that 'Sectors and constituencies can be regarded ... as homogenous systems that allow for the aggregation of vulnerability. Furthermore, sectors and constituencies are also the lowest scale of development planning in Namibia. Most climate change adaptation programs in Namibia are usually implemented at sector and constituency levels' (Government of Namibia, 2020: 131).

10.5   THE GROWTH OF SYSTEM-BASED ADAPTATION

Alignment of Local Threats and Local Actions

Chapters 3 and 4 aimed to establish that, by applying systems thinking to ecological and social systems, the sources of various kinds of strength within those systems could be understood more clearly. This also clarified the kinds of things that undermine them and make them vulnerable to stresses. It was then argued that preserving or restoring certain features of these systems at the local and landscape level would make them more robust to any kind of stress, and that this is important because of the chaotic nature of climate change at the local and landscape level. In other words, an alignment is possible between climate chaos at the local and landscape level and the opportunity to strengthen ecological and social systems at that level, primarily by meeting the needs of the people living there for secure tenure over the ecosystem resources that safeguard their livelihoods, for validation of their traditional ways of managing those resources, and for enrichment of those knowledge resources with information and lessons from elsewhere.

It is this package of local empowerment and environmental education, supplemented by help with mapping, planning and networking with other communities, that is the key to strengthening systems at the local and landscape level against climate chaos and therefore, since every country comprises localities and landscapes, a large part of the solution to climate change adaptation for all countries. This approach is also aligned with the principle of adaptive thinking, since it makes diverse local societies living in intimate contact with diverse local ecosystems responsible for coming up with their own solutions to their own problems – deciding what to conserve and what to change in the face of new challenges. They may be encouraged and enabled by outsiders, but this approach fundamentally requires local people to think realistically and creatively about their own circumstances, rather than relying upon the insights of outside experts and officials.

That local people are often entirely capable of rising to the challenge, thinking in new ways and absorbing and using new information if they need to is quite evident from the case studies in Chapters 5–8 (and the past studies referenced in Chapter 1). This is to be expected, bearing in mind the extraordinary behavioural flexibility of our species and the ability of new generations to question old ways and invent new ones at need. In any case, Table 10.3 summarises evidence that some governments are thinking systemically, and that many are now in favour of using ecosystem-based and community-based adaptation (i.e. local system-based) solutions as central parts of their adaptation strategies. To this evidence can be added the signs in Chapter 9 that Danida was, by 2019, fully aware that local- and landscape-level capacity and processes are key to adaptation, and was actively exploring the implications for aid programming.

10.6   CHANGES IN SYSTEM-BASED APPROACHES BETWEEN 2015 AND 2020

The impression from the adaptation communications is of rapid growth in systems thinking from the first generation NDCs in 2015–2016 to the later National Communications and NAPs. As a scoping exercise, Table 10.4 summarises key features of how the subject of adaptation is treated in five paired NDCs: Suriname 2015 and 2020; Moldova 2015 and 2020; Chile 2015 and 2020; the Marshall Islands 2015 and 2018; and Vietnam 2016 and 2020. Over this period:

- All five countries greatly increased the total volume of their NDCs, but the share of pages dedicated to adaptation varied little between the years for

Table 10.3 *Signs of system-based approaches in the adaptation communications*

Aspect	Feature recognised by stakeholders
Systems thinking	*Mali*: that 'combating the threat of climate change means increasing the resilience of ecological systems, production systems and social systems'. *EU*: European Research and Innovation for Food and Nutrition Security programme, Regional Knowledge Network on Systemic Approaches to Water Resources Management.
Ecosystem-based adaptation (EbA)	*NDCs*: 49 per cent contained substantive references to 'ecosystem', including EbA. *Afghanistan, Botswana, Fiji, Mexico, Seychelles*: that EbA supports the delivery of ecosystem services in an equitable way. *Costa Rica, Fiji, Mexico*: that EbA promotes holistic management of human ecology systems. *Afghanistan, Botswana, Dominican Republic, Fiji, Lao PDR, Mexico, Morocco, Namibia, Palestine, Seychelles, South Sudan, Vietnam*: that EbA acts through biodiversity conservation to enhance adaptive capacity or reduce adverse impacts. *Grenada*: that EbA tends always to have a strengthening effect. *Afghanistan, Dominica*: that EbA tends to be cost-effective. *EU*: that EbA can be used to protect infrastructure. *Dominica, Dominican Republic, Germany, Nepal, Saint Lucia, South Sudan, Uruguay*: that EbA can be used to manage risk. *Ghana*: that EbA can be used to reduce poverty. *Jamaica, Myanmar*: that EbA can be used to promote urban resilience.

Table 10.3 (*cont.*)

Aspect	Feature recognised by stakeholders
	Norway, Madagascar, Namibia, Seychelles: that EbA can be used in monitoring and research. *UAE*: that EbA generates carbon sequestration benefits.
Community-based adaptation (CbA)	*NDCs*: 30 per cent contained substantive references to 'community', including CbA. *Costa Rica, Suriname, Namibia*: that CbA combined with technical solutions is becoming the dominant approach and/or is an essential implementation strategy. *Australia, Rwanda, Solomon Islands, UK, Vietnam*: that CbA can be used in managing disaster risks. *Belize, Palestine*: that CbA can be used in managing health risks. *Dominica, Fiji, Ghana, Kiribati, Suriname, South Sudan, Tonga*: that CbA can be used for managing protected areas and community lands and harvesting areas. *Ethiopia*: that CbA can be used in managing forests and degraded lands. *Ireland*: that CbA can be used in managing wastes. *Guinea-Bissau, Nepal, Nigeria*: that CbA can be used in managing flood and drought risks. *Lesotho*: that CbA can be used in managing water catchments. *Tonga (and others)*: that CbA can be used for scaling down major grant financing to an effective level. *Fiji, Suriname, South Sudan*: that CbA contributes synergies with EbA, traditional knowledge-based and human rights/gender-based approaches.

Source: Text, and Box 10.1 (for EbA), Box 10.2 (for CbA).

Table 10.4 *Five pairs of adaptation communications in the first and second generations*

Government of Suriname (2015)	
Total pages	12 pages
Adaptation pages	1.7 pages (14 per cent): Trends, impacts, vulnerabilities (0.2); Unconditional contribution (0.9); Conditional contribution (0.6).
Substantive mention of 'ecosystem' in adaptation context	7 *in total*: conservation contributions (2); climate risk to coastal ecosystems (2); sustainable ecosystem management (1); conservation investments (1); ecosystem adaptability (1).
Substantive mention of 'community/ies' in adaptation context	*Nil.*
Government of Suriname (2020)	
Total pages	37 pages
Adaptation pages	4.7 pages (13 per cent): Trends, impacts, vulnerabilities (0.3); Introduction, planning (2.4); Sustainable forest management projects (2.0).
Substantive mention of 'ecosystem' in adaptation context	6 *in total*: sustainable forest management (2); biodiversity conservation (1); preserving ecosystem services (1); PES (1); vulnerability of agricultural ecosystems (1).

Table 10.4 (*cont.*)

Substantive mention of 'community/ies' in adaptation context	*2 in total*: resilience of farming systems (1); inclusion of marginal communities (1).
Government of Moldova (2015)	
Total pages	30 pages
Adaptation pages	20.7 pages (69 per cent): Projections (0.5); Trends, impacts, vulnerabilities (1.5); Vision, goals, targets (0.5); Current and planned undertakings (9.0); Gaps, barriers, needs (4.7); Recent external support (2.5); Monitoring and reporting (2.0).
Substantive mention of 'ecosystem' in adaptation context	*4 in total*: adaptive capacity of forests (2); value of ecosystem services (1); vulnerability research (1).
Substantive mention of 'community/ies' in adaptation context	*5 in total*: risk-reducing behavioural change (1); autonomous adaptation of communities (1); protection against extreme weather (1); piloting flood risk management (1); health sector adaptation (1).
Government of Moldova (2020)	
Total pages	74 pages
Adaptation pages	58.1 pages (79 per cent): Institutions: (1.4); Projections (6.2); Trends, impacts, vulnerabilities (3.8); Vision, goals, targets (1.1); Priorities (20.2); Current and planned undertakings (9.9); Gaps, barriers, needs (4.8); Recent external support (3.7); Progress on implementation (4.7); Monitoring and reporting (2.3).

| Substantive mention of 'ecosystem' in adaptation context | 26 *in total*: mitigation co-benefits from adaptation action (2); resilience vision (1); resilience indicators (1); vulnerabilities (6); ecosystem services (3); aquatic system management (4); agriculture sector monitoring (1); forest adaptability (2); investment priorities; (2) sectoral planning (1); ecosystem approach (1); DRR (1). *Plus*: 'Managing the interactions between climate change, land use, and terrestrial ecosystems is still a challenge at sector and national levels, and it requires efficient policy instruments that can help in ensuring conservation and sustainable use of forests, implementation of ecosystem-based adaptation strategies and ecosystem services programme' (page 65). *Plus*: 'The main systemic impediments for an increased political commitment in addressing climate change adaptation include: (i) insufficient prioritization of climate change adaptation in national political agenda, with the focus of politicians on the immediate needs for economic growth; (ii) insufficient knowledge of high-level decision makers on the magnitude of the climate change impacts and the threat to economic growth and ecosystem services; (iii) insufficient statistical data and climate impact studies on health and well-being through a gender perspective' (page 66). |
| Substantive mention of 'community/ies' in adaptation context | 31 *in total*: gender analysis of adaptation (3); integrating community adaptation in the agriculture sector (4); awareness raising (2); pilot |

239

Table 10.4 (cont.)

projects (2); inter-sectoral priorities (1); rural vulnerability (1); impact of extreme weather events (2); ensuring participation (3); vulnerability assessments (1); urban communities (1); rural community access to finance and technology (4); climate vulnerability and health care (5); climate vulnerability and water (1); awareness raising (1).

Plus a section (1.0 page) on community-based adaptation (priorities: Promote resilient development of urban communities; Promote community-based adaptation action in rural areas; Strengthen the role of vulnerable groups and local actors in planning processes that affect their own lives).

Government of Chile (2015)

Total pages

32 pages

Adaptation pages

3 pages (9 per cent), including 0.6 pages on capacity building.

Substantive mention of 'ecosystem' in adaptation context

4 in total: climate impacts on ecosystems (2); ecosystem services (1); adaptability of ecosystems (1).

Substantive mention of 'community/ies' in adaptation context

3 in total: ecosystem services of value to communities (1); adaptive capacity of communities (2).

Government of Chile (2020)

Total pages

96 pages.

Adaptation pages

11 pages (12 per cent), including 50 per cent (7 pages) of integration component.

Substantive mention of 'ecosystem' in adaptation context

14 in total: climate impacts on ecosystems (4); ecosystem services (3); water security (1); nature-based solutions (1); co-benefits of mitigation (2); ecosystems as resources (2); sustainable forest management (1).

Plus extensive reference to mitigation in forestry and peatlands.

Plus a section (1.5 pages) on 'Cross-cutting to ecosystems' (degradation of ecosystems; restoration of ecosystems; ecosystem goods and services).

Plus a section (3.5 pages) on 'Ocean' (changes in marine ecosystems and ecosystem services; marine protected areas and ecosystem recovery; establishing coastal protected areas; co-benefits of marine ecosystems).

Substantive mention of 'community/ies' in adaptation context

9 in total: marine protected area management (1); communities central to climate action (2); value of communities' traditional knowledge (2); equity and communities (1); community co-benefits (1); restoring community ecosystems and ways of life (1); ecosystem services of value to communities (1).

Government of the Marshall Islands (2015)

Total pages

10 pages.

Adaptation pages

1.9 pages (19 per cent), including 0.2 pages on support needs.

Table 10.4 (cont.)

Substantive mention of 'ecosystem' in adaptation context	1 in total: protecting traditional culture and ecosystem resources (1).
Substantive mention of 'community/ies' in adaptation context	3 in total: enhancing livelihoods and resilience (1); ensuring consultation (1); climate impacts on livelihoods and infrastructure.
Government of the Marshall Islands (2018)	
Total pages	79 pages, including Tile Til Eo ('Lighting the Way') annex.
Adaptation pages	14.0 pages (18 per cent); includes 100 per cent of Headline recommendations (0.9); Summary (4.0); Potential measures and next steps (0.2); Trends, impacts, vulnerabilities (3.8); Methods (0.2); and 50 per cent of Gender, human rights and climate (1.5); Health and climate (0.5); Education and climate (0.7); Monitoring and reporting (1.2); 2050 Strategy (1.0);
Substantive mention of 'ecosystem' in adaptation context	Nil.
Substantive mention of 'community/ies' in adaptation context	14 in total: community resilience building (1); vulnerability/impacts on communities (3); priority support for vulnerable groups (4); ensure community participation (3); protection from drought (1); resettlement of communities (1); awareness and education (1).

Government of Vietnam (2016)

Total pages

11 pages.

Adaptation pages

3.8 pages (35 per cent), including 0.5 pages on support needs.

Substantive mention of 'ecosystem' in adaptation context

2 *in total*: natural ecosystems and biodiversity as 'most vulnerable areas' (1); 'ecosystem-based adaptation through the development of ecosystem services and biodiversity conservation, with a focus on the preservation of genetic resources, species at risk of extinction, and important ecosystems' (1).

Substantive mention of 'community/ies' in adaptation context

5 *in total*: need to replicate community adaptation solutions (1); need to invest in community-based climate change adaptation and disaster management (1); need for community-based adaptation, 'including using indigenous knowledge, prioritizing the most vulnerable communities' (1); need to conserve 'biodiversity associated with livelihood development and income generation for communities and forest-dependent people' (1); need to resettle communities from areas affected frequently by 'storm surges, floods, riverbank and shoreline erosion, or areas at risk of flash floods and landslides' (1).

Government of Vietnam (2020)

Total pages

40.5 pages.

Adaptation pages

9.7 (24 per cent), including 0.6 on support needs.

243

Table 10.4 (cont.)

Substantive mention of 'ecosystem' in adaptation context	9 *in total*: natural ecosystems at high risk (4); need to restore natural and social systems (1); need for 'ecosystem-based adaptation and nature-based solutions to minimise damage associated with climate change in each sector' (1); need to increase resilience and the adaptive capacity of ecological systems (1); need to scale up ecosystem-based adaptation models (1); need to preserve and increase forests and coastal protection forests such as mangroves (1).
Substantive mention of 'community/ies' in adaptation context	8 *in total*: need to raise community awareness and community-based disaster risk management (DRM) (3); need to develop 'community-based models to respond to natural disasters and climate change' (1); need to increase resilience and the adaptive capacity of ecological systems (3); need to pilot and scale up community-based, ecosystem-based climate change adaptation models (1).

four of them, the exception being Vietnam where it declined because of the country's quickly growing mitigation actions in the energy sector. It did vary greatly between countries, however, with Suriname, Chile and the Marshall Islands all dedicating a constant 10–20 per cent to adaptation, whereas Vietnam gave 24–35 per cent, and Moldova gave four times more to adaptation in 2020 than in 2015, mainly because by then it had access to much more information on climate trends and projections, it had elaborated a larger portfolio of planned actions, it had received more support for them and it had more to report on their implementation.

- The number of substantive references to ecosystems remained at low and constant levels in Suriname and the Marshall Islands, but increased somewhat in Vietnam, hugely in Moldova and particularly in Chile. This might reflect that Suriname and the Marshall Islands had always prioritised ecosystem-based adaptation, while Vietnam, Moldova and Chile had learned a new style of ecological thinking to support their adaptation proposals.
- The number of substantive references to community was low and constant in Suriname (where Amerindian and Maroon communities had long been seen as important stakeholders), increased somewhat in Chile, Vietnam and the Marshall Islands (where local and community stakeholders had always been seen as among the top priorities for adaptation) and increased greatly in Moldova, which may be a sign of new thinking.

This sample suggests that the knowledge resources available for adaptation planning are increasing rapidly. It also suggests that in some cases adaptation thinking had deepened and broadened in the direction of ecosystem- and community-based approaches, as well as the cross-sectoral integration and mainstreaming of adaptation into sustainable development and rights-based strategies. These directions of travel may become clearer as the next generation of NDCs and other adaptation communications becomes available for the global stocktake in 2023.

11 Adaptation in Specific Geographies

11.1 ADAPTATION IN THE EUROPEAN UNION

Trends and Impacts

The EU possesses the world's most diverse aggregation of wealthy nations, ecosystems and environmental research facilities, all of them working towards common standards of reporting and all in communion with one another. Not surprisingly, therefore, the collective understanding of climate trends and impacts at the regional level of the EU member states and its immediate neighbours (some of which, like Norway, are active contributors to the knowledge resource) is extraordinarily detailed. It is therefore worth summarising the position as seen by the EC on behalf of the member states (EC, 2017). This is done in Table 11.1 for observed climate trends and Table 11.2 for some of their expected impacts. This is a baseline which will be updated in future, and in the absence of significant mitigation all trends mentioned will have continued to intensify, and all impacts will have worsened over the last few years.

EU Adaptation Strategy

The *EU Strategy on Adaptation to Climate Change* (EC, 2013) had three key objectives. First, it promoted action by member states, encouraging them to adopt comprehensive adaptation strategies and confirming that funding would be provided to help them build up their adaptation capacities and take action. Second, it promoted 'climate-proofing' action at EU level by further promoting adaptation in key vulnerable sectors such as agriculture, fisheries and cohesion policy, ensuring that Europe's infrastructure is made more resilient, and advocating the use of insurance against natural and man-made

Table 11.1 *Climate trends in and around Europe*

Temperature. The average annual temperature of the European land area for the decade 2006–2015 was around 1.5°C above pre-industrial levels, making it the warmest decade on record; 2014 and 2015 were the warmest years in Europe since instrumental records began and 500-year-old temperature records were broken in over 65 per cent of Europe in the period 2003–2010. Climate reconstructions show that summer temperatures in Europe in the three decades from 1986 to 2015 have been the warmest for at least 2,000 years, and were outside the range of natural variability. The strongest warming has been observed over the Iberian Peninsula, particularly in summer, and across central and north-eastern Europe. Winter warming has been strongest over Scandinavia.

Precipitation. Average annual precipitation for all of Europe has not changed significantly since 1960, but significant changes have been observed at subcontinental scales. North-eastern and north-western Europe show an increasing trend of up to 70 mm per decade (in western Norway) since 1960, whereas some parts of southern Europe show a decrease of up to 90 mm per decade (in central Portugal). At mid-latitudes, no significant changes in annual precipitation have been observed. Mean summer (June to August) precipitation has decreased by up to 20 mm per decade in most of southern Europe, while increases of up to 18 mm per decade have been recorded in parts of northern Europe.

Fresh waters. In general, river flows in Europe have increased in winter and decreased in summer since the 1960s, but with substantial regional and seasonal variation. Water flows have generally increased in western and northern Europe, while decreasing in southern and parts of eastern Europe, particularly in summer. These trends have led to an increase in the number of floods since 2000, and an increase in the severity and frequency of droughts in south-western and central Europe in particular, impacting water quality and freshwater ecosystems.

Oceans. All coastal regions have experienced an increase in absolute sea level, and most of them also an increase in sea level relative to land. As well as sea level rise, other primary climate change impacts observed in European seas are acidification, increased ocean heat content and increased sea surface temperature. For example, in the north-east Atlantic Ocean sea surface temperatures and ocean heat content are increasing, at different rates across all regions.

Table 11.1 (*cont.*)

The cryosphere. A general loss of glacier mass since the beginning of the measurements has occurred in all glacier regions in the EU. The Alps have lost roughly 50 per cent of their ice mass since 1900.

Extremes. Observations from the European Environment Agency indicate that several weather patterns are becoming more extreme with more intense and frequent events. Large parts of Europe have experienced intense and long heat waves since the 1950s, most of which occurred since 2000 (2003, 2006, 2007, 2010, 2014 and 2015) with notable impacts on society. Heavy precipitation events have increased in northern and north-eastern Europe since the 1960s. An analysis of the timing of river floods in Europe over 50 years found clear patterns of changes in flood timing that can be ascribed to climate effects. These variations include earlier spring snow-melt floods in north-eastern Europe, later winter floods around the North Sea and parts of the Mediterranean coast owing to delayed winter storms and earlier winter floods in western Europe caused by earlier soil moisture maxima. The number of very severe flood events, in terms of socioeconomic impact, in Europe increased in 1980–2010, but with large interannual variability. Storm surge heights along the Estonian coast of the Baltic Sea have increased significantly during the twentieth century, a trend associated with increasing mean sea levels.

Source: adapted from EC (2017: 86–93).

disasters. And third, it sought better-informed decision-making by addressing gaps in knowledge about adaptation and further developing the European climate adaptation platform known as Climate-ADAPT.

The strategy also supported adaptation in cities through the Covenant of Mayors, which had been set up in 2008 and was repurposed in the 'Mayors Adapt' initiative (from 2014) and the Mayors' Climate and Energy initiative (from 2015). This spin-off approach has proved fruitful, with close to a thousand cities committed to action on mitigation and adaptation. An evaluation of the strategy (EC, 2018) found that it had delivered on its objectives (Table 11.3), but that

Table 11.2 *Climate change impacts in and around Europe*

Food supplies. The stress imposed by climate change on agriculture is likely to aggravate disparities between European countries, with some regions experiencing positive and others only negative impacts. Observed changes in crop phenology include the advance of flowering and harvest dates in cereals. Water demand is expected to increase most in southern and central Europe, where crop deficit and irrigation needs are projected to increase. Expansion of a range of agricultural pests not previously found in Europe can be expected due to increased temperatures allowing them to survive wintertime and to have multiple generation cycles per year, and by increasing the susceptibility of crops and trees to new dangerous pests of plants from other continents. The expected main effect of climate change in the coming decades will be to shift food production from southern to northern Europe but without much curtailing of overall production. Europe relies increasingly on imports to meet demand for food and feed supply, and climate impacts on agriculture outside Europe are having an effect on the supply of these commodities within Europe. The Mediterranean area is the most vulnerable to shocks in the flow of agricultural commodities, primarily due to its high dependence on food imports from outside the EU as well as the prominent role food plays in its economy.

Forests. Climate change impacts on forests and on the ecosystem services they provide include shifts of tree species towards higher altitudes and latitudes, an increased risk of forest fires – particularly in southern Europe – as well as an increased incidence of forest pest insects. Forests cover around 215 million hectares across Europe, which is around 33 per cent of the total land area. In recent years large forest fires have repeatedly affected Europe, in particular Mediterranean countries, and the danger of forest fires is expected to increase with climate change. Cold-adapted coniferous tree species are expected to lose large areas of their ranges to broadleaf species, and forest growth is projected to decrease in southern Europe but increase in northern Europe.

Fresh waters. Vulnerability to climate change is intimately linked to the impact on water resources through floods and droughts, but also through the impact on fisheries and low river flows on aquatic ecosystems. The Mediterranean region is expected to be increasingly affected by severe impacts on its water resources, due to extreme high temperatures and droughts, and negative effects on water resources in mountain regions

Table 11.2 (*cont.*)

will impact hydropower production, winter tourism and ecosystems. Physical risk to infrastructure and settlements from slope instability may also increase. Under a 2°C climate future the south-west Mediterranean is a region of concern, with extreme droughts projected for much of southern Europe. River floods already affect around 216,000 Europeans and cost about €5.3 billion annually, and these numbers could triple under a 3°C scenario. Coastal floods already affect around 100,000 Europeans and cost nearly €1 billion annually, and these numbers could increase by a factor of 35 by the end of the century.

Marine resources and fisheries. Increased water temperature and reduced oxygen can result in marked changes in species composition, nutritional value and size, and the functioning of aquatic ecosystems. Climate change has made marine ecosystems more vulnerable to other intense ecological stressors such as overfishing, pollution and introduction of alien species. Of the commonly observed demersal fish species (fish living and feeding on or near the bottom of seas or lakes), 72 per cent have experienced changes in abundance and/or distribution in response to warming waters. This has had important impacts on fisheries in the Atlantic region. Elevated sea temperatures have triggered a major northward retreat of colder-water plankton in the north-east Atlantic, estimated at 1,100 km over the last 40 years. This trend has accelerated since 2000 and is expected to shift the distribution of fisheries. Sub-arctic species are receding northwards as a result, and more subtropical species are appearing in European fisheries. Continued changes in fisheries distribution will affect the livelihoods of fishing communities and impact current international agreements on the exploitation of straddling and highly migratory stocks.

Energy. The energy sector faces multiple threats from climate change, from changing patterns of demand, increasing stress on water resources and extreme weather events. Long-term changes in average energy demand are expected to be accompanied by more acute stress on energy infrastructure as a result of extreme events. Increasingly common droughts and heat waves may mean that cooling water is unavailable and thermal generating capacity is forced offline. Changing patterns of precipitation will continue to impact upon the output of hydropower plants.

Table 11.2 (*cont.*)

Cities. Climate change will have a direct impact on cities, including increasing health problems due to heat, and flooding damage to buildings and infrastructure. With a higher proportion of elderly people, cities will also be more sensitive to heat waves and other climatic hazards. Increased incidences of heavy rainfall can cause flooding along coastlines, within river catchments and also from poor urban drainage. This has an indirect impact on homes, business and critical infrastructure. The urban heat island effect is making heat waves worse, and this is increasingly affecting cities in central and north-western Europe. High soil sealing and urban sprawl, in combination with more extreme precipitation events and sea level rise, increase the risk of urban flooding. Many cities have expanded into floodplains, thus increasing their exposure to floods, and low-density housing built on previously untouched land has increased the risk of forest fires in many residential areas over the last decades, in particular around cities in southern Europe. Damage from heat waves, droughts in southern Europe and coastal floods shows the most dramatic rise, but the risks of inland flooding, windstorms and forest fires will also increase in Europe, with varying degrees of change across regions. Economic losses are expected to be highest for the industry, transport and energy sectors, but southern and south-eastern European countries will be most affected and, as a result, will probably require higher costs of adaptation.

Transport. Climate-related impacts on transport are primarily the result of extreme events. Transport systems in mountainous regions, coastal areas and regions prone to more intense rain and snow are generally expected to be most vulnerable to future climate change. Available projections suggest that rail transport will face particularly high risks from extreme weather events, mostly as a result of increased heavy rain. By the end of the century 196 airports and 852 seaports across the EU could face the risk of inundation due to higher sea levels and extreme weather events.

Biodiversity and ecosystems. The relative importance of climate change as a major driver of biodiversity and ecosystem change is expected to increase. Efforts to mitigate and adapt to climate change can both positively and negatively affect ecosystems, biodiversity and ecosystem services. Europe's marine and alpine ecosystems are currently the most sensitive to climate change. Observed climate change impacts on

Table 11.2 (*cont.*)

terrestrial ecosystems include changes in soil conditions, phenological changes, and altitudinal and latitudinal migration of plant and animal species (the general trend is northwards and upwards), as well as changes in species interactions and composition within communities. In Europe, 14 per cent of habitats and 13 per cent of species of interest are already known to be under pressure because of climate change. The number of habitats threatened by climate change is projected to more than double in the near future. Changes in soil moisture, in particular in the Mediterranean region, will have a direct effect on terrestrial ecosystems. Climate change is also anticipated to exacerbate the spread of invasive species, which is already being experienced across Europe. Climate change therefore significantly affects the capacity of ecosystems to provide services for human well-being and may have already triggered shifts in ecological regimes from one state to another. The Mediterranean region is home to almost half of the plant and animal species and more than half of the habitats listed in the EU Habitats Directive. However, this reservoir of biodiversity is threatened by climate-driven habitat loss because the Mediterranean climate zone is at risk of becoming smaller.

Health. Heat waves were the deadliest extreme weather events in 1991–2015 in Europe, causing tens of thousands of premature deaths. An expected increase in the length, frequency and intensity of heat waves will lead to greater mortality, which will be most pronounced among vulnerable population groups. Mortality effects are observed even for small differences from seasonal average temperatures. Existing vectors and vector-borne diseases are expected to alter their geographic and seasonal distributions. It is widely suspected that climate change has played (and will continue to play) a role in the expansion of disease vectors such as the spread of the Asian tiger mosquito (*Aedes albopictus*), which can disseminate several diseases including dengue, chikungunya and Zika, and *Phlebotomus* species of sand flies which transmit leishmaniasis. Ticks capable of carrying the bacteria that cause Lyme disease and other pathogens will show earlier seasonal activity and a generally northward expansion. Furthermore, climate change has been found to have an impact on food safety hazards throughout the food chain. Increases in water temperatures due to climate change will alter the seasonal windows of growth and the geographic range of suitable habitat for dangerous algae and bacteria.

Table 11.2 (*cont.*)

Economic activity and employment. Rising temperatures and erratic weather patterns have the potential to reduce agricultural productivity across many European regions. Extreme weather events can severely disrupt economic activity. Sea level rise will put physical capital assets at increasing risk. On the other hand, climate change may also offer new business opportunities in the form of new products and services to help people to adapt. The projected damage costs from climate change are distributed very heterogeneously across Europe, with notably higher impacts in southern Europe. The largest economic costs relate to sea level rise and tourism. Countries such as Bulgaria, Croatia, Estonia, Latvia, Lithuania, Greece and Romania, which have high climate change damage costs and a relatively high share of people employed per unit of output, are more likely than others to experience the negative effects of climate change on their agriculture and tourism sectors.

Social issues. Climate change impacts are expected to affect people's daily lives in terms of employment, housing, health, water and energy access as well as the furthering of gender equality and human rights efforts. Populations in some European areas are at a higher risk from climate change than others, depending on their exposure to climatic hazards and their vulnerability. The currently prevailing spatial distribution across Europe of a higher capacity in central and north-western Europe and a lower capacity in southern and in particular in (some of) eastern Europe is expected to continue.

Source: adapted from EC (2017: 93–112).

Europe remained vulnerable to climate impacts within and outside its borders. The evaluation report notes that, 'As the strategy's scope was to focus on the climate change impacts on the EU's territory, it did not address the potential interrelations with climate change adaptation outside the EU' (EC, 2018: 7), but the role of the EU in promoting adaptation is mentioned below, particularly with reference to Africa.

Table 11.3 *Implementation of the EU Adaptation Strategy*

Overview. The strategy defines the EU's main role as supporting the public and private sector at the national, regional and local levels by providing comprehensive information on adaptation. This is mainly provided through the European information platform, Climate-ADAPT, and the Covenant of Mayors for Climate and Energy. The EU also provides guidance on coherent adaptation approaches, in addition to funding adaptation actions (e.g. through the LIFE programme). The EU also strengthens and mainstreams adaptation into those sectors that are closely integrated at the EU level through the single market and common policies. Research on the impacts of climate change, vulnerability and adaptation options has become a high priority for Europe. DRR and new insurance products are two adaptation-related research areas that have been prioritised, to address the impacts of more frequent extreme weather events.

Promoting action by member states. By April 2017, 23 of the 28 EU member states had adopted national adaptation strategies and/or action plans, and the total LIFE Climate Action envelope for 2014–2017 had made €190 million available to support adaptation through four annual calls for proposals. A new Natural Capital Financing Facility was introduced to the LIFE programme in 2015, and was being implemented by the European Investment Bank to support projects promoting the conservation, restoration, management and enhancement of natural capital for biodiversity and climate adaptation benefits.

Better-informed decision-making. The EC recognised that research is key for effective adaptation, as practical adaptation actions and measures must be based on sound, scientific, technical and socioeconomic information. Horizon 2020 is the key EU mechanism to support research in Europe, and was spending €80 billion in 2014–2020 in support of smart, sustainable and inclusive economic growth, with about 35 per cent of the funding reserved for improving understanding of the causes and impacts of climate change and coordinating efforts to address them.

Climate-proofing EU actions. The European Structural and Investment Funds have about 43 per cent of the EU budget and about a quarter of them contribute to EU climate policy objectives. These funds help deliver the EU's Regional/Cohesion, Common Agricultural, Integrated Maritime, Common Fisheries, and Social and Employment policies. Infrastructure

Table 11.3 (*cont.*)

projects in particular, which have long lifespans and high costs, need to withstand the current and future impacts of climate change. The aim is to improve the market penetration of natural disaster insurance and to unleash the full potential of insurance pricing and other financial products for risk awareness prevention and mitigation and for long-term resilience in investment and business decisions.

Source: EC (2017, 2018).

EU Adaptation Scorecards

In 2014 the EC developed key indicators for measuring member states' levels of readiness for climate change, and every EU member state prepared an 'adaptation scorecard' in 2018. These were reviewed in a 'Horizontal assessment of the adaptation preparedness country fiches' as an annex to EC (2018). The chief conclusions are summarised in Table 11.4, showing that progress had been rather patchy, as might be expected from the diversity of the member states. This is a good example of 'experimentalist governance' in action, since: (1) the agreed overarching goal of 'adaptation' had yielded diverse local attempts to make progress in ways that each member state considered appropriate – their freedoms being illustrated by the statement that not all of them 'wish to coordinate sectoral adaptation actions under a single strategy' (EC, 2018: 23); and (2) these efforts were then reviewed by the EC so that each could learn from the others.

11.2   ADAPTATION IN SMALL ISLAND STATES

The SIDS group of countries was first recognised as having common circumstances, needs and interests in sustainable development by the UN at the Rio Conference in 1992. Many of them had been in dialogue with each other over climate change before then, and in 1990 they had established the Alliance of Small Island States (AOSIS) as a lobby group within the UN (where they represent about a quarter of

Table 11.4 *EU member state adaptation scorecards*

Overview. The adaptation cycle has five steps (see below), and while most member states had made good progress with the first three, many had yet to implement adaptation actions and undertake monitoring and reporting. Larger member states and those that had already adopted a National Adaptation Strategy made most progress, but this was also influenced by administrative culture and geography. For instance, not all member states wish to coordinate sectoral adaptation actions under a single strategy, and the need for detailed transboundary arrangements is less relevant for more isolated member states.

Step A: Preparing the ground for adaptation. All member states had a basic governance structure for adaptation policymaking. Although some degree of vertical coordination was in place in almost all member states, to enable subnational stakeholders to influence policy development and implementation, this did not seem to have a sectoral focus. Nevertheless, most country fiches indicated that a wide range of stakeholders had been consulted in the preparation of adaptation policies. While the extent of transboundary cooperation varied between member states, almost all were planning to address common challenges with relevant countries; invariably with regard to water. It was clear that international initiatives (e.g. the International Commission for the Protection of the Danube River, and the Alpine Convention), EU initiatives (e.g. the macro-regional strategies) and EU-funded projects were important in helping to prepare the ground for cooperation.

Step B: Assessing risks and vulnerabilities. Climate change scenarios and projections were widely available at national level. They were being used in most member states to undertake sound, centrally coordinated assessments of climate vulnerabilities, risks and future economic, social and environmental impacts. While most member states had included actions related to knowledge in their national plans and had identified adaptation knowledge gaps, there seemed to be limited activity to address these gaps in almost half of the member states. Adaptation-related data (e.g. climate projections, vulnerability and risk assessments, adaptation tools) were available to at least some stakeholders in almost all member states, and disseminated by a majority of them via a national web-based platform. However, coordination of associated capacity-building activities was established in fewer than half of member states.

Table 11.4 (*cont.*)

Step C: Identifying adaptation options. Most member states had used detailed vulnerability and/or risk assessments in combination with robust methods (e.g. multicriteria analyses and/or stakeholder consultations) to prioritise sectoral adaptation options. However, notably, fewer than half of member states had mechanisms in place to coordinate DRM and climate adaptation. EU funds played an important role in enabling funding to be made available nationally for implementation of adaptation actions in at least a few sectors in almost all member states. Nevertheless, there was a lack of reliable funding, with only half of member states having budgets attached to their national adaptation strategies or plans.

Step D: Implementing adaptation action. Although most member states had begun implementing their national adaptation strategies or plans, around half were yet to ensure: (1) that climate adaptation was considered in Strategic Environmental Assessments; (2) that synergies with DRR were progressed; (3) that land use, spatial, urban and maritime planning policies encouraged adaptation; (4) that adaptation was integrated into insurance policies; (5) that cooperation mechanisms were established to foster local and subnational action; (6) that there was appropriate consideration of potential climate impacts on major projects or programmes and of alternative options, including green infrastructure and (7) that stakeholders were involved in implementing adaptation policies.

Step E: Monitoring and evaluation. While most member states had planned a periodic review of their national adaptation strategies or plans, their monitoring and reporting was not yet robust and there was a need to develop stakeholder involvement (including at subnational levels) in their assessments.

Source: adapted from EC (2018), Annex on 'Horizontal assessment of the adaptation preparedness country fiches'.

members), primarily to promote action on climate. The AOSIS comprises eight countries in the 'African, Indian and South China Seas' (of nine SIDS there), 16 in the Caribbean (of 23 SIDS), 15 in the Pacific (of 20 SIDS) and five observer states (all of them SIDS).

The collective influence of AOSIS is key to mobilising support, since many of the individual members are relatively small and isolated both geographically and economically. They also tend to be exceptionally vulnerable to climate change (Box 11.1), to the point (e.g. in Kiribati, Nauru, Singapore and the Marshall and Solomon islands) where they perceive an existential risk. The issue of forced relocation of the population is being discussed in several places, along with questions like: 'How will sea level rise impact RMI's [Republic of the Marshall Islands] claim to its sovereign territory, exclusive economic zone, and the resources within its current boundaries?' (Government of the Marshall Islands, 2018: 46). This last is a challenging question in many ways, among them the issue of whether an archipelagic country that no longer exists above sea level can continue in law to claim revenues from its exclusive economic zone, in order to sustain its citizens in their new homes, and indeed whether new homes can be found for them without such a claim being recognised.

The Placencia Ambition Forum

Several forums exist among AOSIS members, including those focused on promoting common actions around UNFCCC implementation. The Placencia Ambition Forum, for example, was hosted in April 2020 by the government of Belize, where sessions focused on adaptation and resilience issues (AOSIS, 2020a), and the AOSIS Pacific Regional NDC Hub developed the following ambitious goals (AOSIS, 2020b):

- *fossil fuel-free Pacific* – significant progress towards the shift to 100 per cent renewable energy and decarbonisation of the transport industry, among other sectors;
- *resilient communities* – identification of the region's most vulnerable population groups to climate change impacts and phased implementation of effective adaptation responses;
- *atolls for the future* – creation of safe locations for eventual retreat from the most vulnerable coastal areas, in a dignified and planned manner;

BOX 11.1   **Adaptation challenges in small island states**

- *Antigua and Barbuda.* 'The country's economy is heavily dependent on natural resources, low-lying coastal zones, and favorable climate conditions to support the tourism sector, which accounts for about 80% of output gross domestic product, about 70% of direct and indirect employment and 85% of foreign exchange earnings. Antigua and Barbuda is exposed economically, environmentally and socially to projected climate change impacts. . . . Analysis of climate change for the island also projects accelerated coastal erosion and inundation, lower average annual rainfall, increased rainfall intensity causing flooding and a likely increase in tropical storm intensity' (Government of Antigua and Barbuda, 2015: 6).
- *Guinea-Bissau.* 'Due to its island characteristics, with an archipelago consisting of over 88 islands and islets, Guinea-Bissau environment is an exceptional ecosystem and one of the weakest in the world. The main environmental challenges revolve around deforestation/soil erosion and the coastal area, biodiversity conservation and quality of water resources' (Government of Guinea-Bissau, 2015: 1).
- *Kiribati.* 'Low atolls, isolated location, small land area separated by vast oceans, high population concentration, and the costs of providing basic services make Kiribati, like all Small Island Developing States, especially vulnerable to external shocks including the adverse impacts of climate change. The results of sea level rise and increasing storm surge threaten the very existence and livelihoods of large segments of the population, increase the incidences of water-borne and vector-borne diseases undermining water and food security and the livelihoods and basic needs of the population, while also causing incremental damage to buildings and infrastructure. The Climate Change in the Pacific Report (2011) describes Kiribati as having a low risk of cyclones. However, in March 2015 Kiribati experienced flooding and destruction of seawalls and coastal infrastructure as the result of Cyclone Pam, a Category 5 cyclone that devastated Vanuatu. Thus Kiribati remains exposed to the risk that cyclones will strip the low lying islands of their vegetation and soil' (Government of Kiribati, 2016: 4–5).
- *Maldives.* 'The country's main economic sectors are tourism and fisheries, both of which are extremely climate-sensitive. . . . The challenges Maldives faces in the context of climate change and development are similar to other small island nations. These challenges include, but are not limited to, the low

BOX 11.1 **(cont.)**

lying nature of the islands, high population density, high levels of poverty, and a dispersed geography. Because Maldives is a small low lying island nation, its vulnerability to climate change impacts and associated extreme weather events and disasters are significantly greater due to limited ecological, socio-economic, and technological capacities. Maldives' geography also makes communication difficult and transport expensive. Maldives' small, physically isolated economy is highly susceptible to global influences and shocks' (Government of the Maldives, 2015: 1).

- *Marshall Islands.* 'The Republic of the Marshall Islands is one of the world's lowest-lying and climate vulnerable countries. It is a coral atoll nation comprising 1,156 individual islands/islets and 29 different atolls with an average elevation of just six feet above sea level, dispersed across nearly two million square kilometers of the Pacific Ocean. ... The country is experiencing increasingly damaging effects from climate change and seeing more frequent and intense events, such as drought, floods and swells, and tropical cyclones and storms. ... As king tides become more frequent and intense, salt water is increasingly seeping into fresh water lenses, creating urgent challenges for the islands' (Government of the Marshall Islands, 2018: 6–7).

- *Saint Lucia.* 'Like all Small Island Developing States, Saint Lucia faces an uncertain future as a consequence of both the emerging and anticipated impacts of global climate change on all aspects of its development. These include, but are not limited to, threats to coastal infrastructure and economic assets from sea level rise; the impacts of more intense and possibly more frequent extreme weather events; negative impacts on human and ecosystem health, water, food production and financial services sectors; changes in rainfall distribution and intensity, resulting in both floods and droughts; degradation of coastal resources; and saline intrusion into aquifers' (Government of Saint Lucia, 2015: 1).

- *Singapore.* 'Sea level rise presents an existential challenge to Singapore, posing threats to Singapore's long-term future. Along with fellow members of the AOSIS, Singapore, as a low-lying country, is particularly exposed to the adverse effects of rising sea levels. The dangers are compounded by the fact that Singapore is located in the tropics, since it is predicted that sea level rise in tropical areas could be up to 30% higher than the global average' (Government of Singapore, 2020: 21).

> BOX 11.1 **(cont.)**
>
> - *Solomon Islands*. 'For Solomon Islands, as with other small islands developing States and Least Developed Countries, where climate change threatens the very existence of the people and the nation, adaptation is not an option – but rather a matter of survival' (Government of the Solomon Islands, 2016: 11).
> - *Trinidad and Tobago*. 'As a Small Island Developing State (SIDS), the country is vulnerable to temperature increases, changes in precipitation and sea level rise. Other vulnerabilities include increased flooding, increased frequency and intensity of hurricanes, hillside erosion and loss of coastal habitats' (Government of Trinidad and Tobago, 2018: 2).
>
> *Sources*: adaptation communications to UNFCCC Secretariat, as indicated.

- *resilient ecosystems* – management of marine, coastal and upland ecosystems to sequester carbon, protect assets and livelihoods, and sustain biodiversity;
- *leadership and communication* – implementation of strategic, whole-of-island approaches for preparedness, response and recovery; and
- *climate-proofed infrastructure* – all future infrastructure to be based on climate projections, with retrofitting of infrastructure where justified.

The Marshall Islands and Kiribati

Adaptation challenges and action priorities are listed in Box 11.2 for the Marshall Islands and Box 11.3 for Kiribati. In both cases the determination is clear to consider the problems realistically, holistically and with maximum participation, based on the idea of these nations being whole ecological and social systems, from sea bed to hill crest, and from deep history to the present and future, while not losing sight of immediate, practical concerns. The long list of policy aims by the government of Kiribati in Box 11.3, in particular, illustrates the extent of the challenge posed by climate change in a tropical small island context, and also the diversity of interlocking measures that are needed for such a country to cope even with moderate change.

BOX 11.2   **Adaptation in the Marshall Islands**

The Marshall Islands' adaptation efforts have focused on policies and measures to combat drought, which has already had a significant impact on communities. But attention in coming years will also need to focus more systematically on other aspects of adaptation, including the climate resilience of women, men and youth. Particular immediate areas of focus for adaptation are likely to include coastal resilience and coastal vulnerability assessment. Prioritisation will be critical to attract necessary private, public and international funding to implement policies, and in a way that is consistent with the country's strategic interests. Particular areas of focus for resilience could include: (1) DRM (including working through the Chief Secretary's Office to build further capacity nationally, and improving communications with outer islands) and (2) contingency and emergency response (including putting in place a financial mechanism to allow rapid response and emergency funding). It is already clear that the country's adaptation pathway will need to consider at least the following four areas.

- *Protection*: including coastal protection, infrastructure climate proofing, community and household resilience building, food security, water security, health security, and broader adaptation and DRR investments, including the development of financial mechanisms to finance investments and response and recovery efforts.
- *Elevation*: all new construction of all types of structures should be elevated, and as conditions worsen, particularly with sea level rise, new policies and plans for constructing elevated settlements for future consolidation of the population will be critical.
- *Consolidation*: in the event that the majority of communities become increasingly vulnerable and experience more frequent and dire effects of climate change, including flooding and inundation, there may need to be a policy to consolidate the population onto elevated settlements, possibly rather soon.
- *Relocation*: while relocation should be considered the last-resort option, the country may not be able to accommodate all 60,000 or more people who are

BOX 11.2   **(cont.)**

expected to be living there by 2030, so the government will need to assure the right of citizens to remain in the islands as best they can, while also ensuring continued opportunities for migration.

Sources: adapted from Government of the Marshall Islands (2018: 44–45).

BOX 11.3   **Adaptation in Kiribati**

Policies and priorities. High levels of political awareness and support on climate change have existed in Kiribati since the early 1990s, and the country has been outspoken at the international level about the impacts of climate change. The government takes a 'whole-of-government' approach to climate change, with a focus on mainstreaming and coordination across sectors and scales, aiming for an integrated, multisectoral, programmatic approach to climate change and DRM. Twelve key national adaptation priorities are used, each with consistent aims as follows.

- *Governance, policies and laws.* (1) All policies, strategies, sector operational plans, ministry annual workplans, ministerial plans of operations, project proposals and monitoring and evaluation systems aim to enable proactive and inclusive adaptation and DRR. (2) National and sector legislation is providing an enabling environment for DRR. (c) Coordination must be enhanced between adaptation and DRM programmes and legislation, among government departments, island councils, NGOs, faith-based organisations and the private sector, with the aim of collaboration across sectors and links between these and national development aspirations.
- *Managing knowledge.* (1) An integrated and up-to-date national database providing all relevant information for resilient development should be available and accessible for all. (2) Capacities to communicate science and best practices are to be strengthened by developing and disseminating effective and relevant information, communication and awareness products for decision making and awareness raising across sectors and at all levels. (3) Capacities for data collection, assessment, analysis, interpretation, monitoring and reporting are to be strengthened across sectors.

BOX 11.3 **(cont.)**

- *Greening the private sector.* (1) Investment is to be promoted by businesses, including small and medium-sized enterprises and women in value-adding marine and agricultural products for the domestic and export niche markets, ensuring that they benefit women and men equally. (2) Private sector concerns will be encouraged to implement greening and risk management initiatives (in areas such as tourism, trade, transport, import and export), while incorporating climate change and disaster risks into their strategic and business plans (and feasibility of insurance is to be assessed).
- *Water and food security.* (1) Communities with their island councils should manage and implement adaptation and DRR measures as an integral part of their development efforts and inclusive of vulnerable groups. (2) Salt-, drought-, rain- and heat stress-resilient crops, fruit, vegetables and livestock breeds are to be identified and promoted, and communities encouraged to preserve traditional knowledge of cultivation, preparation and preservation techniques for traditional food crops, fruit trees and seafoods. (3) Communities should manage coastal fisheries by taking into consideration sustainability of marine resources as well as climate change and disaster risks. (4) Communities should have constant access to local produce and basic food commodities. (5) Communities should manage their water resources, including during extreme events such as drought, heavy rain and storm surges.
- *Delivering health services.* (1) Public awareness of water safety is to be increased and the public encouraged to reduce the spread of vector-, water- and food-borne diseases. (2) Surveillance systems for environmental health hazards and climate-sensitive diseases are to be strengthened and the capacity of national and local health systems, institutions and personnel to manage climate change- and disaster-related health risks enhanced. (3) The I-Kiribati population's general health status is to be enhanced so that people are more resilient to climate-related diseases and health impacts. (4) A national climate change, disaster risk, outbreak preparedness governance framework, response plan and a sectoral environmental health plan, which incorporate surveillance and response to climate-sensitive diseases and disaster risks, are to be put in place. (5) Support is to be allocated for retrofitting medical facilities and health infrastructure adversely affected by, or susceptible to, the impacts of climate change. (6) Chemical waste management and alternatives to reduce contamination and pollution are to be encouraged.

BOX 11.3   (cont.)

- *Infrastructure and land management.* (1) Public buildings, infrastructure and utilities are to be well maintained and made as resilient as possible to climate change and disasters. (2) Land and marine planning and management systems for all islands will provide clear regulations on land development, and the capacity of planning authorities strengthened to support enforcement. (3) Coastal resilience will be built through strategic coastal protection initiatives, (4) Water reserves will be protected, and communities must have access to sufficient and adequate fresh water at all times, as well as to improved sanitation facilities. (5) Financial mechanisms to address the risks facing community and public assets will be established, with a focus on climate risk insurance and building on existing initiatives and programmes.
- *Education, training and awareness.* (1) It is recognised that students and professionals have special capacity to act on adaptation, DRM and emission reduction. (2) The I-Kiribati population is to be well informed and all stakeholders will have access to up-to-date and accurate, contemporary and traditional information on climate change and DRM. (3) Formal as well as technical and vocational education and training will be used to ensure that the I-Kiribati population (including vulnerable groups) is well qualified for employment inside and outside Kiribati.
- *Early warning and disaster and emergency management.* (1) There will be greater focus on strengthening disaster risk preparedness (through innovative technology), response and recovery across all sectors including at the island and the community levels. (2) Effective enforcement is needed at Kiribati's ports of entry to safeguard its fragile environment from external threats. (3) Principles of humanitarian assistance and 'building back better' will be enshrined in disaster responses. (4) Data collection and vulnerability analysis will be used to enhance understanding of loss and damage and improve Kiribati's position in engaging with regional and international actors.
- *Renewable sources of energy and energy efficiency.* (1) The transition towards renewable energy will be promoted and enhanced. (2) Energy conservation and energy efficiency will be increased. (3) Policy, legislation and regulation will be used to support renewable energy and energy efficiency.
- *Capacity to access finance, monitor expenditures and maintain partnerships.* (1) In-country coordination will be enhanced on climate finance, climate change and DRM. (2) The Climate Finance Unit and broader national capacity, including NGOs, will be used to engage with key multilateral

BOX 11.3   **(cont.)**

sources and climate finance mechanisms to provide efficient support for
adaptation, mitigation and DRM. (3) Efforts will be increased to mobilise and
scale up sources of financing to implement climate change adaptation,
mitigation and DRM needs and priorities. (4) Monitoring, evaluation and
performance measures for adaptation and DRM will be strengthened,
including budgeting, expenditure, institutional capacity and internal systems
to increase Kiribati's access to, and engagement with, various sources of
climate finance.

- *Sovereignty, identity and cultural heritage.* (1) The rights of Kiribati over its
  existing exclusive economic zone and the resources within it are to be
  protected forever for the people of Kiribati. (2) The cultural heritage of Kiribati
  is to be protected, preserved and promoted.
- *Strategic partnerships for community participation, and engagement,
  ownership and inclusion of vulnerable groups.* (1) Community partnerships
  and members of vulnerable groups are to be increasingly engaged in climate
  change and DRM initiatives. (2) Good governance, sustainability and
  empowerment are to be the basis of long-term partnerships with
  communities. (3) Locally driven resilience programmes will be used in
  identifying issues, strength and opportunities. (4) Community participation
  and engagement will be a priority in all measures to address climate change
  and DRM.

Sources: adapted from Government of Kiribati (2019: 69–76).

11.3   ADAPTATION IN AFRICA

As the most culturally and biologically diverse of the inhabited continents, and the ancestral home of our species, Africa and the relationship between Africans and the changing climate needs a library to
itself, and is fast developing one (e.g. Yaro and Hesselberg, 2016; Leal
Filho et al., 2017; Berck et al., 2018; Chirisa et al., 2018; Leal Filho
et al., 2021). Even so, samples of government thoughts and actions
can illustrate key challenges that are having to be addressed, and
how this is being done, for example in Ghana (Box 11.4), Eswatini

BOX 11.4 **Adaptation in Ghana**

More than 30 years of climate records show that the climatic conditions in Ghana have severely deteriorated and are likely to worsen in the future. The uncertainties in the future climate will be far more significant than in the past. Rainfall variability will be higher in the forest regions than the rest of the country. Ghana will continue to be warm and even get worse by 2080. Temperatures are likely to increase by at least 3°C by 2080 nationwide. The savannah regions are likely to record temperature above 30°C. The high likelihood of wet spells may lead to more floods across the country. The projected increases in dry spells may exacerbate drought conditions, especially in the savannah.

Climate change is likely to bring unbearable disruptions to the electricity system, cash crop production, urban migration, livelihoods of smallholder farmers and the coastline. The changing climate is a threat to energy and food security and rural livelihoods. The already harsh ecological conditions coupled with adverse climate change impacts in Northern Ghana could accelerate migration towards the southern cities. Most migrants tend to live in slums and urban-stressed areas that are prone to recurrent floods. The projected future severe drought conditions may affect cash crop production. Prolonged drought conditions, will increase the population of variegated grasshoppers which destroy cassava. There is an increasing reduction in marine fish catch as well as freshwater landings due to rising sea temperatures and to high fishing intensity, leading to generally low incomes in fishing communities. An increase in extreme temperatures could lead to increased demand by competing water users which could reduce water availability for power generation. Climate change-induced sea level rise is very likely to exacerbate the erosion taking place along vulnerable sections of the 560 km coastline.

Adaptation policies and investments are producing positive returns and must be scaled up at all levels. The principal government flagship policy initiatives support climate change adaptation measures under the 'modernising agriculture' agenda:

BOX 11.4   **(cont.)**

- The *One Village One Dam* policy has increased access to water all year round to vulnerable farmers. So far, 570 small dams and dugouts are under different stages of construction in the Northern, Upper East and Upper West regions.
- The *One District One Warehouse* policy has contributed to reverse post-harvest losses. The construction of 300 warehouses of 1,000 tonnes capacity is underway.
- The *Planting for Food and Jobs* programme involves the supply of improved seeds, fertiliser and extension services to farmers and has contributed to increased food production (maize 75 per cent, rice 24 per cent, soybean 39 per cent and sorghum 100 per cent); 557,000 farmers have been supplied with high-quality seeds; 7,600 tonnes of seeds and cassava planting materials have been distributed to farmers.
- The *Rearing for Food and Jobs* programme supports climate-smart livestock production; 53,500 livestock (sheep, pigs, cockerels and guinea fowls) had been distributed as of June 2019.
- *Planting for Export and Rural Development* (PERD) is a decentralised tree crop programme to promote rural economic growth and farmer incomes. The plan is to develop nine commodity value chains, namely cashew, coffee, cotton, coconut, citrus, oil palm, mango, rubber and shea, through a decentralised system. PERD seeks to support one million farmers in 170 districts with certified free planting materials to cover over one million hectares of farmlands and engage 10,000 young graduates as crop specialised extension officers.

Source: adapted from Government of Ghana (2020: xxxvi–xxxvii).

(Box 11.5) and Namibia (Box 11.6).One thing to note is the realistic attitude, for example, of the government of Burkina Faso (2015: 10):

> The greatest concern for Burkina Faso, as for any other country, is that the climate changes foreseen for the next 50 years are now inevitable. Hence, the primary interest of Burkina Faso, which is not a large GHG emitter, must necessarily be improvement of the people's capability to adapt to the conditions that will exist from now to 2025, 2030 or 2050: a significant rise in the average

BOX 11.5 **Adaptation in Eswatini**

Evidence of climate change is already visible in Swaziland through dwindling crop yields, violent storms and persistent drought. This is exacerbating the country's existing challenges which include chronic poverty, food insecurity and the highest prevalence of HIV/AIDS in an adult population in the world. Adaptation is of utmost importance in Swaziland, particularly in four key sectors of the economy that will form the foundation of the adaptation contribution of Swaziland's INDC: the biodiversity and ecosystems, water, agriculture and health sectors. The health sector will be affected by climate change impacts, with groups such as households with members living with HIV expected to be particularly vulnerable. Health is a cross-cutting sector, however, and is dependent on adaptation in the agriculture, water and biodiversity and ecosystems sectors, where goals include the following:

- *Biodiversity and ecosystems sector.* (1) Scale up investments in restoring and maintaining ecological infrastructure, with a focus on the priority ecological assets. (2) Establish effective long-term biodiversity conservation, landscape management and natural resource management programmes. (3) Strategically plan and manage the ecological infrastructure, which includes healthy grasslands, rivers, wetlands, woodlands and natural forests. (4) Enhance biodiversity and promote ecotourism with benefit sharing for the surrounding communities. The possible actions that have been identified to achieve these contributions include: agro-forestry; ecological pest management; flood mapping; grazing land management; degraded land rehabilitation; fire management; and erosion control through terracing.
- *Water sector.* (1) Develop water pricing structures to encourage efficient water use. (2) Implement measures to reduce water consumption throughout the value chain. (3) Strengthen the capacity of early warning centres, for improved emergency preparedness, disaster risks and response capacities. (4) Develop systems to integrate water resource management across all the sectors of human endeavour, land use and the environment. The possible actions that have been identified to achieve these contributions include: artificial groundwater recharge; integrated river basin management; leakage detection; rainwater harvesting; sand dams; solar pumps borehole water pumping; water recycling and reuse; and wetland restoration.

BOX 11.5   **(cont.)**

- *Agricultural sector.* (1) Increase the contribution of agriculture to economic development, to support both food security and exports. (2) Reduce poverty and improve food and nutrition security through sustainable use of natural resources, improved access to markets and improved disaster and risk management systems. The possible actions that have been identified to achieve these contributions include: conservation tillage; crop diversification; greenhouse farming; hydroponics; livestock selective breeding; micro irrigation; organic farming; and solar dryers.

Source: adapted from Government of Eswatini (2016: 2–3).

BOX 11.6   **Adaptation in Namibia**

Namibia is an arid, water-deficient country, among the largest in Africa, in which agricultural output is extremely sensitive to climatic conditions since it depends on an erratic and low rainfall that feeds dams, ephemeral rivers and aquifers. Water is a dominating priority, and issues in the water sector include scarcity and high prices for irrigation, competition from mining/industry, commerce and an expanding, urbanising population within Namibia, as well as with neighbouring countries, plus increasing levels of pollution and overexploitation of aquifers. It has extraordinary biodiversity (including an important 'hotspot' in the Succulent Karoo) protected by an elaborate community conservation system which is mainly in the form of 115 legally recognised communal conservancies and forests that cover a total of about 165,000 km^2 (20 per cent of the country). This communal tradition of wildlife and forest management influences all other aspects of adaptation strategy since vulnerability assessments and planning are all founded on the circumstances and needs of 122 local constituencies or settlements. The government considers that this local level of society is likely to be the most hard-hit by climate change, so it is where the climate response must be most focused.

BOX 11.6 **(cont.)**

Historically, Namibians have adopted several coping mechanisms, but, despite their adaptation efforts, many communities have low levels of adaptive capacity due to challenges such as marginalisation, underdevelopment, poverty, inequality, maladaptive policies and increasing population. There are also problems of migration, most visible in the spread of informal settlements that accommodate low-income families in shacks on the edges of towns. These informal settlements are becoming more vulnerable and progressively less able to deal with climate change risks, and this adds to existing levels of vulnerability of communities in Namibia which are likely to intensify with climate change. The government's goal is therefore to identify constituencies that are relatively more vulnerable to climate change impacts. This is done to make adaptation planning spatially explicit, thus ensuring that the focus is emphasised on constituencies most in need of resilience building by using an index-based approach in the assessment of the vulnerability of constituencies.

The idea here is that vulnerability is a function of exposure, sensitivity and adaptive capacity, framed in the context of the IPCC definition of vulnerability. Meanwhile, the government and its partners have supported 32 recent adaptation programmes and projects, mostly directed to agriculture, fisheries, sustainable land management, government, climate information and research, ecosystems and biodiversity, forestry and energy. All nationally implemented projects, however, have supported community-based adaptation, and the government considers that adaptation initiatives are now evolving from a sectoral approach to assume a community-based and settlements approach. This has put Namibia ahead of the curve in using highly inclusive community-based initiatives to target vulnerable hotspots identified through local vulnerability analyses (coupled with other findings on livelihoods and resource sustainability), and in seeking replicable, scalable and innovative adaptation strategies.

Source: adapted from Government of Namibia (2020: 1–9).

temperature, more severe dry seasons, strongly and less predictable rainy seasons, a growing problem of drought, lowering of the groundwater table and an increase in the frequency of certain diseases. The only scenario to be prepared for is the trend situation, 'business as usual', because the climate effects which Burkina must confront have already begun and the positive effects of the possible mitigation actions to be envisaged from this point forward, either at the local or global level, will not be felt until after the period of applicability of the INDC (2030).

There is also a confidence in the African population's ability to cope, in the value of traditional knowledge-based solutions to climate problems – for example, Burkina Faso mentions its wood- and metal-free architecture, known as 'Nubian vaults' – and the overwhelming importance given to water. As Burkina Faso again puts it: 'One molecule of water is just as vital to the soil as carbon dioxide for food security and the life cycle chain. For adaptation, the conservation of water is an adaptation indicator just as carbon dioxide is in the case of mitigation' (Government of Burkina Faso, 2015: 35).

EU Support for Adaptation In Africa

A synthesis evaluation of EU climate aid to Africa (Caldecott, Clark et al., 2019) found support for the involvement of African countries in UNFCCC processes, while sector-wide programmes had helped build institutional capacity and promote knowledge sharing. Sectoral support had helped to improve forest governance via two flagship programmes focusing on reducing emissions from deforestation and forest degradation, by supporting the UNFCCC Secretariat (on REDD+) and through the Forest Law Enforcement, Governance and Trade programme (Particip, 2015). These contributed to mitigation through carbon conservation and to adaptation through maintenance of ecosystem services and biodiversity. The mitigation theme was complemented by investment in clean energy, where EU support has raised awareness and promoted policy dialogue around regulatory reforms, institutional strengthening, sector strategies and clean energy delivery, with a strong cumulative impact in preparing the ground for future progress on clean energy while some real progress was made on clean energy access.

The adaptation theme had been complemented and advanced by various initiatives supported by the EU's Global Climate Change Alliance (GCCA). Since so many African farmers are at risk of drought, sharing adaptive solutions among locations is a key theme. The EU-supported programme Monitoring for the Environment and Security in Africa provided satellite and land-based Earth observation monitoring and analysis in support of environment, climate and food security policies, programming and decision-making, in partnership with African institutions (Particip, 2016). It enabled research and innovation, offering technology and data for use by African research organisations, and also by government services in forecasting and planning in meteorology, agriculture, fisheries, transport, environment and climate change mitigation.

The utility of remote imagery analysis was also multiplied in other continent-wide EU initiatives, including the Climate Change, Agriculture and Food Security Programme, an umbrella for climate change-related activities led by various national centres of excellence. Its focus was on engagement with local communities, which identified their own priorities and provided widely shared knowledge on adaptation techniques (Particip, 2016). It also used local research to inform national policy development and influence global climate negotiations. Other EU initiatives enabled knowledge sharing on climate-resilient agriculture in Africa and beyond, including the Global Forum on Agricultural Research, its Alliance for Climate Smart Agriculture, and the Smallholder Innovation for Resilience project of the Global Programme on Agricultural Research for Development (Particip, 2016).

This transfer of knowledge, technology and innovative solutions is a core priority for the GCCA, which has also supported pilot programmes (e.g. in Ethiopia and the Central African Republic) that are providing models of worldwide significance for the integration of community forestry, agriculture, water conservation, climate-smart farming and local capacity building (Particip, 2015; Topper and Pallen, 2014, 2015). The GCCA was established in 2007 and relaunched in 2014 as GCCA+, with additional resources up to 2020 to act as the EU's flagship in mainstreaming the climate change response in development, strengthening societies against climate-related stresses and shocks through adaptive change, DRR and forest protection, and promoting holistic adaptation and mitigation strategies (EC, 2020). Compared with the first phase, GCCA+ had more emphasis on the roles of knowledge management and communication, and was designed to respond more directly to the adaptation communications of developing countries, especially LDCs and SIDS, while also enhancing cooperation with civil society organisations and other non-state actors (Cabinet Office, undated).

11.4   ADAPTATION IN THE AMERICAS

There is no question of summarising the circumstances, experiences and aspirations of two whole continents, not least because two large and influential American countries, the USA and Brazil, have had recent governments headed by individuals who seem deeply sceptical of human-caused climate change and hostile to long-established arrangements for international cooperation. Nevertheless, as was evident from Chapter 6 for Bolivia (and will be from Chapter 13 for Costa Rica), the Americas also contain countries that are exploring some valuable new solutions to adaptation problems, and it is in Latin America alone where particular interest has been expressed in global south–south cooperation and networking for joint learning about adaptation (Government of Chile, 2015; Government of Brazil, 2016; Government of Mexico, 2016).

The value of this networking instinct has been responded to by the EU, at least at the level of regional cooperation through EUROCLIMA, which promoted dialogue on the climate response among Latin American countries as well as supporting their partici-pation in UNFCCC processes (Caldecott, Clark et al., 2019). Among EU best practices was EUROSOLAR, which synergised with EU sup-port to governments through UNFCCC and EUROCLIMA to build capacity on solar energy in Bolivia, Ecuador and Peru, and gave rise to the Inter-American Development Bank-financed 2014–2019 Mass Programme for Solar Energy (Canessa et al., 2015). More directly relevant to adaptation, there are also resonances between Latin America and some of the strategies that countries in the Pacific and Africa are striving towards, for example in recognising, as Chile does as much as Namibia, that it is the local level of society where the impacts of an increasingly chaotic climate are likely to fall most heavily, so this is the level where systems must be strengthened most decisively.

Also from South America, the sentiment in such leadership statements as the following, from an Argentinian Pope in the Paraguayan National Adaptation Plan, and two Chilean presidents, would certainly be appreciated by leaders in Kiribati, the Marshall Islands and elsewhere:

- 'I urgently appeal, then, for a new dialogue about how we are shaping the future of our planet. We need a conversation which includes everyone, since the environmental challenge we are undergoing, and its human roots, concern and affect us all' (Pope Francis, 2015: §14, quoted, in Spanish, in Government of Paraguay, 2017: 2).
- 'Climate change deepens inequalities and multiplies threats. It is our obligation to address this problem before its consequences become irreversible. Future generations will judge us not only for the growth of our economy and its social impacts, but also for our capacity to face the climate change challenge' (Michelle Bachelet, president of Chile, in Government of Chile, 2015: 4).
- 'Every generation faces unique challenges, but none has had to face such an urgent and formidable challenge as the one our generation is facing today in the environment: climate change and global warming; that is the mother of all battles, as it is a fight for humankind's survival. Scientific evidence is

absolutely overwhelming and conclusive. Even by fulfilling all the commitments of the Paris Agreement now, the temperature would significantly exceed the target established, reaching an increase of almost 3.4 degrees, which is disastrous. We need much more demanding and ambitious commitments and measures to limit temperature rise to 1.5 degrees at most. That is Chile's position' (Sebastian Piñera E., president of Chile, in Government of Chile, 2020: 5).

PART V **Conclusions**

12 Designing and Evaluating Adaptation Investments

Designing for Outcomes

Judgements about the design and performance of adaptation investments must be based on transparent definitions and evidence. Aid objectives can be defined, agreed among stakeholders and articulated as policies, under such headings as increased respect for human rights, sustainable inclusive economic and human development, biodiversity conservation and climate change mitigation and/or adaptation. These overarching priorities are periodically refreshed into lists of global aims, most recently the SDGs, while a similar but more limited role is also taken by the many national, sectoral and thematic policy dialogues and strategic plans that are developed each year by the officials of partner countries and other stakeholders. All specific actions at all other levels should contribute to these aims; how they are expected to do so is usually explained in the design documents for each investment; and how well they actually did so is usually related in the evaluation documents.

Theories of Change

Design quality assessment targets the question of how well the action was formulated so as to achieve its specific objectives, whatever they may be. Here design quality is considered high if there is evidence that an intervention was based on a rational theory of change supported by plausible assumptions. A theory of change states what the designers hoped to achieve, why and by what means. It depends on assumptions about cause and effect. These assumptions are often linked, and all contribute to the theory of change, so the plausibility of each must be assessed using evidence and reasonable inference.

279

A chain is only as strong as its weakest link, so design quality depends on the defensibility of all the assumptions and also the links between them. A project's design should thus provide a convincing analysis of the context, problems, needs and risks upon which it is founded, sufficient evidence that its approach can deliver useful results and sustainable impacts, and a complete sense of how and why the project is to be implemented. It is also necessary to confirm that key stakeholder groups were meaningfully consulted in the design process, to give confidence that participants understand and agree with the aims.

Chains of Causality

To assess the quality of a project design involves first understanding its theory of change, as explained by the designers or reconstructed from documents on context and purpose. Reconstructing the theory of change can be hard, since the systems involved are complex and not all designers fully explain their reasoning. The next step is to identify and state clearly the assumptions that appear to underlie the theory of change, so their plausibility can be assessed, and also establish the logic linking one to another, so the strength of these links can also be assessed.

The purpose is to spell out the chain of causality, why it is expected to operate and any uncertainties, beliefs or hopes that may be folded into the reasoning. This is done in Table 12.1 for the community forestry component of NARMSAP in Nepal (Chapter 5), both as a generic example but also because the content is relevant to strengthening ecological and social systems. It shows how considering the theory of change, the assumptions and the reservations about the assumptions can help evaluators form an opinion about the design. In this case the participatory history of the programme was accepted as evidence of adequate consultation with stakeholder groups, and the confidence of the evaluators was expressed as a high score of six for design quality, on a seven-point scale explained further below.

Table 12.1 *Design quality of the community forestry component of NARMSAP*

Theory of change. The livelihoods of many rural people in Nepal depend on goods and services provided by natural forest ecosystems. These livelihoods can be safeguarded and potentially improved if groups (communities, or, in principle, companies or individuals) possess the accountable authority, ecological knowledge and managerial skills to control specific forest areas and use them exclusively and permanently in their own interests. Since 1978, first with World Bank and later with Danida support, the government of Nepal accepted this logic and established a framework of policy and law to encourage and enable the process of appointing CFUGs among traditional forest users as exclusive and autonomous managers of community forests within the national forest estate. The expectation was that community forest owners would have an interest in maintaining and improving forest condition and in inventing ways to generate revenues and other livelihood benefits, so they would be willing to accept, adopt and apply new ideas and skills that facilitate these outcomes. The community forestry component thus focused on facilitating the legal establishment of community forests, the CFUGs to be responsible for managing them and the operational plans to guide their management, while also offering and delivering training to CFUG members and government forest staff.

Assumptions underlying the theory of change	Judgements on the validity of assumptions
Assumption 1. Community integrity and policy/legislative support is sufficient to allow inclusive and accountable CFUGs to be established and empowered over local forests.	*Strongly plausible*. Community forestry may be vulnerable to civil discord if it is seen as a priority of one faction and is opposed by another. In practice, however, both Maoist and government groups were willing to accept the strategy (although government staff were progressively excluded from rural areas).
Assumption 2: CFUGs and government forest staff (as trainers, co-managers and supervisors)	*Strongly plausible*. The assumption was plausible both initially (based on local lobbying

Table 12.1 (*cont.*)

would be willing to participate in actions and accept training in ecology and forest management to maintain and improve forest condition and seek livelihood improvements.	and international experience), and later (based on local experience).
Assumption 3: Stakeholders will use the existence of a secure forest under their own control as a resource to plan more sustainable ways to meet demand for fuel wood and other forest products, and to invent and build new livelihood strategies that will contribute to reduced poverty without undermining the integrity of the ecosystems concerned.	*Strongly plausible*. Forests are vulnerable to conflict and competitive overexploitation of particular resources in the course of individual enterprises, so requiring strong, informed and accountable governance (e.g. by CFUGs) to prevent and resolve conflict.

Overall conclusion. Accepting the validity of assumptions 1–3, it was reasonable to expect that the legal establishment and training measures would be welcomed, and would result in forest conservation and livelihood benefits. *Score for design quality*: 6.

Source: adapted from Caldecott et al. (2017), annex E.11: evidence on design quality, NARMSAP.

12.2   DESIGNING FOR ADAPTIVE PERFORMANCE

Recognition of Local System Values

It seems clear from the foregoing chapters that some countries, and perhaps many, are starting to recognise:

- that climate chaos is most threatening and requires most action at the local and landscape level, where stronger ecological and social systems can best be built to resist it;
- that community-based and ecosystem-based approaches offer valid, potent and cost-effective strategies with multiple kinds of usefulness in the context of climate adaptation;

- that strengthening local community tenure over resources is often a key feature of such approaches, so should be sought as a potential way to rebuild the strength of local systems;
- that unique locally adapted cultures, the traditional knowledge they contain and the wild species and ecosystems to which they relate, are valuable in diverse ways;
- that unique locally adapted ecosystems, the biodiversity they contain and the goods and services they offer to human society are also valuable in diverse ways; and
- that in all this, the proper role of government is not to impose but to empower, by encouraging and enabling local action but also by responding to local needs and requests with knowledge and other resources.

These ideas have practical consequences for the climate response, and are consistent with the high ideals of environmentally sustainable human development. But another point is that these system-based ideas also require the abandonment of entitlement myths that can block common-sense solutions, for example by regarding people far from the centres of power as inadequate and in need of direction. Such biases have no place in our efforts to adapt to climate change. Moreover, adaptive thinking requires new concepts of authority and leadership, ones that challenge everyone to think for themselves in dialogue with everyone else, rather than giving orders or prescriptions based on prior knowledge. All these changes go to some extent against the grain of much that has made the modern world so structured and so destructive, so wealthy yet so unequal. Such approaches are therefore still far from being universally accepted, although their time may well have arrived.

'Perfect' Design for a Generic Project

It is now possible to describe what a 'perfect' design might look like, in the same format as in Table 12.1 but for a generic project intended to strengthen local ecological and social systems against climate chaos. This is done in Table 12.2, which is based on a combination of the principles listed above with the following system needs identified in Chapter 4.

- Strong social systems need:
  - *Social homeostats* – forums where members can agree to recognise and relate to each other through a common system of values, arrangements for conflict resolution, and a shared understanding of roles, rights responsibilities.
  - *Social foundations* – clear tenure over useful territory, and locally accountable management of all the things that are important to local livelihoods, security and well-being.
  - *Strengthening social systems* – can involve shielding the original system (e.g. in a 'homeland'), or validating and preserving parts of it (e.g. language and sacred myth), but mostly means renewing the system with a forum to clarify local identities, values, roles, rights, knowledge and education, and with the authority and capacity to safeguard local livelihoods through healthy ecosystems under clear ownership and locally accountable management, so that people can seek their own new equilibrium.
- Strong ecological systems need:
  - *Ecological homeostats* – whole webs of life, which can be preserved only through ecological integrity and a degree of area protection.
  - *Ecological foundations* – the physical structure of biomass, as well as soils, microclimate and flows of water, nutrients and energy.
  - *Strengthening ecological systems* – can involve shielding the original system (e.g. in a 'nature reserve'), or validating and preserving parts of it (e.g. 'flagship species'), but mostly means renewing the system with targeted protection and restoration of soils, water, structure and microclimate, and against alien invasives and overharvesting, so that the ecosystem can seek a new equilibrium.

12.3   SCORING DESIGN AND PERFORMANCE USING EVIDENCE

Previous studies (Caldecott et al., 2010, 2014, 2015, 2017, 2020; Caldecott, Hawkes et al., 2012; Caldecott, van Sluijs et al., 2012; Caldecott, Valjas et al., 2012; Caldecott, 2017a) have shown that the performance of different projects at different times can be compared, and strengths and weaknesses highlighted for lesson-learning purposes, by using a system in which judgements on performance are

Table 12.2 *'Perfect' design quality in a local system-strengthening programme*

Theory of change. By assigning to local people clear, long-term rights to own, manage and benefit from ecosystems, they will become more interested in the condition of those ecosystems. By also establishing ecosystem management groups or forums where people can agree how to manage the ecosystems for which they are responsible, the ecosystems will tend to be managed in line with agreed locally perceived priorities. Because of long-term tenure, community responsibility and forum accountability, these priorities will usually include an interest in the long-term supply, and hence sustainability, of a selection of ecosystem goods and services, for collective rather than private benefit. The forum itself will also preserve, restore or introduce practices of debate and consensus building within the empowered community. Some of these effects (e.g. biodiversity conservation, catchment services, carbon storage, conflict reduction, poverty avoidance) will also be appreciated as benefits by national society, creating an incentive among national stakeholders to support and replicate the community empowerment process. Some national benefits may be seen by outsiders, because of their different priorities, as unlikely to be adequately addressed or developed by local people, creating an incentive to improve local ecological knowledge by offering technical support, environmental education and networking with other communities where similar issues have been resolved. The net result of secure local tenure, local forums and outside support offered but not imposed (i.e. without consent or compensation), will be healthier ecosystems and more engaged and confident local people (i.e. stronger systems) at landscape level, and improvements to governance and environmental security at the national level, as a result of which the whole country will be better able to cope with climate chaos.

Assumptions underlying the theory of change	*Judgements on the validity of assumptions*
Assumption 1: Local interest in residential sustainability. There is sufficient interest among local people in the future of their society in its present location to support a	*Strongly plausible.* If this interest is absent because the community has been resettled in the past, it might be rekindled if the community is restored to its

Table 12.2 (*cont.*)

community-based ecosystem management arrangement.	traditional location. Also, if the community is deeply affected by fatalism and alienation (expressed, for example, in drug and alcohol problems or religious manias), then special measures may be needed to support the community while more strategic solutions are found. Also, if the community is in a location that cannot be saved and must be evacuated, then relocation support will be needed instead.
Assumption 2: Local interest in process and outcome. There is sufficient interest in the ecosystems in their present state, which may be highly degraded and unproductive, to motivate local people to take an interest in them.	*Strongly plausible.* But sometimes additional support (e.g. grant-supported employment, proactive education) will be needed to create an interest in the future value of regenerated ecosystems. Sometimes, too, security of tenure has value to local people regardless of the condition of the lands and waters returned to them.
Assumption 3: Participation and accountability. That local people are sufficiently represented by the appointed ecosystem management group or forum (or the appointed group is sufficiently accountable to all those with a claim to or other interest in the ecosystems concerned) for a collective system to be possible.	*Strongly plausible.* But this may require anthropological research to clarify, since some systems of participation and accountability may be unrecognisable to members of other cultures.
Assumption 4: Social trust and conflict. That there is sufficient social trust among local people, and/or mild, few or soluble enough	*Strongly plausible.* But sometimes these conflicts will need to be resolved through special measures (e.g. arbitration, out-of-court

Table 12.2 (*cont.*)

conflicts over the ecosystems concerned, to allow an ecosystem management group or forum to be appointed and to operate effectively.	settlement, legal ruling, facilitated negotiation, expropriation), especially where competing or outside interests have become entrenched, before the ecosystem management group or forum can be established.
Assumption 5: Permissive policies and laws. That national policies and laws allow for ecosystem management groups or forums to be established in a way that allows them to take charge of ecosystems accountably on behalf of communities, and that areas can be assigned to this sort of management.	*Strongly plausible.* But sometimes it will be necessary to find ways around or to amend national policies and laws to make these things legally possible.
Assumption 6: Inter-level cooperation. That there is sufficient overlap between the values and intentions of local people and outsiders in relation to local ecosystems, and that local people have adequate knowledge and skills for the task, to allow for an empowerment or co-management agreement.	*Strongly plausible.* But if preliminary studies and dialogue have established that the values and intentions of local people differ from those of outsiders, and/or that their capacities and skills are inadequate, then an agreement will need to be negotiated to allow for both sides (i.e. local ecosystem managers and the staff of interested outside agencies) to train and work together on these aspects of management.
Assumption 7: Political stability of the arrangement. That national partners will not renege on their agreements with local people before or after the national-level benefits of the arrangement (e.g. from the restoration of ecosystem services) have been documented.	*Strongly plausible.* Participatory research and monitoring should be built into all such arrangements to provide the data to support economic analysis and lobbying for political protection, as a safeguard against high-level political change.

Table 12.2 (*cont.*)

Overall conclusion. Accepting the validity of assumptions 1–7, it is reasonable to expect that the legal establishment of an ecosystem management group or forum and the assignment to it of responsibility for managing an ecosystem, would be welcomed by local people, that the necessary agreements on training and technical support would be secured, and that as a result local and national systems would be made more robust to climate chaos while generating additional co-benefits in terms of good governance, poverty avoidance and biodiversity conservation. *Score for design quality*: 7.

---

scored, from 7 (best) to 1 (worst). In this, if the evidence suggests *perfection* then a score of 7 is awarded; if there are *any doubts* it is scored 6; and if the project is *basically good but with some flaws* it is scored 5. If the intervention has *no merits at all* then it is scored 1; if there are *some possible merits*, it is scored 2; and if it is *basically weak but with some good points* it is scored 3. The remaining score of 4 is used for those that are *moderate in value and have good and bad points*.

These scores must be supported by evidence, and the scores for each criterion and project are only individually meaningful to the extent this is done. Evidence of performance is gathered by searching project reports and interview notes for statements and data that are relevant to each performance criterion (see Section 12.6), and assembling these in a table for each component of the project. These are the raw materials upon which judgements are based, and typically amount to 100–300 words in each cell after editing. It is then easy to award preliminary scores for each criterion, which are reviewed by other members of the evaluation team who will be looking at the material from different points of view. When all are satisfied with the score, it can stand to represent all the evidence collected, offering a powerful way to summarise the strengths and weaknesses of the project as a whole.

12.4   TELLING 'BEFORE AND AFTER'
AND CONTRIBUTION STORIES

When all this has been done, it should be possible to approach the question of what a project or programme actually contributed, by setting out the conditions before and after, and hopefully identifying causal links between them (Box 12.1). Some of these 'before and after' accounts, if clear enough, might be used as 'contribution stories' to communicate achievements to a wider audience, and an example of such a story was given in Chapter 8. This process can be repeated separately for all of the components of all the projects that constitute the subject of evaluation – in the case of Danish cooperation with Nepal, all 48 of them (Chapter 5).

BOX 12.1   **The situation before and after the community forestry component of NARMSAP**

Major changes to governance in Nepal were underway during NARMSAP, especially including administrative decentralisation and the promotion of local control over forests. The design and implementation of NARMSAP responded to this and was 'on the right side of history' in the sense of pushing in the same direction as government, people and ecological need. The result was that 'The programme has contributed positively to improved physical conditions of forests and formation of new partnership between the government and local people in the management and utilisation of natural resources' (NARMSAP completion reports, Danida, 2005a, 2005b).

There are few baseline data against which forest recovery can be measured objectively, but the completion reports note that NARMSAP worked with over a million households and 'as several of the supported interventions by their nature are long term there are grounds to believe that NARMSAP also will have some future impact on the overall national key development objectives: full benefits from well managed young forest patches will only materialise in 10–20 years time;

BOX 12.1   **(cont.)**

successful sustained social change processes at the community level will have far-reaching long-term impacts; the effects of support to gene base preservation and distribution of more than 10,000 kilos of improved tree seeds will only show its full value in 10–50 years time'.

It is impossible to say whether NARMSAP might have been substituted by other initiatives given the encouraging policy environment, but it is clear that the programme contributed strongly to the multiplication, empowerment and enlightenment of community forest management enterprises at a large scale, and that this had a beneficial environmental and social impact that is likely to have continued indefinitely. The fact that similar models are still being used, for example in Nepal's 2014 Emission Reduction Programme Idea Note to the Carbon Fund of the Forest Carbon Partnership Facility (MoFSC, 2014), suggests that the approach is still seen as valid – and there are hundreds of similar processes around the world that are equally relevant.

The NARMSAP completion reports do make the point that the programme was 'less successful in contributing effectively to social equity. Promoting equitable access to resources and benefits in the forestry sector means addressing imbalances and barriers for access of women, poor, Dalits ['low caste'] and other excluded groups. Although the programme has started to respond to these imbalances by introducing measures such as increased access to fuel-wood, fodder, training, and other livelihood improvements, there is still much room for improvement. Lessons from programme implementation demonstrate the need for better governance systems to ensure that access and benefits are equitably distributed among the diverse members of society.' The main research question for the evaluation was therefore the long-term impact and sustainability of the socioecological changes introduced, amplified and facilitated by NARMSAP over the following 15 years, and the potential for knowledge sharing and networking among community resource

BOX 12.1 **(cont.)**

management initiatives in the context of the global response to mass extinction, desertification and climate change.

Source: Caldecott et al. (2017), annex F.12: evidence on performance, NARMSAP.

12.5 COMPARING PROJECTS AND PORTFOLIOS

One advantage of using design quality and particularly performance scores in this way is that they can be added and averaged across components to yield mean scores for performance criteria, and for the performance of multiple interventions if the aim is to do a synthesis evaluation or to compare overall performance between aid actions. Without giving excessive weight to the quantitative appearance of qualitative judgements, these methods offer an easy way to draw attention to consistent strengths and weaknesses in aid performance, between institutions, places, portfolios, methods and themes, and over time. Some of these uses are illustrated in Box 12.2, where two of the Danish-funded programmes in Nepal (PRG and education) were found to be particularly strong performers, while relevance stood out overall. Legacy effects were visible long after projects were completed, but could not be detected by the scoring system (as sustainability within the projects) and had to be reported in other ways.

A key point in Box 12.2 is that good design quality leads to higher performance, an effect that had also been seen in the Finnish and Swiss aid programmes by Caldecott et al. (2010, 2014). The implication is that effort invested in design is likely to more than pay for itself through better performance. An example of how findings in one country or programme can be put to use in a later evaluation is provided by the study in Tanzania of which the work in Zanzibar in Chapter 7 was a part:

The mean performance scores for the three Institutional Cooperation Instrument (ICI) projects were 4.4 ('moderate/strong')

BOX 12.2   **Comparing design and performance scores among project portfolios**

Among the five main themes [of the Danish partnership with Nepal], the *strongest performers* were: (a) the 2003–2018 peace, rights and governance interventions (including tax reform), with a mean score across the eight scorable criteria for the various interventions ranging from 4.5 to 5.3, which in a seven-point scale are all equivalent to 'strong'; and (b) the 1992–2012 education sector reform and development interventions with a score of 5.1 (also 'strong'). The others (renewable energy, environment, renewable natural resource management) scored on average in the range 3.8–4.4, all equivalent to 'moderate' performance.

The mean performance score across all 48 components was 4.4, which is equivalent to 'moderate' (but only 0.1 away from being rounded to 'strong', and the score is indeed 4.6 or 'strong' among the 43 components for which both design quality and performance scores are available). This is based on such a large sample that it is considered very robust, is a good performance score for any country programme, and should be a source of satisfaction to all its participants. Moreover, according to the criterion of *relevance* its mean score was 5.6 ('very strong'), suggesting very close targeting on correct and high-priority issues, and for the crucial criterion of effectiveness the mean score was 4.7 ('strong').

Many subtleties are visible in the data that will not be explored further here. As an example, though, the criterion of *sustainability* might be mentioned. This describes a judgement on the anticipated legacy of an intervention, beyond its impact over a year or two, based partly on what irreversible or long-acting changes might have been introduced by it. It is only possible to assess a legacy long after the event, which is a luxury not available to most evaluations. Here, however, interviews were being done more than a decade after the end of the first peace, rights and governance and energy assistance programmes, and after the conclusion of the environment sector support and community forestry programmes. In these cases, strong

BOX 12.2 **(cont.)**

legacy effects were noted that are not captured by the scoring system and can only be described, but they do help to inform answers given in the Evaluation Matrix.

It would be reasonable to expect *design quality* to affect performance, even though other factors will also do so. If the scores for design quality and performance are compared across all the 43 components for which both are available, they are indeed positively and significantly correlated ($\sum d_i^2$ = 2,661.5, r_s = 0.995, t = 11.137, p > 0.001). The same relationship has previously been found in the Finnish aid programme and the Swiss climate change portfolio. These findings help confirm that the scores represent real phenomena, and that the performance of aid portfolios can be enhanced by applying sound design principles. The latter implies that it is feasible through better design to improve aid performance per unit cost to the public.

Taken together with design quality, the key finding is that the Danida country programme was on average and with few exceptions *well designed, well-targeted and strongly effective*. This is consistent with the judgement that it made a significant contribution to Nepal's development over many years, even though some errors were made and the actual drivers of that development have primarily been the Nepalese themselves. This conclusion is about as good as it gets for any country aid programme of this diversity and duration subjected to this intensity of scrutiny.

Source: Caldecott et al. (2017): 44.

for INFORES [Implementation Support of Results and Data of first National Forest Resources Monitoring and Assessment at Regional and Local Level in Tanzania], and 5.1 ('strong') for ZAN-SDI and GST-GTK [Mineral Resources Potential and Small-scale Mining in Nachingwea Area and a General Nation-wide Geochemical Map of

Tanzania], similar to the successful 2003–2018 governance and 1992–2012 education programmes funded by Danida in Nepal.

(Caldecott, Killian et al., 2019: §3.4 page 19).

It can also be noted that all three of the Tanzanian projects had mean scores higher than the 3.5 ('weak/moderate') mean score for 50 aid programmes analysed by Caldecott (2017a), and that 'since all these studies used the same methods, the ICI projects seem to have performed relatively well as development cooperation actions'. A consistently applied, evidence-based scoring system is thus able to offer a useful point of reference for comparing performance among aid investments.

12.6 CRITERIA FOR EVALUATING GENERIC PERFORMANCE

Evaluation requires evidence-based judgements to be made on performance as well as design. This depends on the delivery and influence of the project considered from various points of view that correspond to the evaluation criteria, each of which offers a different way to examine and think about the project. Box 12.3 describes these in a way that indicates the signs that are looked for in evaluation and that might support a judgement of high performance in each case. The evaluation questions in the terms of reference should indicate the priorities of the client in the search for evidence to support a conclusion on the project from each of these points of view. Not all of them will ordinarily be a stated priority, although relevance, efficiency, effectiveness, impact and sustainability are almost always included, and interest in the cross-cutting themes, partner satisfaction (a proxy for 'aid effectiveness') and coherence is generally assumed.

12.7 SPECIAL FACTORS IN EVALUATING ADAPTIVE PERFORMANCE

The conceptual gap between design and evaluation is not great, since what an evaluator would like to see is a (near-perfect) design put into

BOX 12.3   **Criteria for evaluating performance of aid projects**

- *Relevance* is considered high if there is evidence that the intervention responded in a balanced way to the needs of the national partners, while also being consistent with national policies and strategies. It can often most usefully be assessed separately for the social system whose resources are being invested (e.g. the donor country and its aid institutions) and the social system where the investment is undertaken (e.g. the recipient country and its beneficiary institutions).
- *Efficiency* is considered high if there is evidence that the intervention contained measures that through elegance and accountability promoted sound management and value for money, including consistent patterns in management, governance, capacity or relationships, and in difficulties that arose and how they were overcome.
- *Effectiveness* is considered high if there is evidence that results contributed to achieving the specific purpose. Evidence for direct effectiveness may be quantitative or qualitative, depending on the subject of study. Evidence for indirect effectiveness includes information on side effects and expected or unexpected consequences. Reasons to expect this kind of intervention to be effective can also be based on other knowledge (e.g. of similar kinds of intervention elsewhere), especially if reasons for consistent outcomes can be identified. Of special interest would be consistent patterns in management, governance, capacity or relationships that may have affected results.
- *Impact* is considered high if there is evidence that the intervention had effects that were wider and longer term than its results, including strategic changes attributable directly or indirectly to the intervention, which may be more or less subtle, beneficial or sustainable in nature. Impacts might include changes in skills, education, relationships, institutions, legislation and administration. Negative impacts should also be noted, and could include unintended economic externalities, perverse incentives, population movements and ecological deterioration. Effectiveness, impact and sustainability are connected ideas, and respectively stress immediate (short-term), systemic (strategic) and transformational (irreversible) changes.
- *Sustainability* is considered high if there is evidence that the intervention had effects that continued after it ended, due to induced changes: (1) in policies, laws and regulations, systems and working practices, establishment of new forums or creation of new permanent staff positions; (2) to fiscal arrangements

BOX 12.3   **(cont.)**

and budget allocations, or creation of thriving businesses with local participation in benefits; (3) in trends in environmental deterioration and ecosystem restoration, or introduction of incentives and resource management systems that reward sustainable use of ecosystems; or (4) in the introduction of new ideas, groups and activities that contributed to environmental or social protections.

- *Partner satisfaction* is considered high if there is evidence that the intervention promoted ownership, accountability and enthusiasm in partner organisations. The existence of a partnership can be confirmed using records of activities (visits, joint workshops, reports, etc.) which show the exchange of goods, services and knowledge, and the quality of the partnership and its value for 'aid effectiveness' depends partly on the frequency and content of these exchanges, but mainly on the extent to which participants are enthusiastic about them, which can be assessed using interviews. An additional factor is that partnerships must adjust themselves to remain relevant to changing priorities on both sides, and require both sides to see enough mutual advantage to be willing to invest in overcoming challenges.

- *Coherence* is considered high if there is evidence that the intervention has ways to promote synergy with, and to manage interference from, the plans and actions of other actors, including other donors and the impact of one donor's actions on another. Factors include: *compatibility* (i.e. how well the goals of all participants are taken into account and where necessary reconciled); *coordination* (i.e. the existence and likely use of forums to sustain dialogue among stakeholders); and *complementarity* (i.e. how well participants' policies, plans, actions and choices support one another, and the degree of harmony among partners in achieving desired outcomes). There is often insufficient evidence to examine different aspects of coherence separately, and coordination arrangements such as forums for stakeholder dialogue are then used as a proxy.

- *Replicability* is considered high if there is good reason to expect that the intervention will yield lessons that can be used to improve actions in the future or elsewhere, based on the expectation that previous choices, policies or planning approaches will be effective against new but similar challenges. Knowing that this has actually occurred would be strong evidence for high replicability.

- *Connectedness* is considered high if there is evidence that the intervention was designed and implemented to anticipate and mitigate external factors and

BOX 12.3 **(cont.)**

influences to which it may be vulnerable but over which it has little or no control, such as climate change, macroeconomic pressures or civil discord.

- *Cross-cutting themes* (CCTs) include human rights (i.e. as set out in the UN Charter and the Universal Declaration of Human Rights), good governance (i.e. stable, lawful and effective governance maintained by accountability to an informed electorate), gender equity and social inclusion (i.e. ensuring due attention to groups who are disadvantaged because of landlessness, caste, poverty, ethnicity, gender, age, faith or other reasons) and environmental sustainability (maintaining the full integrity of ecosystems and hence their ability to nurture and protect human interests). These CCTs are common to the different sectors and face the identical challenge of ensuring that they become part of standard institutional procedures (i.e. through 'mainstreaming'). The issue for evaluation is therefore to assess to what extent each of the CCTs has been mainstreamed within the intervention, and to identify possible barriers to respecting the CCTs that prevailed in the culture where the intervention occurred.

Sources: Caldecott et al. (2017); Caldecott, Killian et al. (2019).

(near-perfect) practice, and in the process having conformed smoothly to unexpected events without losing its relevance, potency (effectiveness, impact and sustainability) or coherence (with other projects, programmes, donors and policies). For adaptation purposes, however, as understood here, the strategic purpose of an investment is always to enhance the strength of social and ecological systems, and the aims are always to improve things under the general headings of governance, ecology and knowledge.

Qualitative Indicators of System Strength

Designing for adaptive performance leads naturally to the issue of evaluating projects that either aim to deliver it or that might deliver it as an additional benefit of whatever else they are trying to do.

The 'six principles of adaptation' in Chapter 3 are that understanding systems and changes, envisioning outcomes, exchanging knowledge, choosing good leaders and leaving no one behind are all important to strengthening systems. Ways to describe and measure the strength of systems include their resilience, resistance and flexibility, indicating respectively their ability to bounce back, fight change and bend rather than break. But there are other important qualities that are even harder to describe and measure, such as fairness (a degree of equity that is considered acceptable by the society concerned, and which therefore promotes solidarity) and wisdom (a capacity for collective judgement based on discussion or *talanoa*, and embedded knowledge). These qualities are all too abstract to use as planning and evaluation tools, but they do have a useful role in highlighting desirable strengths in systems that must adapt to challenge, risk, danger and distress. A system that is strong in these ways is more likely than others to survive *any* challenge, which is vital considering the present uncertainties in which it is known that there will be severe challenges, but not what form they will take in any given place and time. Thus generic indicators of system strength will often be more useful in judging adaptation capacity than formulae based on particular technologies or business models.

Key Features of Social and Ecological Systems

The critical systems for adaptation are social and ecological ones. They are complex because they contain many distinctive parts (species, families, communities, classes, peoples, etc.) which are both active and interactive, and which relate to one another in many different ways. Neither kind of system can be understood outside the places where they exist, which may be at any scale from local to global. The two kinds of system in each place are linked by so many dependencies and impacts that they cannot meaningfully be considered separately, but they are also very different:

- Social systems comprise all the human attributes of culture (how people think, feel and express their group identities), economics (how people understand and organise their transactions) and governance (how decisions are made and decision makers held accountable to those affected by them), all of which are accessible, at least in principle, to the consciousness and reason of their participants.
- Ecological systems comprise the attributes of non-human organisms and how they relate to one another and to the physical environment, none of which are capable of consciousness or reason in ways that are easily recognisable to us.

Dimensions of Adaptive Capacity

If society and ecology are distinct but linked, and equally necessary for understanding climate change and adapting to it, then it also becomes important to know how knowledge is obtained and managed by people to support this understanding of society and ecology, and of how they fit together. Thus, adaptive capacity can be considered in terms of three 'dimensions' that describe a society's form of organisation, its relations with its environment and its collective understanding of itself and its place in the world. These can be thought of respectively as *governance, ecology* and *knowledge*. The aim of evaluating climate change adaptation investments is to find out how they affect governance, ecology and knowledge, separately and together in an *adaptive synthesis*, and thereby strengthen societies. 'Investments' here include all the expenditures and efforts associated with or driven by projects, programmes, policies and treaties, and also relevant co-benefits (and co-costs) that arise from other investments. Adaptation is not an end point but a process, and can only succeed within limits, but making adaptation-friendly arrangements in all three dimensions and the synthesis should allow as much adaptation as possible to occur, while also maintaining as much human progress as possible in the face of new environmental conditions. The adaptation-friendly features of each dimension are identified in Box 12.4.

BOX 12.4 **Features of adaptive dimensions**

Governance. The extent to which decision makers are accountable to those affected by their decisions. Where accountability (and therefore, potentially, 'good governance') exists, it depends both on transparency to stakeholders and on their power to recall decision makers or veto their decisions, so leaders must be compelled to explain their actions to stakeholders, and stakeholders must be able to understand those explanations.

Ecology. The realistic description of an environment derived from an understanding of how living systems work, the goods and services they offer to people when they work properly, the harm that can befall them under conditions of abuse, overuse and disturbance, and the goods and services that can be lost when this occurs. Such an understanding must include but cannot be exclusively based on information derived from scientific study, since there are many aspects of an environment, including aesthetic, social and spiritual values, that are meaningful to people but are hard to access through scientific investigation.

Knowledge. Reliable and meaningful information about social and ecological systems and livelihoods that is organised in useful ways and accessible to all.

Adaptive synthesis. High levels of accountability, understanding and knowledge all brought together in a sustained way. This requires that powers of decision, examination, recall and veto are exercised, within reason, by all sides, creating a need for wise leadership and informed stakeholders. It also requires that scientific study of the environment is not practised only by professional scientists, since citizen science has proved its worth and public involvement is needed to capture an appreciation of all important aspects of the environment. Knowledge must also be managed and shared through formal science, citizen science, public involvement and education. There must also be due consideration of the interests of all stakeholders, whether born or unborn, old or young, man or woman, rich or poor, powerful or weak, more or less educated, majority or minority and human or non-human.

BOX 12.4 **(cont.)**

This inclusiveness will flow automatically from paying attention to governance, ecology and knowledge, but it is easy for weaker stakeholders to fall through the cracks so attention must deliberately be given to those who are relatively voiceless and powerless.

Adaptive Qualities and Synthesis

Climate change and adaptation to it are both system-wide issues that are intimately related to complex social and ecological phenomena, and neither can be addressed only in a top-down and technical way. A more inclusive social response is needed, but there is much that can be done within existing structures without exceeding their capacity to change quickly and safely. To proceed within these bounds, and drawing on the evaluation approach described in Sections 12.1–12.3 and 12.6, the *dimensions* of governance, ecology, knowledge and adaptive synthesis (Box 12.4) can be presented alongside the *qualities* of relevance, potency and coherence (Box 12.5). These can all be used as criteria for judging the effect of investments and how they are designed, ranging from the investments involved in overseas aid to those upon which is based the management of nations, subnational regions and localities – cities, towns, municipalities, rural districts, etc. The aim is a consistent set of dimensions and criteria, and similar indicators, that can be used for investments at all social scales and in all geographies, and against which an investment might be assessed (Table 12.3).

12.8   CO-BENEFITS OF LOCAL SYSTEM STRENGTHENING

Given the range of benefits listed in Table 10.3 that governments recognise as flowing from ecosystem-based and community-based adaptation, and the synergies between them, projects could have any number of targets while still strengthening local systems against climate chaos, and a locally focused adaptation project could yield

---

BOX 12.5 **Adaptive qualities as evaluation and performance criteria**

Relevance. Because climate change is best understood primarily (though not exclusively) as an ecological phenomenon, and adaptation cannot work without addressing climate change primarily (though not exclusively) at an ecological level, relevance here requires the possibility (which is usually to be sought in education or governance reform) of at least some change in human behaviour or relationships in a way that promotes the integrity of natural ecosystems, or that encourages and enables their use without harm to the environmental services they perform, or that restores or enhances the ability of any natural, planted or built ecosystem to deliver environmental services, or any combination of these. Relevance is considered high where there is evidence suggesting a realistic chance of such changes resulting.

Potency is considered high if there is evidence that the investment has had or will have an effect whose consequences are relevant to promoting adaptation in the short, medium or long term, including direct or indirect immediate effects, and strategic and/or continuing and/or irreversible changes attributable directly or indirectly to it. Adaptive potency is an amalgam of the conventional criteria of effectiveness, impact and sustainability, reflecting both the continuity between these criteria over time, and the uncertain duration over which any given investment in adaptation may act.

Coherence is considered high if there is evidence that the effects of the investment are enhanced by, or at least not neutralised by or competitive with, the effects of other investments.

---

any number of other benefits. Entire aid portfolios could be formulated as adaptation measures while also delivering progress on all the SDGs, and entire portfolios could be designed to deliver the SDGs while also delivering adaptation benefits. The two approaches can go together, so a clear adaptation criterion for design and evaluation in all aid programming would be a powerful gain.

Table 12.3 *Dimensions and qualities in adaptation investments*

	Qualities		
Dimensions	Relevance	Potency	Coherence
Governance	Considers barriers and incentives that affect the quality and influence of local decisions and local accountability for them.	Encourages behaviour and new relationships that promote more influential local decisions and better accountability for them.	Promotes synergy and reduces conflict among policies and actions that affect local decision-making and accountability.
Ecology	Considers barriers and incentives that affect the management of ecosystems and livelihoods.	Encourages a fair and rewarding distribution of benefits from protection and sustainable management of ecosystems.	Promotes synergy and reduces conflict and actions that affect environmental quality and the integrity of natural, planted or built ecosystems.
Knowledge	Considers barriers and incentives that affect the understanding of social and ecological systems and livelihoods.	Encourages validation, collection, organisation and use of all forms of knowledge concerning the nature and sustainability of ecosystems and livelihoods.	Promotes reconciliation among points of view through appeal to common understanding of how systems work, and shared values of how they should work.
Adaptive synthesis	Results in an integrated, holistic view of	Results in more influential, better informed and	Results in synergy and reduced

Table 12.3 (*cont.*)

| Dimensions | Qualities | | |
	Relevance	Potency	Coherence
	behaviour, relationships and knowledge that affect the integrity of natural, planted or built ecosystems, and the management and sustainability of ecosystem qualities and services.	more accountable local decision-making oriented to safeguarding and enhancing the integrity of natural, planted or built ecosystems, and the sustainability of ecosystem qualities and services in the short, medium or long term.	interference among decisions that affect the adaptation-friendliness of issues concerning governance, ecology and knowledge.
Evidence	Documented content of dialogue, analyses and plans.	Quantitative (i.e. meaningfully measurable and measured), qualitative (i.e. indicated by proxies or the opinions of informed observers) or inferential (i.e. plausibly suggested by reason and examples or case studies) evidence of real effectiveness, impact and sustainability.	Documented content of dialogue, analyses and plans.

This ground has in the past been covered by the 'sustainability' criterion, or else by calls to consider climate change as a 'cross-cutting issue'. The problem with the first is that sustainability is mostly treated as concerning the survival of the changes introduced by the project (as in *financial* or *institutional sustainability*), and 'sustainability' in general (as in *sustainable development*) is too vague and too general to be much use. The problem with the second is that climate change then falls into a long list of open-ended topics like human rights, gender equality, good governance and environmental sustainability, which are common to multiple sectors or to whole systems and are so hard to address in a project context that they are often merely touched on in project documents, unless they happen to concern the central purpose of the investment.

I have tried to correct for some of this in my own work, by adding 'connectedness' as an evaluation criterion (Box 12.3), but it would still be better to be more specific about what a local adaptation project should be doing, or what another kind of project should be doing for adaptation. The demand for this will only increase as the urgency of adaptation needs and the scale of adaptation investments continue to grow. The nature of adaptation as a process of system strengthening means that it is process indicators that are most relevant. The establishment of forums and inclusive participation can be documented, but most of the other important aspects are qualitative and require judgements and multiple lines of often indirect evidence to assess.

This can be frustrating to those who seek measurable quantities of input and output, and there are sometimes conflicts as well as synergies between adaptation and other investments (Chapter 9). Pending the spread of a more systemic way of thinking about ecological and social systems, the best that can be done at the moment is to table the arguments in favour of systems thinking (Chapters 3 and 4), examples of what community-based and ecosystem-based adaptation look like in practice (Chapters 5–8), evidence that increasing numbers of government stakeholders understand the principles involved and the benefits of the approach (Chapters 10 and 11, and this chapter) and a checklist of features that should be designed into and looked for in a project with adaptive purpose and significance (this chapter).

It is important to add that much the same case can be made for seeking adaptive co-benefits in the implementation of a range of international treaties. The kinds of actions that can reasonably be said to contribute to mitigation and adaptation are listed in Boxes 9.1 and 9.2, but there is a high degree of overlap between these and also with the actions called for in the biodiversity and desertification conventions (Caldecott, 2017a), including:

- that ecological mitigation largely concerns finding ways to encourage and enable people to conserve ecosystems as carbon stores, even though those same ecosystems will also be 'storing' biodiversity and providing many other goods and services;
- that many of the actions needed for adaptation (e.g. those listed under 'resilience' in Box 9.2) are identical to those required in combating desertification or conserving biodiversity, with the details depending largely on ecological context;
- that living ecosystems can help regulate water supplies, protect soils from depletion and erosion, absorb the energies of storm impacts and perform other services that we can rely on to take the edge off some of the effects of climate change; and
- that biodiversity in all its forms underpins agriculture, which when adapted to harsh conditions can make the difference between a desertified and a verdant landscape.

These various aims can synergise, especially when they are bundled together around a common process like land use planning at local level, since the actions that safeguard water catchments, soils, farms and forests can also contribute to reducing GHG emissions. The practical adaptation dimension is often far more comprehensible and motivating to local people and governments than a mitigation dimension explained in terms of physics and global systems. The Danish-funded Locally Appropriate Mitigation Actions in Indonesia project, for example, mainstreamed mitigation priorities in land use planning at provincial and district level (Caldecott et al., 2021). It succeeded because the Indonesian branch of the World Agroforestry Centre, in implementing it, promoted an inclusive, integrative and informed approach that focused on ecosystem services and their role in local economies (Sonya Dewi, personal communication, November 2020). In short, it is a hallmark of good ecological mitigation projects that they can also function and be presented as adaptation projects, or indeed as sustainable low-carbon, biodiversity-friendly, 'green growth' projects.

13 Adaptive Thinking, Feeling and Acting

13.1 MITIGATION AS A TYPE I ISSUE

'Technically Simple and Easy to Solve'

We have known for decades that we must reduce net GHG emissions and how to do it. The mitigation challenge is thus a 'Type I' issue in the medical terminology of Heifetz (1994), in which diagnosis is certain, a cure is possible with known techniques and the authority to prescribe the cure and compliance by the patient are all that are required for a likely resolution. In medicine, authority comes from an expectation that the physician is competent, as guaranteed by peers, laws and institutions, and compliance depends on the patient's wish to minimise suffering and postpone death. In the case of climate change, now that the public is informed and alarmed we naturally look to government to tell us what to do in order to save our lives and livelihoods, the well-being of our grandchildren and a healthy and beautiful biosphere. Once brought to consciousness, fear of losing these is enough to mobilise almost everyone behind a fair and effective leadership agenda for life-saving change. Hence neither leadership nor compliance should be a problem for the mitigation agenda, yet it is.

13.2 MITIGATION AS A TYPE II ISSUE

'Technically Simple but Very Hard to Solve'

The mitigation problem has not yet been solved for several reasons. The first is that there is no one authority in any single country that has fully understood what needs to be done and has stayed in power long enough to do it. Authoritarian rulers and one-party states seldom give priority to environmental issues, and in the very few cases where a stable democracy has been able to follow a consistent environmental trajectory over three decades or more, as in Costa Rica with its 'Peace with Nature' agenda (see Section 13.5), and the UK with its climate mitigation agenda (Bowen and Rydge, 2011), this has depended upon sustained elite consensus along with public understanding and/or indifference.

But the mitigation problem cannot be solved by any one country, and there is no all-knowing, global 'hegemon with the power to impose a single set of rules' (Chapter 1). Only nature could have that role, in wielding infinite power and universal rules, but this is seldom acknowledged. International cooperation is therefore essential, and this has fallen foul of multiple sources of reluctance to cooperate, rooted in historical, cultural and economic experiences, rivalries and reasons for distrust, ranging from slavery, trade and religion to imperialism, ideology and war. Even where cautious and conditional arrangements to promote global cooperation have been made, first in the League of Nations (1920–1946), then in the UN, the Association of Southeast Asian Nations and the EU, there are structural issues with the distribution of powers, and above all the problem of willingness to pay.

The latter manifests continually at all levels, within and between countries, where it becomes entangled with all the other sources of rivalry and distrust among groups, classes and nations, all of which are exploitable politically and any of which, thus exploited, can derail agreements on cost sharing. The slow and painful process of the UNFCCC and its gradual, erratic progress towards approximate agreements on fixing the mitigation problem is the compromise solution that has gradually been squeezed out of the tension in the UN system between everyone knowing that something must be done and no one being willing to pay for it.

Things are a bit different in the EU, at least since 2010 when the 'hegemonic' approach was abandoned in favour of an 'experimentalist' one (Chapter 1). This allowed continent-wide climate mitigation strategies to be agreed and funded, and also the facilitation of UN agreements such as the SDGs and the Paris Agreement itself.

But more generally, in the absence of well-explained, fair and consistent instructions from their governments, citizens everywhere are at a loss in relation to climate mitigation. They know that there is a problem, cannot fix it on their own, yet are bombarded by messages from parties with various agendas that they must personally 'do something'. The obvious thing to do, which is to vote for 'green, social democratic' leaders and a 'Green New Deal' at every opportunity, is not a consistent part of the messaging that we receive in practice. Instead, and meanwhile, there exists a messy power contest more or less in every country, and in every international institution to which we might look for solutions.

Again framing it in terms used by Heifetz (1994), the net result of all this since 1992 has been to transform mitigation from a Type I issue, in which effective solutions are known and require only expertise and compliance to implement, into a Type II issue. In these, the problem is definable but solutions are less certain because the systems involved are complex and not fully understood, and most importantly because the likeliest path to recovery relies on a carefully orchestrated regime involving multiple changes in behaviour. In the case of a patient with heart disease, for example, these changes may include new patterns of exercise and diet, which the physician and patient must orchestrate through dialogue.

In the case of a global population faced with a biosphere crisis, however, such changes would have to involve new patterns of energy generation, consumption, travel, farming, waste management, etc., and a near-universal willingness to comply with many new rules over at least several decades. Not only may there not be decades left to make these changes, due to imminent tipping points in the biosphere (Chapter 2), but there are no collective bodies equivalent either to a physician or a patient who could do the orchestrating. In their absence, until 2016 the issue of mitigation lay becalmed in a Type II ocean with barely a breeze of leadership to provide headway. Whether the EU-inspired Paris Agreement can stir up an adequate wind remains to be seen.

13.3 ADAPTATION AS A TYPE III ISSUE

'Technically Complex but Soluble through Imaginative Cooperation'

The challenge of adaptation is quite different to that of mitigation. The point was made in Chapter 12 that separating the two agendas at a treaty level would have allowed us to think more clearly about how to adapt to climate change, rather than being distracted by the question of who would pay to prevent or mitigate it. Now that the two agendas have been separated by the Paris Agreement, at least partially, we can see that adaptation is akin to a Type III issue.

To explain the significance of this in rather robust terms, Heifetz describes a patient with terminal cancer. Here, he argues, the problem is not the cancer itself, but rather the approaching death, which requires the patient to adapt their thinking, to stop worrying about why they have cancer or what might miraculously cure it, but rather to focus on 'making the most out of his life; considering what his children may need after he is gone; preparing his wife, parents, loved ones, and friends; and completing valued professional tasks' (Heifetz, 1994: 6).

Some have concluded that adaptation is *exactly* like this – that civilisation is doomed because we cannot solve the mitigation issue in time, or because climate change is only one sign of a global system that is now so unstable that it can only disintegrate, so we should learn to think in new ways about the utter transformation of all things (e.g. Scranton, 2015; Ahmed, 2017; Bendell, 2018; Read, 2018; Bologna and Aquino, 2020). Many who have been trying to conserve ecosystems and biodiversity in the past few decades may be drawn to a similar conclusion.

But this is not the only possible response. This book contains much evidence that all is not necessarily lost, including the multiple signs of a *Zeitgeist* shift in favour of a galvanised mitigation response in Chapter 2, the increasingly well-informed and well-thought-out adaptation plans of many national governments reviewed in

Chapter 10 and the examples of aid efforts with high adaptation potential in many locations described in Chapters 5–8. Thus, 'terminal cancer' may not yet be the correct diagnosis for our collective circumstances in the face of climate change.

But there is another way in which the diagnosis of a Type III issue *does* apply to adaptation, in the sense that in order to adapt we must learn to think adaptively. This can be hard to do since it requires creativity, a pragmatic willingness to accept reality, an ability to endure the pain of abandoning past certainties and the bravery to relinquish and think beyond entitlement myths that may be deeply entrenched and have never before been questioned. These myths convey the idea that those who belong to apex power structures must naturally be in charge, which could work in a Type I situation like mitigation if there were an agreed power structure and if those at its apex knew what they were doing. But it seems much less appropriate in a Type III situation like adaptation, where rigid hierarchies and obedience to traditional authority can only take you so far in a changing world.

Correcting Biases

Many people still think in maladaptive ways, and retrograde political movements that exploit entitlement myths and propagate conspiracy theories have become influential in recent years (Applebaum, 2020). Yet, the overall trend is for people to learn to think more adaptively. In the nineteenth century the organised slave trade was suppressed and the principle of slavery renounced; in the twentieth century racially constructed empires and their surviving fragments were replaced by majority self-rule; and in the twenty-first century leftover biases are being challenged one by one – police impunity and structural racism here, gender pay gaps and structural sexism there.

In July 2020, the name of Dundas Street in Edinburgh, which commemorates Henry Dundas (1742–1811), a man of many works but who is now blamed for resisting the abolition of slavery, was challenged on every corner by new signs saying 'Emancipation Street'.

These were soon removed, but the point was made. Meanwhile, the National Trust began reinterpreting its properties, asking where the wealth had come from to build the stately homes and gardens in its charge, and finding that colonial exploitation and slavery were often the answer (Huxtable et al., 2020).

These were small items among many in a global wave of action against racism. Even the principles of ecology have attracted new interest with the resurgence of environmental activism through Extinction Rebellion and other movements (Chapter 2). So it is clear that there is nothing truly inevitable about anything that people think or believe, even though there may also be strong conservative forces at work to resist or reverse cultural change (Chapters 2 and 3).

13.4 HOPE AND PURPOSE

My impression is that the most contented and energetic people have hope and purpose, and that people can do without one of these but not both. One or other of them will prevent feelings of desolation, and the degradation of self harm. Hopeless loses its sting to purpose; purpose-lessness loses its sting to hope. Being responsible for a local, familiar ecosystem (a garden, allotment, farm, sacred site, traditional land-holding or nature reserve) and local, familiar people (a group of friends, a community, a small electorate) is an excellent way to nurture feelings of purpose. The sense of security that comes from being surrounded by healthy ecosystems and in contact with trusted acquaintances is an excellent way to nurture feelings of hope.

The implication is that while countries must do many things to adapt, the best way to enhance the contentedness, energy and security of their people, and to strengthen their own ecological and social fabric, is to ensure that throughout their territories local people feel responsible for local ecosystems and trust their neighbours. In prac-tice this means reforming land tenure in favour of local people, build-ing on traditional land claims or creating new ones based on secure tenure, and encouraging and enabling local people to organise them-selves into mutual-support groups – which they will tend to do naturally, but are often prevented or discouraged from doing by barriers that must first be removed. It also means doing nothing in the name of adaptation or development, without a very good reason and many compensating advantages, that undermines or blocks these arrangements to nurture hope and/or purpose among people.

13.5 PEACE WITH NATURE

The 'hope and purpose' issue is directly relevant to individuals living in small communities, urban or rural, but there is another sense in which it is of the gravest concern to people far more widely. After a brief introduction to the issue of climate change in Chapter 1, a more alarming note was introduced on the state and trajectory of the biosphere, including that our collective behaviour almost suggests that we are 'at war' with nature. If so, it is a war that we cannot possibly win, but meanwhile the damage being inflicted on all sides, and the destabilising effects of the conflict, are causing many to lose hope entirely. Even the sense of purpose is at risk, as calamity follows calamity and nothing seems to be being achieved despite so many efforts over so long. Although there are in fact reasons to keep hope alive, a renewed sense of purpose would also be welcome. And here the idea of 'grand strategy' is useful, since it addresses the 'goals of the peace' – what the world will look like when we end the war and opt instead for peace with nature.

Costa Rica is a country in Central America, between Panama and Nicaragua. It abolished its armed forces in 1948, and redirected its military budget to healthcare, education and protecting the environment. During the 1970s and 1980s, however, rapid deforestation convinced Costa Rican conservationists that virtually all private lands in the country risked being cleared of natural ecosystems. Their lobbying led in 1989 to a National Biodiversity Planning Commission, which started with the premises that biodiversity was economically valuable, and should therefore be preserved and used for public benefit, and that forest protection could not succeed unless the people living around each protected area were willing to help protect it.

The commission proposed new laws to consolidate a National System of Conservation Areas, with all the units being managed locally and for local benefit. To help pay for this, the commission recommended a national biodiversity inventory, to find out exactly what made up Costa Rica's biological richness and what it might be used for, and also the creation of a National Biodiversity Institute to manage the inventory. This led in the early 1990s to the birth of bioprospecting as a high-profile strategy for tropical developing countries to use and conserve their own living resources for their own long-term benefit (Caldecott and Lovejoy, 1996).

By the mid-1990s the logic of using ecosystems creatively to pay for their own conservation and contribute to national development priorities had resulted in a PES programme in Costa Rica (Caldecott, 2017a). This is a national system to manage payments for carbon storage, hydrological services and the protection of biodiversity and landscapes. It has been credited with turning deforestation into net reforestation by the early 2000s. In 1997–2004, some US$200 million was invested in PES to protect over 460,000 hectares of forests, to establish forestry plantations and to provide additional income to more than 8,000 forest owners. The PES system is managed by the National Fund for Forestry Financing, and is mostly financed by a 3.5 per cent share of revenues from a sales tax on fossil fuels, but the aim is that all beneficiaries of environmental services will eventually pay for those they receive.

These environmentally based economic programmes, together with debt-for-nature swaps and ecotourism, helped to transform the country's self-image and future, and by 2007 Costa Rica was ready for the next logical step, which was to declare peace with nature. This is like a super-imaginative version of the 'climate emergency' declarations that so many governments have made more recently (Chapter 2). In its own declaration, Costa Rica pledged to abolish all forces that destroy nature: phasing out and banning net GHG emissions and single-use plastics; promoting environmental action planning by all state institutions; investing in the protected area system and biodiversity; making arrangements for users of ecosystem services to pay for their conservation; and ensuring environmental education in all schools (Caldecott et al., 2008). Twelve years later Costa Rica was named 'UN Champion of the Earth' (UNEP, 2019d).

Thus, Costa Rica's 30-year process of change is built on consistent ecological reasoning, appropriate technologies and the sharing of costs and benefits to involve multiple aspects of society and the economy. The factors that made such an approach feasible in Costa Rica (with historical starting positions, luck and leadership among them) need to be understood, but there are lessons to be learned here that are certainly applicable in all other countries. Three factors are particularly relevant:

- First, that consensus for a peace with nature declaration by a country or group of states might be achieved far more quickly, starting now, than it was in Costa Rica starting in 1989, since we have all been wrestling with and learning from similar issues since then.
- Second, that with 'experimentalist governance' as a breakthrough concept (Chapter 1), peace with nature can provide an overarching goal to which all countries aspire and which they can all compete and cooperate to achieve.
- And third, that peace with nature need not stop at an inspiring declaration and a set of government programmes, but could also be the basis for a process of constitutional reform or reinvention.

In late 2020, the UN secretary general stated that, 'Making peace with nature is the defining task of the 21st century. It must be the top, top priority for everyone, everywhere' (Guterres, 2020). Earlier in the year a public petition in Scotland was calling for a Peace with Nature Constitution to acknowledge the supremacy of ecological reality and our dependence on nature (Change.org, 2020; Scotland's Climate Assembly, 2020b). Its key practical aim was to establish a Court of Ecology to which citizens would have the right of appeal for any law to be examined for ecological safety, and potentially struck down if it failed the test. This would provide an essential protection for citizens, future generations, non-human species and nature as a whole, against unsafe decisions by politicians. The effect of this would be to place ecological law at a higher level than human law, and establish that the people are sovereign while nature is supreme.

This is a wholly new constitutional idea, since all other modern constitutions make either the people or parliament both sovereign and supreme. By accepting it, Scotland would set a new standard for other countries to follow, affirming in the process that the world is not there just to be exploited by humanity. It is this kind of practical but idealistic revisioning of the future – joined by others mentioned in Chapter 2 – that can allow a sense of purpose to re-enter the whole climate response. And with purpose as well as hope and realism, we can resist chaos and perhaps survive the transition to an Anthropocene climate, where further challenges await.

13.6   TAKE-HOME MESSAGES

My initial motivation to write this book was to understand Danida's decisions in 2005–2010 to abandon three excellent programmes – two in Nepal and one in Bolivia – and to explain why these decisions seemed so wrong in hindsight, yet a decade earlier they had seemed right to those making them (if not to local stakeholders). At least for the community, land and forest projects, the answer now seems to be that 2015 was the 'Paris moment', an historic watershed between 2010 and 2020 which changed the context for all such decisions. Before Paris these projects might have seemed complex, difficult and slow, with rather obscure outcomes; but after Paris they look more like effective safeguards of development gains through adaptive investment as part of the climate response, with huge advantages for mitigation, biodiversity and human rights as well. That change of perspective means that this book is essentially an exploration of the forward-looking significance of the Paris Agreement in the context of the global ecological crisis and its local manifestations. While writing it, I noted that there are a number of different kinds of stakeholders who may have different interests in the content and conclusions. By way of final remarks, the following messages can be identified for these various potential readers.

For Staff of the UNFCCC Secretariat

- Community-based and ecosystem-based approaches have multiple advantages, and are increasingly prominent in national adaptation efforts. This is so significant that the parties may wish to consider amending the global adaptation goal to endorse these approaches more explicitly. For example, the goal could state that national adaptation plans should contain substantive sections on ecosystem-based adaptation and a clear explanation of how cooperation and capacity at the grassroots community level will be built, and how action at that level will be encouraged and enabled.

- The parties might also wish to consider amending the global mitigation goal to include stabilising and then reducing GHG concentrations in the atmosphere, since at least for CO_2 this is easier to measure than mean surface temperature and would provide a more direct and immediate indicator of humanity's collective progress.

- Outreach to the secretariats of other conventions could be a way to begin forging the overarching goal of a sustainable human–biosphere relationship, perhaps flagged as 'peace with nature' and supported by the amended adaptation and mitigation goals and the LPI to keep track of biodiversity and ecosystem integrity.

- For the first global stocktake in 2023 it would be particularly helpful to analyse the adaptation communications for further signs of the practical growth of system-based adaptation approaches, as well as the cross-sectoral integration and mainstreaming of adaptation into sustainable development and rights-based strategies.

For Government Officials in Developing Countries

The attitudes of Burkina Faso, Namibia and Grenada are relevant to many others:

- Burkina Faso observes that major climate change is now inevitable, so everyone should prepare for rising temperatures, more severe dry seasons, less predictable rainy seasons, reduced groundwater and altered disease profiles. This attitude is also taken by at least Eswatini and Ghana, as well as by Pacific countries such as Kiribati, Nauru and the Marshall and Solomon islands, faced by rising sea levels and more violent storms and storm surges.

- Namibia assesses vulnerability and plans, designs and implements adaptation programmes at the local and landscape (constituency) levels, always prefers a community-based and ecosystem-based approach and sees that adaptation initiatives are now evolving to assume a community-based and settlements-oriented approach. This attitude is also taken by at least the Seychelles, Suriname and Chile.
- Grenada takes the view that ecosystem-based solutions will probably always be rewarded by positive social impacts regardless of how exactly the climate changes, an attitude apparently shared by Mali but rarely spelled out. Many countries, however, seem aware that these solutions – like community-based ones – can generate a wide variety of economic, financial, social and environmental benefits (Table 10.3).
- These opinions show that aspects of a single emerging global pattern, one with a common underlying ecological and social logic, are being detected and articulated by different stakeholders in advance of an emerging new consensus. It would be wise for developing countries to insist that these ideas are reflected in all policy and aid programming dialogue from now on.

For National and Local Government Officials Generally

All countries have an interest in becoming better able to resist stresses and threats, including those arising from climate chaos, especially if this can be done at low cost. The model described in Chapter 8 offers a useful way forward, since community networks on the neighbourhood scale form very easily and can transform local abilities to organise collective discussion and certain kinds of action. Once organised, which often seems to need one or two motivated individuals and an issue of common interest to form around, a neighbourhood group can:

- make proposals for improving the local environment, which local government can support, or take collective action to do so, which local government can applaud;
- discuss and plan for minor or major emergencies, which local government can enrich with information on the UN Awareness and Preparedness for Emergencies at Local Level or EU European Civil Protection and Humanitarian Aid Operations models;

- federate with other local groups to obtain a larger voice in managing the landscape as a whole, which simplifies the job of local government in developing policies and plans with wide support; and can also
- network with other groups and federations elsewhere to share insights, experiences and solutions to all manner of challenges, thus enhancing the knowledge and competence of the people within the local government's area of responsibility.

All these rewards are available for trivial levels of expenditure, if any. A country made up of thousands of small, self-organised community networks will be likely, on average and compared with others, to be more robust, responsive, prepared and capable of dealing with whatever a changing climate throws at it. And since there is no country to which this logic does not apply, whatever its ecological and economic circumstances, a desirable goal would be a world comprising hundreds of thousands or even millions of such groups. To this aim any country can contribute through its own system of organisation, and through its own international partnerships. Such an arrangement would be likely, on average, to enhance the effectiveness, impact and sustainability of all government expenditure, and it would greatly facilitate the global adaptation process.

For Institutional Aid Professionals

Suggestions for designing and evaluating investments that will strengthen local ecological and social systems against climate chaos are offered in Chapter 12, with relevant examples and discussion throughout this book. The benefits and synergies available from ecosystem-based and community-based adaptation mean that whole aid portfolios can be formulated as adaptation measures while also delivering progress on all the SDGs, and that they can be designed to deliver the SDGs while also delivering adaptation benefits. The two aims are easily unified by adopting a clear adaptation criterion for design and evaluation in all aid programming:

- The case for governments to promote ecosystem- and community-based adaptation and to facilitate the emergence of climate-aware community groups everywhere applies equally to the donors involved in any partnership.
- It may well be that the quickest and most cost-effective way to build global capacity to resist the impacts of climate chaos would involve the mobilisation of hundreds of thousands of local community action groups in all countries.
- The synergies between community-based, ecosystem-based, traditional knowledge-based and human rights/gender-based approaches that are recognised by Fiji, Suriname, South Sudan and others provide a strong steer towards measures that can address multiple key priorities at least cost.
- Moreover, the lead of Tonga and others in favour of community-oriented small grants schemes, as a way to scale down major financing to an effective level, should be recognised and followed widely.
- And finally, the lead on international networking by the GCCA across Africa and with South American advocacy draws attention to south–south collaboration as a vital accelerator of adaptation performance, if only donor institutions could find ways to accommodate it within their budget lines and portfolios.

For Students, Researchers and Teachers

Writing this book turned into a bigger research project, and a different one, than I had expected. This was mainly due to the volume of adaptation communications that have poured into the UNFCCC Secretariat since 2015. Although adaptation research has been described as at an 'impasse' (Chapter 9), this really just means that the specialisms into which researchers are artificially divided by their institutions are wrong for the subject matter – but expectations of governments for easy and simple answers are unhelpful as well. Three points can be made:

- The real potential of adaptation research lies in understanding complex systems through studies where all the life and social sciences overlap and inform one another, and in applying the principles, practices and technologies of governance and networking to the task of accelerating adaptive thinking, learning and cooperating at all levels of society and in all social and ecological contexts.

- The NDCs in particular are often discussed only in terms of their conditional and unconditional commitments to reduce GHG emissions, and the question of whether they add up to keeping mean global temperature rise below 1.5 or 2.0°C. But for adaptation the more important point is that the national reports are an unprecedentedly huge resource for the study of how the human world sees what is happening to the biosphere at a unique moment in geological time.

- Never before has an animal species single-handedly destabilised its own habitat at a global level, and in the process brought to a halt the evolutionary trajectories of millions of other species. We are doing this while also applying the full force of investigative and analytical technology to document and explain events and foresee outcomes, and also while in full possession of enough advanced ecological and anthropological knowledge to make sense of what we are seeing. The adaptation communications are an ever-deepening ocean of knowledge to which any student, researcher or teacher should be drawn at this, the most interesting and desperate time in human history.

For People Who Live in Localities and Landscapes

As people become motivated to form or join a community group to do anything important, like protect and restore their local environment or prepare for a dangerous emergency, news from those that have taken the same path becomes more meaningful. The people of Ekuri village in Nigeria have several messages that seem relevant to anyone living anywhere.

The British government had been helping the community build a forest management partnership based on secure resource tenure, traditional forest-related knowledge and some technical advice:

> Some of the lessons learned were revealed when the Ekuri people, upon being asked to advise another village trying to solve its own problems of forest depletion, said that the people there should: (a) Be united and prepared to work hard; (b) Believe in themselves and start self-help projects after full discussion of their own problems

and opportunities; (c) Ensure prudent and realistic management of all the village's resources; and (d) Work with government departments and other outside groups to obtain help with transport and marketing, training and technical advice, financing, and monitoring and evaluation.

(CSD, 1996: 21)

In a nutshell, therefore, we have the ideas that a small group can be potent if it wants to be, and that partnerships are vital. There is nothing to stop everyone on Earth becoming an active part of the climate response by joining with others to make a difference. We are social animals and can achieve nothing outside of social systems, including the saving of the ecological systems on which we all depend. And one of the ways that a small local group can make a real difference is by joining forces with others to enact change and to insist on it. The more groups there are, and the more they talk with each other, the more valuable they all become, to each other, to local and national governments, and to the future.

Abbreviations and Acronyms

ALBA	(Bolivarian) Alliance for the Peoples of Our America (*Alianza Bolivariana para los Pueblos de Nuestra América*)
AOSIS	Alliance of Small Island States
AQMS	air quality monitoring system
CbA	community-based adaptation
CBD	Convention on Biological Diversity
CBRM	community-based resource management
CFUG	Community Forest User Group
CO_2	carbon dioxide
COLA	Commission for Lands (Zanzibar)
CoP	Conference of the Parties (to UNFCCC)
DFD	Department of Fisheries Development (Zanzibar)
dmv	dated mitigation value, a modifier of net physical GHG savings to reflect the year in which the savings occur relative to mid-century climate breakdown, on the basis that earlier savings are more valuable for mitigation than later ones, and expressed as $tCO_2edmv2021$, $tCO_2edmv2022$, etc.
DoURP	Department of Urban and Rural Planning (Zanzibar)
DRM	disaster risk management
DRR	disaster risk reduction
EbA	ecosystem-based adaptation
EC	European Commission
EIA	environmental impact assessment
ESPS	Environment Sector Programme Support
EU	European Union
EV	electric vehicle
GCCA	Global Climate Change Alliance

GEF	Global Environment Facility
Geo-ICT	geospatial and ICT capacities in Tanzanian higher education institutions
GHG	greenhouse gas
ICI	Institutional Cooperation Instrument (*Instituutioiden välisen kehitysyhteistyön instrumentti*) (Finland)
ICT	information and communication technology
IEP	Integrated Environmental Programme
INDC	Intended Nationally Determined Contribution
IPBES	Intergovernmental Science-Policy Platform on Biodiversity and Ecosystem Services
IPCC	Intergovernmental Panel on Climate Change
LDC	least developed country
LPI	Living Planet Index
MAS	Movement Toward Socialism - Political Instrument for the Sovereignty of the Peoples (*Movimiento al Socialismo – Instrumento Político por la Soberanía de los Pueblos*)
MCCN	Mwambao Coastal Community Network (*mwambao* = 'coast' in KiSwahili)
MoFSC	Ministry of Forest and Soil Conservation (Nepal)
NAP	National Adaptation Plan
NARMSAP	Natural Resource Management Sector Assistance Programme
NDC	Nationally Determined Contribution (to Paris Agreement objectives)
NEPAP	Nepal Environmental Policy and Action Plan
NGO	non-governmental organisation
PDR	People's Democratic Republic
PES	payment for ecosystem services
PM10	particulate matter with a diameter of 10 μm (millionths of a metre) or less
PRG	peace, rights and governance

REDD+	Reducing (GHG) emissions from deforestation and (forest) degradation, with internationally-agreed forestry, biodiversity and social safeguards
RGoZ	Revolutionary Government of Zanzibar
SCC	social cost of carbon
SDG	Sustainable Development Goal
SDI	spatial data infrastructure
SIDS	small island developing state
SUZA	State University of Zanzibar
SWIOFish	Southwest Indian Ocean Fisheries Governance and Shared Growth Programme
TCO	*Tierras Comunitarias de Origen* ('community lands of origin')
tCO_2	tonne of carbon dioxide, a physical unit of carbon dioxide mass
tCO_2e	tCO_2 equivalent, a unit of the total greenhouse effect of mixtures of GHGs, taking into account their different potencies as solar heat trapping agents
tCO_2edmv	tCO_2e saving expressed as 'dated mitigation value', to specify the year in which the saving is delivered and hence its value on the basis that earlier savings are more valuable for mitigation than later ones
UN	United Nations
UNEP	United Nations Environment Programme
UNFCCC	United Nations Framework Convention on Climate Change
WFD	Water Framework Directive
ZAN-SDI	National Spatial Data Infrastructure for Integrated Coastal and Marine Spatial Planning in Zanzibar (ICI project)
ZanSea	Zanzibar Social Environmental Atlas for Coastal and Marine Areas
ZEMA	Zanzibar Environmental Management Authority

References

Aboulnaga, M.M., Elwan, A.F. and Elsharouny, M.R. (2019) *Urban Climate Change Adaptation in Developing Countries: Policies, Projects, and Scenarios.* Springer International (Cham, Switzerland).

Adams, M. (2018) *Aelfred's Britain: War and Peace in the Viking Age.* Head of Zeus (London).

Adaptation Committee (2020) *Outcomes of the 17th Meeting of the Adaptation Committee.* 24–27 March 2020. Adaptation Division, UNFCCC Secretariat (Bonn). www4.unfccc.int/sites/NWPStaging/News/Pages/News.aspx. (Accessed 20 December 2020).

Adaptation Fund (2011) *Results Framework and Baseline Guidance: Project-Level.* Adaptation Fund (Washington, DC). www.adaptation-fund.org/document/results-framework-and-baseline-guidance-project-level/. (Accessed 20 December 2020).

Adger, W.N., Paavola, J., Saleemul Huq and Mace, M.J. (eds) (2006) *Fairness in Adaptation to Climate Change.* MIT Press (Cambridge, MA).

Aerts, J., Botzen, W., Bowman, M.J., Ward, P.J. and Dircke, P. (eds) (2012) *Climate Adaptation and Flood Risk in Coastal Cities.* Routledge (Abingdon and New York).

Agrawal, A. and Ostrom, E. (2001) Collective action, property rights, and decentralization in resource use in India and Nepal. *Politics and Society,* 29: 485–514.

Ahmed, N.M. (2017) *Failing States, Collapsing Systems: BioPhysical Triggers of Political Violence.* Springer International (Cham, Switzerland).

Alberge, D. (2017) UK's north–south divide dates back to Vikings, says archaeologist. *The Guardian,* 16 October 2017. www.theguardian.com/society/2017/oct/16/uk-north-south-divide-vikings-watford-gap. (Accessed 20 December 2020).

Allen, M., Axelsson, K., Caldecott, B., Hale, T., Hepburn, C., Hickey, C., Mitchell-Larson, E., Malhi, Y., Otto, F., Seddon, N. and Smith, S. (2020) *The Oxford Principles for Net Zero Aligned Carbon Offsetting.* Smith School of Enterprise and the Environment, University of Oxford (Oxford). www.smithschool.ox.ac.uk/publications/reports/Oxford-Offsetting-Principles-2020.pdf. (Accessed 20 December 2020).

Alliance of World Scientists (2020) *World Scientists' Warning of a Climate Emergency.* https://scientistswarning.forestry.oregonstate.edu. (Accessed 20 December 2020).

Alverson, K. and Zommers, Z. (eds) (2018) *Resilience: The Science of Adaptation to Climate Change.* Elsevier (Amsterdam).

Alves, F., Leal Filho, W. and Azeiteiro U. (eds) (2018) *Theory and Practice of Climate Adaptation.* Springer International (Cham, Switzerland).

Ambrose, J. (2020) Almost half of thermal coal firms set to defy climate pledge – report. *The Guardian,* 12 November 2020. www.theguardian.com/business/2020/nov/12/thermal-coal-firms-climate-pledge-report-paris-goals. (Accessed 20 December 2020).

Anon. (2017) *33 Theses for an Economics Reformation.* Rethinking Economics and the New Weather Institute. https://neweconomics.opendemocracy.net/33-theses-economics-reformation/. (Accessed 20 December 2020).

AOSIS (2020a) *The Placencia Ambition Forum Breakout Group: Enhancing Action in Adaptation and Resilience.* www.aosis.org/wp-content/uploads/2020/04/Adaptation-and-Resilience-Thematic-Breakout-Concept.pdf. (Accessed 20 December 2020).

AOSIS (2020b) *AOSIS Submission to the Standing Committee on Finance Compilation of Indicative Needs of Small Island Developing States 30 April 2020.* https://unfccc.int/sites/default/files/resource/SCF%20Submission-AOSIS%20Needs%20Survey.pdf. (Accessed 20 December 2020).

Applebaum, A. (2020) *Twilight of Democracy: The Failure of Politics and the Parting of Friends.* Allen Lane (London).

Azeiteiro, U.M., Leal Filho, W. and Aires, L. (eds) (2018) *Climate Literacy and Innovations in Climate Change Education: Distance Learning for Sustainable Development.* Springer International (Cham, Switzerland).

Bäck, L., Visti, M. and Moussa, Z. (2014) *Evaluation of Complementarity in Finland's Development Policy and Co-operation: A Case Study on Complementarity in the Institutional Co-operation Instrument.* Evaluation report 2014:1. Ministry for Foreign Affairs of Finland (Helsinki).

Ball, P. (1999) *The Self-Made Tapestry: Pattern Formation in Nature.* Oxford University Press (New York).

Ballantyne, A.P., Alden, C.B., Miller, J.B., Tans, P.P. and White, J.W.C. (2012) Increase in observed net carbon dioxide uptake by land and oceans during the past 50 years. *Nature,* 488 (7409): 70–72. https://doi.org/10.1038/nature11299. (Accessed 20 December 2020).

Bank of England (2019) *Enhancing Banks' and Insurers' Approaches to Managing the Financial Risks from Climate Change.* Supervisory Statement 3/19. www.bankofengland.co.uk/prudential-regulation/publication/2019/enhancing-banks-and-insurers-approaches-to-managing-the-financial-risks-from-climate-change-ss. (Accessed 20 December 2020).

Basnet, S. (2016) Be smart, go electric. *Nepali Times*, 27 May–2 June 2016. http://nepalitimes.com/article/nation/Be-smart-go-electric,3062. (Accessed 20 December 2020).

Bateson, G. (1972) *Steps to an Ecology of Mind*. Chicago University Press (Chicago).

Bateson, G. (1979) *Mind and Nature: A Necessary Unity*. Wildwood House (London).

BEH (2017) *WorldRiskReport Analysis and Prospects 2017*. Bündnis Entwicklung Hilft (Berlin).

Bendell, J. (2018) *Deep Adaptation: A Map for Navigating Climate Tragedy*. IFLAS Occasional Paper 2. Institute of Leadership and Sustainability, University of Cumbria (Carlisle). www.lifeworth.com/deepadaptation.pdf. (Accessed 20 December 2020).

Berck, C.S., Berck, P. and Di Falco, S. (eds) (2018) *Agricultural Adaptation to Climate Change in Africa: Food Security in a Changing Environment*. RFF and Routledge (Abingdon and New York).

Biskaborn, B.K., Smith, S.L., Noetzli, J., Matthes, H., Vieira, G., Streletskiy, D.A., Schoeneich, P., Romanovsky, V.E., Lewkowicz, A.G., Abramov, A., Allard, M., Boike, J., Cable, W.L., Christiansen, H.H., Delaloye, R., Diekmann, B., Drozdov, D., Etzelmüller, B., Grosse, G., Guglielmin, M., Ingeman-Nielsen, T., Isaksen, K., Ishikawa, M., Johansson, M., Johannsson, H., Joo, A., Kaverin, D., Kholodov, A., Konstantinov, P., Kröger, T., Lambiel, C., Lanckman, J-P., Luo, D-L., Malkova, G., Meiklejohn, I., Moskalenko, N., Oliva, M., Phillips, M., Ramos, M., Sannel, A.B.K., Sergeev, D., Seybold, C., Skryabin, P., Vasiliev, A., Wu, Q-B, Yoshikawa, K., Zheleznyak, M. and Lantuit, H. (2019) Permafrost is warming at a global scale. *Nature Communications*, 10, Article 264. https://doi.org/10.1038/s41467-018-8240-4. (Accessed 20 December 2020).

Bista, D.B. (1991) *Fatalism and Development: Nepal's Struggle for Modernization*. Orient Longman (Kolkata).

Blasetti, N. and Williams, S.A. (2020) *The European Green Deal: A Post-COVID Green Recovery*. Trinità Dei Monti (31 Maggio 2020). http://trinitamonti.org/2020/05/31/the-european-green-deal-a-post-covid-green-recovery/?fbclid=IwAR06opw_JrsyllYZeeJ8MrJhX-gmWv7UOdEL5pBFZyl8S1hL5X-cbtS7HIE. (Accessed 20 December 2020).

Bob, U. and Bronkhorst, S. (eds) (2014) *Conflict-Sensitive Adaptation to Climate Change in Africa*. Berliner Wissenschafts-Verlag (Berlin).

Bolle, M. de (2019) *19–15 The Amazon Is a Carbon Bomb: How Can Brazil and the World Work Together to Avoid Setting It Off?* Policy Brief 19-15. Peterson Institute for International Economics (Washington, DC). www.piie.com/sites/default/files/documents/pb19–15.pdf. (Accessed 20 December 2020).

Bologna, M. and Aquino, G. (2020) Deforestation and world population sustainability: a quantitative analysis. *Scientific Reports*, 10: 7631. https://doi.org/10.1038/s41598-020-63657-6. (Accessed 20 December 2020).

Bonifaz, C.R. (2010) Prologue, pages 10–11 in Parellada et al. (2010) *The Rights of Indigenous Peoples: Cooperation between Denmark and Bolivia (2005–2009)*.

Boulter, S., Palutikof, J., Karoly, D. and Guitart, D (eds) (2013) *Natural Disasters and Adaptation to Climate Change*. Cambridge University Press (Cambridge).

Bours, D., McGinn, C. and Pringle, P. (2014a) *Twelve Reasons Why Climate Change Adaptation M&E Is Challenging*. Guidance Note 1. SEA Change CoP (Phnom Penh) and UKCIP (Oxford). www.managingforimpact.org/sites/default/files/resource/2014_01_sea_change_ukcip_gn1_12_reasons_why_cca_mande_is_challenging.pdf. (Accessed 20 December 2020).

Bours, D., McGinn, C. and Pringle, P. (2014b) *Selecting Indicators for Climate Change Adaptation Programming*. Guidance Note 2. SEA Change CoP (Phnom Penh) and UKCIP (Oxford). www.ukcip.org.uk/wp-content/PDFs/MandE-Guidance-Note2.pdf. (Accessed 20 December 2020).

Bours, D., McGinn, C. and Pringle, P. (2014c) *The Theory of Change Approach to Climate Change Adaptation Programming*. Guidance Note 3. SEA Change CoP (Phnom Penh) and UKCIP (Oxford). www.ukcip.org.uk/wp-content/PDFs/MandE-Guidance-Note3.pdf. (Accessed 20 December 2020).

Bours, D., McGinn, C. and Pringle, P. (2014d) *Design, Monitoring, and Evaluation in a Changing Climate: Lessons Learned from Agriculture and Food Security Programme Evaluations in Asia*. Evaluation Review 1. SEA Change CoP (Phnom Penh) and UKCIP (Oxford). www.ukcip.org.uk/wp-content/PDFs/UKCIP-SeaChange-MandE-ER1-agriculture.pdf. (Accessed 20 December 2020).

Bours, D., McGinn, C. and Pringle, P. (2014e) *International and Donor Agency Portfolio Evaluations: Trends in Monitoring and Evaluation of Climate Change Adaptation Programmes*. Evaluation Review 2. SEA Change CoP (Phnom Penh) and UKCIP (Oxford). www.ukcip.org.uk/wp-content/PDFs/UKCIP-SeaChange-MandE-ER2-donor-agencies.pdf. (Accessed 20 December 2020).

Bowen, A. and Rydge, J. (2011) *Climate Change Policy in the United Kingdom*. Centre for Climate Change Economics and Policy, Grantham Research Institute on Climate Change and the Environment. www.lse.ac.uk/GranthamInstitute/wp-content/uploads/2014/03/PP_climate-change-policy-uk.pdf. (Accessed 20 December 2020).

Brofeldt, S., Argyriou, D., Turreira-García, N., Meilby, H., Danielsen, F. and Theilade, I. (2018) Community-based monitoring of tropical forest crimes and forest resources using information and communication technology: experiences from Prey Lang, Cambodia. *Citizen Science: Theory and Practice*, 3 (2): 4, 1–14. https://doi.org/10.5334/cstp.129. (Accessed 20 December 2020).

Brofeldt, S., Theilade, I., Burgess, N.D., Danielsen, F., Poulsen, M.K., Adrian, T., Nguyen Bang, T., Budiman, A., Jensen, J., Jensen, A.E., Kurniawan, Y., Lægaard, S.B.L., Mingxu, Z., van Noordwijk, M., Rahayu, S., Rutishauser, E., Schmidt-Vogt, D., Warta, Z., and Widayati, A. (2014) Community monitoring of carbon stocks for REDD+: does accuracy and cost change over time? *Forests*, 5: 1834–1854. https://doi.org/10.3390/f5081834. (Accessed 20 December 2020).

Brooks, N., Anderson, S. Ayers, J., Burton, I. and Tellam, I. (2011) *Tracking Adaptation and Measuring Development (TAMD)*. Working Paper 1. International Institute for Environment and Development (London).

Brooks, N., Anderson, S. Ayers, J., Burton, I. and Tellam, I. (2013) *TAMD, an Operational Framework for Tracking Adaptation and Measuring Development*. Working Paper 5. International Institute for Environment and Development (London).

Brooks, R.E. (ed.) (1997) Stillpoints, pages 69–70 in *Life in Motion: The Osteopathic Vision of Rolin E. Becker, D.O.* Stillness Press (Portland, OR).

Bryant-Tokalau, J. (2018) *Indigenous Pacific Approaches to Climate Change: Pacific Island Countries*. Palgrave (London).

Buchner, B., Clark, A., Falconer, A., Macquarie, R. Meattle, C., Tolentino, R. and Wetherbee, C. (2019) *Global Landscape of Climate Finance 2019*. Climate Policy Initiative (London). www.climatepolicyinitiative.org/publication/global-landscape-of-climate-finance-2019/. (Accessed 20 December 2020).

Burke, J. (2020) Tanzanian opposition accuses police of killing nine during protests. *The Guardian*, 27 October 2020. www.theguardian.com/world/2020/oct/27/tanzanian-president-accused-of-repression-on-eve-of-election. (Accessed 20 December 2020).

Burton, I. and van Aalst, M. (1999) *Come Hell or High Water: Integrating Climate Change Vulnerability and Adaptation into Bank Work*. Environment Department, The World Bank (Washington, DC).

Bush, M.J. (2018) *Climate Change Adaptation in Small Island Developing States*. Wiley-Blackwell (Oxford).

C40 Cities (2012) *Why Cities? Ending Climate Change Begins in the Cities*. www.c40.org/ending-climate-change-begins-in-the-city. (Accessed 20 December 2020).

C40 Cities (2020) *C40 Cities Annual Report 2019*. C40 Cities Climate Leadership Group (New York). https://c40-production-images.s3.amazonaws.com/other_uploads/images/2574_C40_2019_Annual_Report.original.pdf?1587634742. (Accessed 20 December 2020).

C40 Cities, Realdania and Nordic Sustainability (2019) *Cities 100 2019: 100 City Projects Making the Case for Climate Action.* C40 Cities Climate Leadership Group (London), Realdania (Copenhagen) and Nordic Sustainability (Copenhagen). www.climaterealityproject.org/sites/default/files/cities100_2019_report.pdf. (Accessed 20 December 2020).

Cabinet Office (undated) *The Plus of GCCA+, An EU Flagship Initiative Supporting Climate Resilience.* http://europeanmemoranda.cabinetoffice.gov.uk/files/2015/12/The_Global_Climate_Change_Alliance_Plus.pdf. (Accessed 20 December 2020).

Caldecott, B. (ed.) (2018) *Stranded Assets and the Environment: Risk, Resilience and Opportunity.* Routledge (Abingdon and New York).

Caldecott, B. (2019) *Net Zero by 2050: A Policy Response, Keynote Address.* Rothamsted Research (Harpenden). 20 September 2019. Summarised at: www.mostimportantthings.org/2019/09/26/net-zero-by-2050-notes-on-the-rothamsted-conference/. (Accessed 20 December 2020).

Caldecott, J.O. (1988) *Hunting and Wildlife Management in Sarawak.* IUCN Tropical Forest Programme Monographs No. 7. World Conservation Union (Cambridge and Gland).

Caldecott, J.O. (1996) *Designing Conservation Projects.* Cambridge University Press (Cambridge).

Caldecott, J.O. (2005) Lessons learned and the path ahead, pages 276–285 in *World Atlas of Great Apes and Their Conservation* (edited by J.O. Caldecott and L. Miles). California University Press (Berkeley, CA).

Caldecott, J.O. (2015) *Review of Emission Reduction Program Idea Notes (ER-PINs) prepared for the Forest Carbon Partnership Facility (FCPF) by the Congo Republic, Ghana, Guatemala, México, Nepal and Vietnam.* Creatura Ltd (Bath), 26 August 2015.

Caldecott, J.O. (2017a) *Aid Performance and Climate Change.* Routledge/Earthscan (Abingdon and New York).

Caldecott, J.O. (2017b) *Historical Context of Danida's Work in Bolivia, Mozambique, and Nepal.* Danida Evaluation Day, 26 September 2017. Ministry for Foreign Affairs of Denmark (Copenhagen).

Caldecott, J.O. (2020) *Water: Life in Every Drop*, 2nd edition. Bladud (Bath).

Caldecott, J.O., Bird, N.M. and Grøn, H.R. (2021) *Evaluation of Danish Funding for Climate Change Mitigation in Developing Countries, Final Report, April 2021.* Particip GmbH (Freiburg) and Ministry for Foreign Affairs of Denmark (Copenhagen).

Caldecott, J.O., Chambers, B. and Kurukulasurya, L. (2008) *Green Breakthroughs: Solving Environmental Problems through Innovative Policies and Law.* United Nations Environment Programme (Nairobi).

Caldecott, J.O., Clark, O., Woel, B., Klaasens, E., Laanouni, F. and Willot, P. (2019) *Lessons Learned from Strategic Evaluations: On (a) EU Cooperation with sub-Saharan Africa, (b) Private Sector Engagement, (c) Gender Equality, and (d) Climate Change and Sustainable Energy.* Evaluation Support Service, Landell Mills (Trowbridge) and European Commission (Brussels).

Caldecott, J.O., Dornau, R., Gollan, J., Halonen, M., Saalismaa, N., Schilli, A., Simonett, O., Stuhlberger C. and Tommila, P. (2014) *Technical Report on Effectiveness of the Swiss International Cooperation in Climate Change Mitigation and Adaptation Interventions 2000–2012.* Mandate of the Swiss Agency for Development and Cooperation (SDC) and the State Secretariat for Economic Affairs (SECO) (Bern).

Caldecott, J.O., Halonen, M., Sørensen, S.E., Dugersuren, S., Tommila, P. and Pathan, A. (2010) *Evaluation of Sustainability in Poverty Reduction: Synthesis.* Evaluation Report 2010:4. Ministry for Foreign Affairs of Finland (Helsinki).

Caldecott, J.O., Hansen, F., Visser, M., White, P., Basnet, G. and Efraimsson, A. (2017) *Evaluation of Danish–Nepalese Development Cooperation, 1991–2016.* FCG (Helsinki) and Ministry for Foreign Affairs of Denmark (Copenhagen).

Caldecott, J.O., Hawkes, M., Bajracharya, B. and Lounela, A. (2012) *Evaluation of the Country Programme between Finland and Nepal.* Evaluation report 2012:2. Ministry for Foreign Affairs of Finland (Helsinki).

Caldecott, J.O., Killian, B., Siltanen, M. and Smit, R. (2019) *Final and Ex-post Evaluation of Three Institutional Cooperation Projects in Tanzania, Final Report, 8 January 2019.* Impact Consulting Ltd in association with Saffron Consulting International Ltd and Ministry for Foreign Affairs of Finland (Helsinki).

Caldecott, J.O., Killian, B., Tommila, P., Rinne, P., Halonen, M. and Oja, L. (2013) *Scoping Mission for a Possible Renewable Natural Resource Economic Governance Programme in Tanzania, Final Report.* Gaia Consulting Oy (Helsinki).

Caldecott, J.O. and Lovejoy, A. (1996) Costa Rica, pages 18–27 in *Decentralization and Biodiversity Conservation* (edited by E. Lutz and J. Caldecott). The World Bank (Washington, DC).

Caldecott, J.O., Robledo, C. and Clarke, M. (2015) *The World Bank: Second Program Evaluation of the Forest Carbon Partnership Facility (FCPF), Inception Report.* Indufor Oy (Helsinki) and Forest Carbon Partnership Facility (Washington, DC).

Caldecott, J.O., Valjas, A., Killian, B. and Lounela, A. (2012) *Evaluation of the Country Programme between Finland and Tanzania.* Evaluation report 2012:3. Ministry for Foreign Affairs of Finland (Helsinki).

Caldecott, J.O., van Sluijs, F., Aguilar, B. and Lounela, A. (2012) *Evaluation of the Country Programme between Finland and Nicaragua*. Evaluation report 2012:1, Ministry for Foreign Affairs of Finland (Helsinki).

Calliari, E., Mysiak, J. and Vanhala, L. (2020) A digital climate summit to maintain Paris Agreement ambition. *Nature Climate Change*, 10: 480. https://doi.org/10 .1038/s41558-020-0794-0. (Accessed 20 December 2020).

Canessa, R., Barnini, M. and Lundsgaard, T. (2015) *Evaluation of EUROCLIMA (2010–2014), Final Report*. B&S Europe for the European Commission (Brussels).

Capra, F. and Luisi, P.L. (2014) *The Systems View of Life: A Unifying Vision.* Cambridge University Press (Cambridge).

CarbonBrief (2017) *Explainers, February 14. 2017, Q&A: The Social Cost of Carbon.* www.carbonbrief.org/qa-social-cost-carbon. (Accessed 20 December 2020).

Carstensen, J. (ed.) (2017) *Climate Change Adaptation and Development.* Routledge (Abingdon and New York).

Castro, P., Azul, A.M. and Leal Filho, W. (eds) (2019) *Climate Change-Resilient Agriculture and Agroforestry: Ecosystem Services and Sustainability.* Springer International (Cham, Switzerland).

CBD (2020) *Global Biodiversity Outlook 5.* Secretariat of the Convention on Biological Diversity (Montreal). www.cbd.int/gbo5. (Accessed 20 December 2020).

CED (2020) *Climate Emergency Declarations in 1,859 Jurisdictions and Local Governments Cover 820 Million Citizens.* Posted 17 December 2020. https:// climateemergencydeclaration.org/climate-emergency-declarations-cover-15-million-citizens/. (Accessed 20 December 2020).

Chang, H.J. (2011) *23 Things They Don't Tell You about Capitalism.* Penguin (London).

Change.org (2020) *Write Peace with Nature into the Scottish Constitution.* www.change .org/p/citizens-assembly-of-scotland-write-peace-with-nature-into-the-scottish-con stitution?utm_source=share_petitionandutm_medium=custom_urlandrecruited_ by_id=b4d1fc20-c88a-11e4-97e5-ddf2108d5632. (Accessed 20 December 2020).

Cheng, L., Abraham, J., Trenberth, K.E., Fasullo, J., Boyer, T., Locarnini, R., Zhang, B., Yu, F., Wan, L., Chen, X., Song, X., Liu, Y., Mann, M.E., Reseghetti, F., Simoncelli, S., Gouretski, V., Chen, G., Mishonov, A., Reagan, J., and Zhu, J. (2021) Upper ocean temperatures hit record high in 2020. *Advances in Atmospheric Sciences*, https://doi.org/10.1007/s00376-021-0447-x (in press). (Accessed 28 January 2020).

Chirisa, I., Matamanda, A. and Matamanda, J. (2018) Africa's dilemmas in climate change communication: universalistic science versus indigenous technical knowledge, pages 1–14 in *Handbook of Climate Change Communication: Vol. 1 – Practice of Climate Change Communication* (edited by W. Leal Filho, E. Manolas, A.M. Azul, U.M. Azeiteiro and H. McGhie). Springer International (Cham, Switzerland).

Chrisafis, A. (2020) Citizens' assembly ready to help Macron set French climate policies. *The Guardian*, 10 January 2020. www.theguardian.com/world/2020/jan/10/citizens-panels-ready-help-macron-french-climate-policies. (Accessed 20 December 2020).

Christoffersen, L. (2018) Amazonian erasures: landscape and myth-making in lowland Bolivia. *Rural Landscapes: Society, Environment, History*, 5 (1): 3, 1–19. https://doi.org/10.16993/rl.43. (Accessed 20 December 2020).

CIF (2012) *Revised PPCR Results Framework*. Climate Investment Funds (Washington, DC).

Climate Assembly UK (2020) *The Path to Net Zero*. House of Commons (London). www.climateassembly.uk/report/read/final-report.pdf. (Accessed 20 December 2020).

Convention Citoyenne (2020) *Les Propositions de la Convention Citoyenne pour le Climat*. Conseil Économique, Social et Environnemental (Paris). https://propositions.conventioncitoyennepourleclimat.fr/pdf/ccc-rapport-final.pdf. (Accessed 20 December 2020).

CSD (1996) *Programme Element 1.3: Traditional Forest Related Knowledge*. Report of the Secretary General. E/CN.17/IPF/1996/13 2 Implementation of forest-related decisions of the United Nations Conference on Environment and Development at the national and international levels, including an examination of sectoral and cross-sectoral linkages. Ad Hoc Intergovernmental Panel on Forests, third session, 9–20 September 1996. Economic and Social Council, Commission on Sustainable Development. https://digitallibrary.un.org/record/221382. (Accessed 20 December 2020).

Cultural Survival (2015) *Observations on the State of Indigenous Human Rights in Denmark in Light of the UN Declaration on the Rights of Indigenous Peoples*. Cultural Survival (Cambridge, MA). www.culturalsurvival.org/sites/default/files/media/uprdenmarkfinal.pdf. (Accessed 20 December 2020).

Dakos, V., Matthews, B., Hendry, A.P., Levine, J., Loeuille, N., Norberg, J., Nosil, P., Scheffer, M. and De Meester, L. (2019) Ecosystem tipping points in an evolving world. *Nature Ecology and Evolution*, 3: 355–362. www.nature.com/articles/s41559-019-0797-2. (Accessed 20 December 2020).

Danida (1999) *Final Sector Programme Support Document: Environment Sector, Nepal.* Ministry for Foreign Affairs of Denmark (Copenhagen).

Danida (2000) *Annex 5 to the Environment Sector Programme Support Document (Industrial and Urban Development), Nepal: Component Brief, Support to Air Quality Management in Kathmandu Valley.* Ministry for Foreign Affairs of Denmark (Copenhagen).

Danida (2004a) *Draft Programme Document: Integrated Environmental Programme, Nepal (Ref 104.Nepal, Nov 2004).* Ministry for Foreign Affairs of Denmark (Copenhagen).

Danida (2004b) *Appraisal of Integrated Environmental Programme, Final Draft Appraisal Report (104.Nepal.806.2, Feb 2005).* Ministry for Foreign Affairs of Denmark (Copenhagen).

Danida (2004c) *Strategy for Danish Support to Indigenous Peoples.* Ministry for Foreign Affairs of Denmark (Copenhagen).

Danida (2005a) *Final Component Completion Report, Community Forestry Field Implementation Component (July 1999–July 2005), Nepal, November 2005.* Ministry for Foreign Affairs of Denmark (Copenhagen).

Danida (2005b) *Final Programme Completion Report, Natural Resource Management Sector Assistance Programme (NARMSAP), February 1998–July 2005, Nepal, November 2005.* Ministry for Foreign Affairs of Denmark (Copenhagen).

Danida (2005c) *Environment Sector Programme Support (ESPS), Nepal: Completion Report 1999–2005, Component No. 2: Promotion of Cleaner Production in Industry.* Ministry for Foreign Affairs of Denmark (Copenhagen).

Danida (2005d) *Environment Sector Programme Support (ESPS), Nepal: Completion Report 1999–2005, Sub-Component No. 2.1 EE: Strengthening Energy Efficiency Aspects of Cleaner Production.* Ministry for Foreign Affairs of Denmark (Copenhagen).

Danida (2005e) *Environment Sector Programme Support (ESPS), Nepal, Completion Report 2001–2005, Component No. 5 – Air Quality Management in Kathmandu Valley.* Ministry for Foreign Affairs of Denmark (Copenhagen).

Danida (2016a) *Evaluation of the Danish-Nepalese Development Cooperation (1991–2016): Terms of Reference.* Ministry for Foreign Affairs of Denmark (Copenhagen), November 2016.

Danida (2016b) *Evaluation of Danish-Bolivian Cooperation 1994–2016: Terms of Reference.* Ministry for Foreign Affairs of Denmark (Copenhagen). www.netpublikationer.dk/UM/evaluation_danish-bolivian_cooperation_1994–2016/Html/kap13.html. (Accessed 20 December 2020).

Danida (2019) *Evaluation of the Danish Support for Climate Change Adaptation in Developing Countries: Terms of Reference.* Ministry for Foreign Affairs of Denmark (Copenhagen), June 2019.

Danielsen, F., Adrian, T., Brofeldt, S., van Noordwijk, M., Poulsen, M.K., Rahayu, S., Rutishauser, E., Theilade, I., Widayati, A., The An, N., Nguyen Bang, T., Budiman, A., Enghoff, M., Jensen, A.E., Kurniawan, Y., Li, Q.-H., Mingxu, Z., Schmidt-Vogt, D., Prixa, S., Thoumtone, V., Warta, Z. and Burgess, N. (2013) Community Monitoring for REDD+: International Promises and Field Realities. *Ecology and Society*, 18 (3): 41. https://doi.org/10.5751/ES-05464-180341. (Accessed 20 December 2020).

Dasgupta, P. and HMT (2020) *The Dasgupta Review: Independent Review on the Economics of Biodiversity, Interim Report* (April 2020). H.M. Treasury (London). https://assets.publishing.service.gov.uk/government/uploads/system/uploads/attachment_data/file/882222/The_Economics_of_Biodiversity_The_Dasgupta_Review_Interim_Report.pdf. (Accessed 20 December 2020).

Davies, P. (1987) *The Cosmic Blueprint: Order and Complexity at the Edge of Chaos.* William Heinemann (London).

DECC (2009) *Carbon Valuation in UK Policy Appraisal: A Revised Approach.* Climate Change Economics, Department of Energy and Climate Change (London). www.gov.uk/government/publications/carbon-valuation-in-uk-policy-appraisal-a-revised-approach. (Accessed 20 December 2020).

Dennett, D.C. (1995) *Darwin's Dangerous Idea: Evolution and the Meanings of Life.* Simon and Schuster (New York).

Dhakal, S. (2010) GHG emissions from urbanization and opportunities for urban carbon mitigation. *Current Opinion in Environmental Sustainability*, 2 (4): 277–283. https://doi.org/10.1016/j.cosust.2010.05.007. (Accessed 20 December 2020).

Dossey, L. (2016) Is friendship limited? An inquiry into Dunbar's number. *Metapsychosis Journal of Consciousness, Literature, and Art.* www.metapsychosis.com/is-friendship-limited-an-inquiry-into-dunbars-number/. (Accessed 20 December 2020).

DoURP (2018a) *Kiwengwa Local Area Plan: Pilot Coastal Zone Planning for the Shehia of Kiwengwa as Part of the North East Unguja Special Area Plan, an Integrated Planning Document Including Marine and Land Spatial Planning.* Department of Urban and Regional Planning, Ministry of Land, Housing, Water and Energy, Revolutionary Government of Zanzibar.

DoURP (2018b) *Pongwe Local Area Plan: Pilot Coastal Zone Planning for the Shehia of Pongwe as Part of the North East Unguja Special Area Plan, an*

Integrated Planning Document Including Marine and Land Spatial Planning. Department of Urban and Regional Planning, Ministry of Land, Housing, Water and Energy, Revolutionary Government of Zanzibar.

DoURP (2019) *North-East Unguja Special Area Plan: Pilot Coastal Zone Planning in an Area Covering Mkokotoni and Including the Hinterland between Mkokotoni and Nungwi.* Department of Urban and Regional Planning, Ministry of Land, Housing, Water and Energy, Revolutionary Government of Zanzibar.

Dunbar, R.I.M. (2010) *How Many Friends Does One Person Need? Dunbar's Number and Other Evolutionary Quirks.* Harvard University Press (Cambridge, MA).

EBA (2018) *Invitation for Proposals: Evaluation of the Swedish Climate Change Initiative 2009–2012.* Expertgruppen för biståndsanalys (Stockholm).

EC (2013) *Communication from the Commission to the European Parliament, the Council, the European Economic and Social Committee and the Committee of the Regions: An EU Strategy on Adaptation to Climate Change: COM(2013) 216 Final.* European Commission (Brussels). https://eur-lex.europa.eu/legal-content/EN/TXT/?uri=CELEX:52013DC0216. (Accessed 20 December 2020).

EC (2017) *C(2017)8511 Report from the Commission: Seventh National Communication and Third Biennial Report from the European Union under the UN Framework Convention on Climate Change (UNFCCC) (Required under the UNFCCC and the Kyoto Protocol), and Its Accompanying Staff Working Documents: SWD(2017)457 Seventh National Communication of the EU and SWD(2017)458 Third Biennial Report of the EU. December 2017.* https://unfccc.int/sites/default/files/resource/459381_European%20Union-NC7-BR3–1-NC7%20BR3%20combined%20version.pdf. (Accessed 20 December 2020).

EC (2018) *Report from the Commission to the European Parliament and the Council on the Implementation of the EU Strategy on Adaptation to Climate Change: COM(2018) 738 final.* European Commission (Brussels). https://eur-lex.europa.eu/legal-content/EN/TXT/?uri=CELEX%3A52018DC0738. (Accessed 20 December 2020).

EC (2019) *The European Green Deal: Communication from the Commission to the European Parliament, the European Council, the Council, the European Economic and Social Committee and the Committee of the Regions.* European Commission, Brussels, 11.12.2019 COM(2019) 640 final.

EC (2020) *GCCA+ Orientation Package: The EU Global Climate Change Alliance Plus Flagship Initiative.* https://gcca.eu/sites/default/files/documents/2020–02/1.%20GCCA+%20Orientation%20Package%20January%202020%20EN%20-%20DESIGN.pdf. (Accessed 20 December 2020).

ECHO (2003) *The DIPECHO programme: Reducing the Impact of Disasters.* European Commission Humanitarian Aid Office (Brussels). https://op.europa .eu/en/publication-detail/-/publication/7844f42f-7aa8–11e9–9f05–01aa75ed71a1. (Accessed 20 December 2020).

ECHO and ADPC (2004) *Community-Based Disaster Risk Management Field Practitioner's Handbook.* European Commission Humanitarian Aid Office (Brussels) and Asian Disaster Preparedness Center (Pathumthani). www.adpc .net/igo/category/ID428/doc/2014-xCSf7I-ADPC-12handbk.pdf.   (Accessed 20 December 2020).

EEA (2018) *Water and Marine Environment: EEA 2018 Water Assessment.* European Environment Agency (Copenhagen). www.eea.europa.eu/themes/ water/european-waters. (Accessed 20 December 2020).

EEA (2019) *Water and Marine Environment: European Bathing Water Quality in 2019.* European Environment Agency (Copenhagen). www.eea.europa.eu/ themes/water/europes-seas-and-coasts/assessments/state-of-bathing-water/ european-bathing-water-quality-in-2019. (Accessed 20 December 2020).

EPI (2020) *Environmental Performance Index 2020: Nepal, Country Scorecard.* https://epi.yale.edu/sites/default/files/files/NPL_EPI2020_CP.pdf.   (Accessed 20 December 2020).

EU Councils (2020) *Tanzania: Declaration by the High Representative on Behalf of the EU on the Elections in Tanzania, 2 November 2020.* European Council and Council of the European Union. www.consilium.europa.eu/en/press/press-releases/2020/11/02/tanzania-declaration-by-the-high-representative-on-behalf-of-the-eu-on-the-elections-in-tanzania/. (Accessed 20 December 2020).

Farand, C. (2019) Asset managers worth $15 trillion make climate risk promise to Macron. *Climate Change News,* 12 July 2019. www.climatechangenews.com/ 2019/07/12/asset-managers-worth-15-trillion-make-climate-risk-promise-macron/. (Accessed 20 December 2020).

FFI (2020) *Year in Review. &ffi,* 2 (2020–2021). Fauna and Flora International (Cambridge).

Field, C.B., Barros, V.R. and Dokken, D.J. (eds) (2014) *Climate Change 2014: Impacts, Adaptation, and Vulnerability. Part A – Global and Sectoral Aspects: Contribution of the Working Group II to the Fifth Assessment Report of the Intergovernmental Panel on Climate Change.* Cambridge University Press (Cambridge).

FPP, IIFB, IWBN, CDILK and CBD (2020) *Local Biodiversity Outlooks 2: The Contributions of Indigenous Peoples and Local Communities to the Implementation of the Strategic Plan for Biodiversity 2011–2020 and to Renewing Nature and Cultures. A complement to the Fifth Edition of the*

Global Biodiversity Outlook. Forest Peoples Programme, International Indigenous Forum on Biodiversity, Indigenous Women's Biodiversity Network, Centres of Distinction on Indigenous and Local Knowledge and Secretariat of the Convention on Biological Diversity (Moreton-in-Marsh). www.cbd.int/gbo5/local-biodiversity-outlooks-2. (Accessed 20 December 2020).

Frankenberger, T.R., Spangler, T., Nelson, S. and Langworthy, M. (2012) *Enhancing Resilience to Food Security Shocks in Africa: Discussion Paper.* TANGO International (Tucson, Arizona).

Friedlingstein, P., Jones, M.W., O'Sullivan, M., Andrew, R.M., Hauck, J., Peters, G.P., Peters, W., Pongratz, J., Sitch, S., Le Quéré, C., Bakker, D.C.E., Canadell, J.G., Ciais, P., Jackson, R.B., Anthoni, P., Barbero, L., Bastos, A., Bastrikov, V., Becker, M., Bopp, L., Buitenhuis, E., Chandra, N., Chevallier, F., Chini, L.P., Currie, K.I., Feely, R.A., Gehlen, M., Gilfillan, D., Gkritzalis, T., Goll, S.S., Gruber, N., Gutekunst, S., Harris, I., Haverd, V., Houghton, R.A., Hurtt, G., Ilyina, T., Jain, A.K., Joetzjer, E., Kaplan, J.O., Kato, E., Goldewijk, K.K., Korsbakken, J.I., Landschützer, P., Lauvset, S.K., Lefèvre, N., Lenton, A., Lienert, S., Lombardozzi, D., Marland, G., McGuire, P.C., Melton, J.R., Metzl, N., Munro, D.R., Nabel, J.E.M.S., Nakaoka, S-I., Neill, C., Omar, A.M., Ono, T., Peregon, A., Pierrot, D., Poulter, B., Rehder, G., Resplandy, L., Robertson, E., Rödenbeck, C., Séférian, R., Schwinger, J., Smith, N., Tans, P.P., Tian, H., Tilbrook, B., Tubiello, F.N., van der Werf, G.R., Wiltshire, A.J. and Zaehle, S. (2019) Global Carbon Budget 2019. *Earth System Science Data*, 11: 1783–1838. https://doi.org/10.5194/essd-11-1783-2019. (Accessed 20 December 2020).

Friedlingstein, P., O'Sullivan, M., Jones, M.W., Andrew, R.M., Hauck, J., Olsen, A., Peters, G.P., Peters, W., Pongratz, J., Sitch, S., Le Quéré, C., Canadell, J.G., Ciais, P., Jackson, R.B., Alin, S., Aragão, L.E.O.C., Arneth, A., Arora, V., Bates, N.R., Becker, M., Benoit-Cattin, A., Bittig, H.C., Bopp, L., Bultan, S., Chandra, N., Chevallier, F., Chini, L.P., Evans, W., Florentie, L., Forster, P.M., Gasser, T., Gehlen, M., Gilfillan, D., Gkritzalis, T., Gregor, L., Gruber, N., Harris, I., Hartung, K., Haverd, V., Houghton, R.A., Ilyina, T., Jain, A.K., Joetzjer, E., Kadono, K., Kato, E., Kitidis, V., Korsbakken, J.I., Landschützer, P., Lefèvre, N., Lenton, A., Lienert, S., Liu, Z., Lombardozzi, D., Marland, G., Metzl, N., Munro, D.R., Nabel, J.E.M.S., Nakaoka, S.-I., Niwa, Y., O'Brien, K., Ono, T., Palmer, P.I., Pierrot, D., Poulter, B., Resplandy, L., Robertson, E., Rödenbeck, C., Schwinger, J., Séférian, R., Skjelvan, I., Smith, A.J.P., Sutton, A.J., Tanhua, T., Tans, P.P., Tian, H., Tilbrook, B., van der Werf, G., Vuichard, N., Walker, A.P., Wanninkhof, R., Watson, A. J., Willis, D., Wiltshire, A.J., Yuan, W., Yue, X. and Zaehle, S. (2020) Global Carbon Budget 2020. *Earth System Science Data*, 12:

3269–3340. https://doi.org/10.5194/essd-12-3269-2020. (Accessed 20 December 2020).

Friend, R., Jarvie, J., Reed, S.O., Sutarto, R., Thinphanga P. and Vu Canh Toan (2014) Mainstreaming urban climate resilience into policy and planning; reflections from Asia. *Urban Climate*, 7: 6–19.

Funder, M., Lindegaard, L.S., Friis-Hansen, E. and Gravesen, M.L. (2020a) *Integrating Climate Change Adaptation and Development: Past Trends and Ways Forward for Danish Development Cooperation.* Danish Institute for International Studies (Copenhagen).

Funder, M., Lindegaard, L.S., Friis-Hansen, E. and Gravesen, M.L. (2020b) *Three Steps to Integrate Climate Change Adaptation and Development: Addressing Resilience in Danish Development Policy. DIIS Impact, 2020.* Danish Institute for International Studies (Copenhagen). www.diis.dk/en/research/three-steps-to-integrate-climate-change-adaptation-and-development. (Accessed 20 December 2020).

Gaveau, D.L.A., Salim, M.A., Hergoualc'h, K., Locatelli, B., Sloan, S., Wooster, M., Marlier, M.E., Molidena, E., Yaen, H., DeFries, R., Verchot, L., Murdiyarso, D., Nasi, R., Holmgren, P. and Sheil, D. (2014) Major atmospheric emissions from peat fires in Southeast Asia during non-drought years: evidence from the 2013 Sumatran fires. *Scientific Reports*, 4 (6112): 1–7. https://doi.org/10.1038/srep06112. (Accessed 20 December 2020).

Geertz, C. (1993) *Local Knowledge.* Fontana/HarperCollins (London).

GEF (2010) *Updated Results-Based Management Framework for the Least Developed Countries Fund (LDCF) and the Special Climate Change Fund (SCCF) and Adaptation Monitoring and Assessment Tool.* GEF LDCF/SCCF.9/Inf.4. Global Environment Facility (Washington, DC). www.thegef.org/council-meeting-documents/updated-results-based-management-framework-ldcf-and-sccf-and-adaptation. (Accessed 20 December 2020).

GEF (2012) *LDCF/SCCF Adaptation Monitoring and Assessment Tool (AMAT) Guidelines and Tracking Tool.* Global Environment Facility (Washington, DC).

Geo-ICT (2020) *Geospatial and ICT Capacities in Tanzanian Higher Education Institutions.* www.geoict.org. (Accessed 20 December 2020).

Ghanbari, S. and Daneshvar, M.R.M. (2020). Urban and rural contribution to the GHG emissions in the MECA countries. *Environment, Development and Sustainability.* https://link.springer.com/article/10.1007%2Fs10668-020-00879-8. (Accessed 20 December 2020).

Gladwell, M. (2000) *The Tipping Point: How Little Things Can Make a Big Difference.* Little, Brown (Boston).

GML (2020a) *Monthly Average Mauna Loa CO₂*. National Oceanic and Atmospheric Administration Global Monitoring Laboratory. www.esrl.noaa .gov/gmd/ccgg/trends/. (Accessed 20 December 2020).

GML (2020b) *How We Measure Background CO₂ Levels on Mauna Loa*. National Oceanic and Atmospheric Administration Global Monitoring Laboratory (Pieter Tans and Kirk Thoning). www.esrl.noaa.gov/gmd/ccgg/about/co2_ measurements.html. (Accessed 20 December 2020).

Goldsmith, Z. (2007) Foreword, page x in *Water: Life in Every Drop* (by J. Caldecott). Virgin Books (London).

Government of Afghanistan (2017) *Second National Communication under the UNFCCC*. National Environmental Protection Agency, Islamic Republic of Afghanistan. ccc.int/sites/default/files/resource/SNC%20Report_Final_20180801% 20.pdf. (Accessed 20 December 2020).

Government of Antigua and Barbuda (2015) *Intended Nationally Determined Contribution*. www4.unfccc.int/sites/ndcstaging/PublishedDocuments/Antigua% 20and%20Barbuda%20First/Antigua%20and%20Barbuda%20First.pdf. (Accessed 20 December 2020).

Government of Armenia (2020) *Fourth National Communication on Climate Change*. UNDP and Ministry of Environment, Republic of Armenia (Yerevan). unfccc.int/sites/default/files/resource/NC4_Armenia_.pdf.   (Accessed 20 December 2020).

Government of Australia (2017). *Australia's 7th National Communication on Climate Change*. Department of the Environment and Energy. https://unfccc .int/documents/69238. (Accessed 20 December 2020).

Government of Belize (2014) *A National Climate Change Policy, Strategy and Action Plan to Address Climate Change in Belize*. Caribbean Community Climate Change Centre and Ministry of Forestry, Fisheries and Sustainable Development (Belmopan, Belize).

Government of Bolivia (2016) *Intended Nationally Determined Contribution from the Plurinational State of Bolivia*. www4.unfccc.int/sites/ndcstaging/ PublishedDocuments/Bolivia%20(Plurinational%20State%20of)%20First/ INDC-Bolivia-english.pdf. (Accessed 20 December 2020).

Government of Botswana (2019) *Botswana's Third National Communication to the UNFCCC*. Ministry of Environment, Natural Resources Conservation and Tourism. unfccc.int/sites/default/files/resource/BOTSWANA%20THIRD %20NATIONAL%20COMUNICATION%20FINAL%20.pdf.   (Accessed 20 December 2020).

Government of Brazil (2016) *National Adaptation Plan to Climate Change. Volume 1: General Strategy*. Ministry of Environment (Brasilia). www4.unfccc.int/sites/

NAPC/Documents/Parties/Brazil%20NAP%20English.pdf. (Accessed 4 April 2021).

Government of Burkina Faso (2015) *Intended Nationally Determined Contribution in Burkina Faso.* www4.unfccc.int/sites/ndcstaging/PublishedDocuments/ Burkina%20Faso%20First/INDC%20Burkina_ENG.%20version_finale.pdf. (Accessed 20 December 2020).

Government of Chile (2015) *Intended Nationally Determined Contribution of Chile Towards the Climate Agreement of Paris 2015.* www4.unfccc.int/sites/ ndcstaging/PublishedDocuments/Chile%20First/INDC%20Chile%20english %20version.pdf. (Accessed 20 December 2020).

Government of Chile (2020) *Chile's Nationally Determined Contribution: Update 2020.* www4.unfccc.int/sites/ndcstaging/PublishedDocuments/Chile%20First/ Chile%27s_NDC_2020_english.pdf. (Accessed 20 December 2020).

Government of Costa Rica (2015) *Costa Rica's Intended Nationally Determined Contribution.* Ministry of Environment and Energy (San José). www4.unfccc.int/ sites/ndcstaging/PublishedDocuments/Costa%20Rica%20First/INDC%20Costa% 20Rica%20Version%202%200%20final%20ENG.pdf. (Accessed 20 December 2020).

Government of Dominica (2015) *Intended Nationally Determined Contribution (INDC) of the Commonwealth of Dominica.* www4.unfccc.int/sites/ndcsta ging/PublishedDocuments/Dominica%20First/Commonwealth%20of% 20Dominica-%20Intended%20Nationally%20Determined%20Contributions %20(INDC).pdf. (Accessed 20 December 2020).

Government of Dominica (2020) *Third National Communication to the United Nations Framework Convention on Climate Change of the Commonwealth of Dominica.* Environmental Coordinating Unit, Ministry of Environment, Rural Modernisation and Kalinago Upliftment, Commonwealth of Dominica (Roseau). unfccc.int/sites/default/files/resource/Dominica%20TNC%20-% 20Final%20%28March%202020%29.pdf. (Accessed 20 December 2020).

Government of Dominican Republic (2015) *Intended Nationally Determined Contribution.* www4.unfccc.int/sites/ndcstaging/PublishedDocuments/Dominican% 20Republic%20First/INDC-DR%20August%202015%20(unofficial%20translation) .pdf. (Accessed 20 December 2020).

Government of Egypt (2015) *Egyptian Intended Nationally Determined Contribution.* www4.unfccc.int/sites/ndcstaging/PublishedDocuments/Egypt %20First/Egyptian%20INDC.pdf. (Accessed 20 December 2020).

Government of Eswatini (2016) *Swaziland's INDC.* www4.unfccc.int/sites/ndcsta ging/PublishedDocuments/Eswatini%20First/Eswatini%27s%20INDC.pdf. (Accessed 20 December 2020).

Government of Ethiopia (2019) *Ethiopia's Climate Resilient Green Economy National Adaptation Plan.* Federal Democratic Republic of Ethiopia. www4 .unfccc.int/sites/NAPC/Documents/Parties/Final%20Ethiopia-national-adapta tion-plan%20%281%29.pdf. (Accessed 20 December 2020).

Government of Fiji (2018) *Republic of Fiji National Adaptation Plan: A Pathway towards Climate Resilience.* www4.unfccc.int/sites/NAPC/Documents/ Parties/National%20Adaptation%20Plan_Fiji.pdf. (Accessed 20 December 2020).

Government of Fiji (2020) *Third National Communication: Report to the United Nations Framework Convention on Climate Change.* Ministry of Economy, Republic of Fiji (Suva). unfccc.int/sites/default/files/resource/Fiji_TNC% 20Report.pdf. (Accessed 20 December 2020).

Government of Germany (2017). *Germany's Seventh National Communication on Climate Change: A Report under the UNFCCC.* Federal Ministry for the Environment, Nature Conservation, Building and Nuclear Safety (Berlin). unfccc.int/sites/default/files/resource/26795831_Germany-NC7-1-171220_7% 20NatCom%20to%20UNFCCC.pdf. (Accessed 20 December 2020).

Government of Ghana (2020) *Ghana's Fourth National Communication to the UNFCCC.* Environmental Protection Agency (EPA). unfccc.int/sites/default/ files/resource/Gh_NC4.pdf. (Accessed 20 December 2020).

Government of Grenada (2017a) *National Climate Change Adaptation Plan (NAP) for Grenada, Carriacou and Petite Martinique 2017–2021.* Ministry of Climate Resilience, the Environment, Forestry, Fisheries, Disaster Management and Information (St George's). www4.unfccc.int/sites/NAPC/Documents/Parties/ Grenada_National%20Adaptation%20Plan_%202017–2021.pdf. (Accessed 20 December 2020).

Government of Grenada (2017b) *Grenada, Carriacou, Petite Martinique: Second National Communication to the United Nations Second National Communication to the United Nations.* unfccc.int/sites/default/files/ resource/Grenada%20Second%20National%20Communication_Final%20% 281%29%20%281%29.pdf. (Accessed 20 December 2020).

Government of Guinea-Bissau (2015) *Intended Nationally Determined Contributions.* www4.unfccc.int/sites/ndcstaging/PublishedDocuments/ Guinea%20Bissau%20First/GUINEA-BISSAU_INDC_Version%20to%20the% 20UNFCCC%20(eng).pdf. (Accessed 20 December 2020).

Government of Guinea-Bissau (2018) *Third National Communication: Report to the UNFCCC.* Ministry of Environment and Sustainable Development, Republic of Guinea-Bissau (Bissau). unfccc.int/sites/default/files/resource/ TCN_Guinea_Bissau.pdf. (Accessed 20 December 2020).

Government of Ireland (2018) *Seventh National Communication Ireland.* Department of Communications, Climate Action and Environment, Republic of Ireland. unfccc.int/sites/default/files/resource/63014825_Ireland-NC7-BR3–1-Seventh%20National%20Communication%20Ireland.pdf. (Accessed 20 December 2020).

Government of Jamaica (2016) *Intended Nationally Determined Contribution of Jamaica.* www4.unfccc.int/sites/ndcstaging/PublishedDocuments/Jamaica%20First/Jamaica%27s%20INDC_2015–11–25.pdf. (Accessed 20 December 2020).

Government of Kenya (2016) *Kenya National Adaptation Plan 2015–2030: Enhanced Climate Resilience towards the Attainment of Vision 2030 and Beyond.* Republic of Kenya, Ministry of Environment and Natural Resources. www4.unfccc.int/sites/NAPC/Documents%20NAP/Kenya_NAP_Final.pdf. (Accessed 20 December 2020).

Government of Kiribati (2016) *Intended Nationally Determined Contribution.* www4.unfccc.int/sites/ndcstaging/PublishedDocuments/Kiribati%20First/INDC_KIRIBATI.pdf. (Accessed 20 December 2020).

Government of Kiribati (2019) *Kiribati Joint Implementation Plan for Climate Change and Disaster Risk Management.* www4.unfccc.int/sites/NAPC/Documents/Parties/Kiribati-Joint-Implementation-Plan-for-Climate-Change-and-Disaster-Risk-Management-2019–2028.pdf. (Accessed 20 December 2020).

Government of Lao PDR (2015) *Intended Nationally Determined Contribution.* Lao People's Democratic Republic. www4.unfccc.int/sites/ndcstaging/PublishedDocuments/Lao%20People%27s%20Democratic%20Republic%20First/Lao%20PDR%20First%20NDC.pdf. (Accessed 20 December 2020).

Government of Lesotho (2017) *Lesotho's Nationally Determined Contribution under the UNFCCC.* Lesotho Meteorological Services, Ministry of Energy and Meteorology. www4.unfccc.int/sites/ndcstaging/PublishedDocuments/Lesotho%20First/Lesotho%20First%20NDC.pdf. (Accessed 20 December 2020).

Government of Madagascar (2016) *Madagascar's Intended Nationally Determined Contribution.* www4.unfccc.int/sites/ndcstaging/PublishedDocuments/Madagascar%20First/Madagascar%20INDC%20Eng.pdf. (Accessed 20 December 2020).

Government of the Maldives (2015) *Maldives' Intended Nationally Determined Contribution.* www4.unfccc.int/sites/ndcstaging/PublishedDocuments/Maldives%20First/Maldives%20INDC.pdf. (Accessed 20 December 2020).

Government of the Marshall Islands (2015) *Intended Nationally Determined Contribution.* www4.unfccc.int/sites/ndcstaging/PublishedDocuments/Marshall%20Islands%20First/150721%20RMI%20INDC%20JULY%202015%20FINAL%20SUBMITTED.pdf. (Accessed 20 December 2020).

Government of the Marshall Islands (2018) *Nationally Determined Contribution, Including Annex 1 – Tile Til Eo: 2050 Climate Strategy 'Lighting the Way'.* www4.unfccc.int/sites/ndcstaging/PublishedDocuments/Marshall%20Islands %20Second/20181122%20Marshall%20Islands%20NDC%20to%20UNFCCC %2022%20November%202018%20FINAL.pdf. (Accessed 20 December 2020).

Government of Mexico (2016) *Intended Nationally Determined Contribution.* www4.unfccc.int/sites/ndcstaging/PublishedDocuments/Mexico%20First/ MEXICO%20INDC%2003.30.2015.pdf. (Accessed 20 December 2020).

Government of Moldova (2015) *Republic of Moldova's Intended National Determined Contribution.* www4.unfccc.int/sites/ndcstaging/PublishedDocuments/Republic%20 of%20Moldova%20First/INDC_Republic_of_Moldova_25.09.2015.pdf. (Accessed 20 December 2020).

Government of Moldova (2020) *Updated Nationally Determined Contribution of the Republic of Moldova.* www4.unfccc.int/sites/ndcstaging/PublishedDocuments/ Republic%20of%20Moldova%20First/MD_Updated_NDC_final_version_EN.pdf. (Accessed 20 December 2020).

Government of Morocco (2016) *Nationally Determined Contribution under the UNFCCC.* www4.unfccc.int/sites/ndcstaging/PublishedDocuments/Morocco% 20First/Morocco%20First%20NDC-English.pdf. (Accessed 20 December 2020).

Government of Myanmar (2015) *Myanmar's Intended Nationally Determined Contribution.* www4.unfccc.int/sites/ndcstaging/PublishedDocuments/Myanmar %20First/Myanmar%27s%20INDC.pdf. (Accessed 20 December 2020).

Government of Namibia (2020) *Fourth National Communication to the UNFCCC.* unfccc.int/sites/default/files/resource/Namibia%20-%20NC4%20-%20Final% 20signed.pdf. (Accessed 20 December 2020).

Government of Nepal (2016) *Nationally Determined Contributions, October 2016.* Government of Nepal, Ministry of Population and Environment (Kathmandu). www4.unfccc.int/sites/submissions/INDC/Published%20Documents/Nepal/ 1/Nepal_INDC_08Feb_2016.pdf. (Accessed 20 December 2020).

Government of Nigeria (2020) *Third National Communication (TNC) of the Federal Republic of Nigeria.* Federal Ministry of Environment (Abuja). unfccc .int/documents/226453. (Accessed 20 December 2020).

Government of Norway (2018) *Norway's Seventh National Communication Under the Framework Convention on Climate Change.* Norwegian Ministry of Climate and Environment. unfccc.int/sites/default/files/resource/321045_Norway-NC7-BR3–2- Norways_seventh_national_communication_2.pdf. (Accessed 20 December 2020).

Government of Palestine (2016) *National Adaptation Plan (NAP) to Climate Change.* Environment Quality Authority, State of Palestine (El-Bireh,

Ramallah).   www4.unfccc.int/sites/NAPC/Documents%20NAP/National%
20Reports/State%20of%20Palestine%20NAP.pdf.   (Accessed   20   December
2020).

Government of Paraguay (2017) *Plan Nacional de Adaptación al Cambio
Climático*. Ministerio de Salud Pública y Bienestar Social, en coordinación:
Secretaría del Ambiente, Secretaría de Emergencia Nacional, Ministerio de
Educación y Cultura, academia e instituciones de investigación, cooperativas,
gobernaciones y municipios (Asunción). www4.unfccc.int/sites/NAPC/
Documents/Parties/Plan%20Nacional%20de%20Adaptación%20al%
20Cambio%20Climático_Paraguay_final.pdf. (Accessed 20 December 2020).

Government of Rwanda (2015) *Intended Nationally Determined Contribution
(INDC) for the Republic of Rwanda*. www4.unfccc.int/sites/ndcstaging/
PublishedDocuments/Rwanda%20First/INDC_Rwanda_Nov.2015.pdf.
(Accessed 20 December 2020).

Government of Saint Lucia (2015) *Intended Nationally Determined Contribution
under the UNFCCC*. www4.unfccc.int/sites/ndcstaging/PublishedDocuments/
Saint%20Lucia%20First/Saint%20Lucia%27s%20INDC%2018th%
20November%202015.pdf. (Accessed 20 December 2020).

Government of Saint Lucia (2018) *Saint Lucia's National Adaptation Plan (NAP):
2018–2028*. Department of Sustainable Development, Ministry of Education,
Innovation, Gender Relations and Sustainable Development. www4.unfccc.int/
sites/NAPC/Documents/Parties/SLU-NAP-May-2018.pdf.   (Accessed   20
December 2020).

Government of Seychelles (2015) *Intended Nationally Determined Contribution
Under the UNFCCC*. www4.unfccc.int/sites/ndcstaging/PublishedDocuments/
Seychelles%20First/INDC%20of%20Seychelles.pdf. (Accessed 20 December
2020).

Government of Singapore (2020) *Singapore's Update of Its First Nationally
Determined Contribution (INDC) and Accompanying Information*. www4
.unfccc.int/sites/ndcstaging/PublishedDocuments/Singapore%20First/Singapore
%27s%20Update%20of%201st%20NDC.pdf. (Accessed 20 December 2020).

Government of the Solomon Islands (2016) *Intended Nationally Determined
Contribution*.   www4.unfccc.int/sites/ndcstaging/PublishedDocuments/Solomon%
20Islands%20First/SOLOMON%20ISLANDS%20INDC.pdf. (Accessed 20 December
2020).

Government of South Africa (2016) *Intended Nationally Determined
Contribution*. www4.unfccc.int/sites/ndcstaging/PublishedDocuments/South
%20Africa%20First/South%20Africa.pdf. (Accessed 20 December 2020).

Government of South Sudan (2018) *Initial National Communication to the UNFCCC*. Ministry of Environment and Forestry (Juba). unfccc.int/sites/default/files/resource/South%20Sudan%20INC.pdf. (Accessed 20 December 2020).

Government of Suriname (2015). *Second National Communication to the UNFCCC*. Office of the President of the Republic of Suriname (Paramaribo). https://unfccc.int/sites/default/files/resource/Surnc2rev.pdf. (Accessed 20 December 2020).

Government of Suriname (2020). *Suriname National Adaptation Plan (NAP) 2019–2029*. www4.unfccc.int/sites/NAPC/Documents/Parties/Suriname%20Final%20NAP_apr%202020.pdf. (Accessed 20 December 2020).

Government of Tonga (2019) *Third National Communication on Climate Change Report*. Ministry of Meteorology, Energy, Information, Disaster Management, Environment, Climate Change and Communications, Kingdom of Tonga (Nuku'alofa). unfccc.int/sites/default/files/resource/Final%20TNC%20Report_December%202019.pdf. (Accessed 20 December 2020).

Government of Trinidad and Tobago (2018) *Intended Nationally Determined Contribution (iNDC) under the UNFCCC*. www4.unfccc.int/sites/ndcstaging/PublishedDocuments/Trinidad%20and%20Tobago%20First/Trinidad%20and%20Tobago%20Final%20INDC.pdf. (Accessed 20 December 2020).

Government of Turkey (2018) *Seventh National Communication of Turkey under the UNFCCC*. unfccc.int/sites/default/files/resource/14936285_Turkey-NC7–2-Seventh%20National%20Communication%20of%20Turkey.pdf. (Accessed 20 December 2020).

Government of UAE (2015) *Intended Nationally Determined Contribution of the United Arab Emirates*. www4.unfccc.int/sites/ndcstaging/PublishedDocuments/United%20Arab%20Emirates%20First/UAE%20INDC%20-%2022%20October.pdf. (Accessed 20 December 2020).

Government of United Kingdom (2017) *7th National Communication*. Department for Business, Energy and Industrial Strategy. unfccc.int/sites/default/files/resource/19603845_United%20Kingdom-NC7-BR3-1-gbr%20NC7%20and%20BR3%20with%20Annexes%20%281%29.pdf. (Accessed 20 December 2020).

Government of Uruguay (2017) *First Nationally Determined Contribution to the Paris Agreement*. Oriental Republic of Uruguay. www4.unfccc.int/sites/ndcstaging/PublishedDocuments/Uruguay%20First/Uruguay_First%20Nationally%20Determined%20Contribution.pdf. (Accessed 20 December 2020).

Government of Vietnam (2016) *Intended Nationally Determined Contribution of Viet Nam*. www4.unfccc.int/sites/ndcstaging/PublishedDocuments/Viet%20Nam%20First/%27S%20INDC.pdf. (Accessed 20 December 2020).

Government of Vietnam (2020) *Updated Nationally Determined Contribution (NDC)*. www4.unfccc.int/sites/ndcstaging/PublishedDocuments/Viet%20Nam%20First/Viet%20Nam_NDC_2020_Eng.pdf. (Accessed 20 December 2020).

Guterres, A. (2020) *Secretary-General's Address at Columbia University: 'The State of the Planet'*. www.un.org/sg/en/content/sg/speeches/2020–12–02/address-columbia-university-the-state-of-the-planet. (Accessed 20 December 2020).

Harvey, F. (2020a) UN chief: don't use taxpayer money to save polluting industries. *The Guardian*, 28 April 2020. www.theguardian.com/environment/2020/apr/28/un-chief-dont-use-taxpayer-money-to-save-polluting-industries. (Accessed 20 December 2020).

Harvey, F. (2020b) Green stimulus can repair global economy and climate, study says. *The Guardian*, 5 May 2020. www.theguardian.com/environment/2020/may/05/green-stimulus-can-repair-global-economy-and-climate-study-says. (Accessed 20 December 2020).

Harvey, F. and Rankin, J. (2020) What is the European Green Deal and will it really cost €1tn? *The Guardian*, 9 March 2020. www.theguardian.com/world/2020/mar/09/what-is-the-european-green-deal-and-will-it-really-cost-1tn. (Accessed 20 December 2020).

Heifetz, R.A. (1994) *Leadership without Easy Answers*. Belknap Press of Harvard University Press (Cambridge, MA).

Hepburn, C., O'Callaghan, B., Stern, N., Stiglitz, J. and Zenghelis, D. (2020) Will COVID-19 and the fiscal recovery packages accelerate or retard progress on climate change? Smith School of Enterprise and the Environment, Working Paper No. 20–02. *Oxford Review of Economic Policy*, 36 (S1).

Hill, R.A. and Dunbar, R.I.M. (2003) Social network size in humans. *Human Nature*, 14 (1): 53–72.

Hinkel, J., Bharwani, S., Bisaro, A., Carter, T. Cull, T., Davis, M., Klein, R., Lonsdale, K., Rosentrater, L. and Vincent, K. (2013) *PROVIA Guidance on Assessing Vulnerability, Impacts and Adaptation to Climate Change: Consultation Document*. Global Programme of Research on Climate Change Vulnerability, Impacts and Adaptation (PROVIA, Nairobi).

Hobsbawm, E.J. (1962a) *The Age of Revolution 1789–1848*. Weidenfeld and Nicolson (London).

Hobsbawm, E.J. (1962b) *The Age of Capital 1848–1875*. Weidenfeld and Nicolson (London).

Hobsbawm, E.J. (1987) *The Age of Empire 1875–1914*. Weidenfeld and Nicolson (London).

Hobsbawm, E.J. (1994) *Age of Extremes: The Short Twentieth Century 1914–1991*. Michael Joseph (London).

Hoffmann, S., Irl, S.D.H. and Beierkuhnlein, C. (2019) Predicted climate shifts within terrestrial protected areas worldwide. *Nature Communications*, 10, Article 4787. https://doi.org/10.1038/s41467-019-12603-w. (Accessed 20 December 2020).

Holdaway, E., Dodwell, C., Sura, K. and Picot, H. (2015) *A Guide to INDCs Intended Nationally Determined Contributions*, 2nd edition (May) CDKN and Ricardo-AEA. https://cdkn.org/wp-content/uploads/2015/04/CDKN-Guide-to-INDCs-Revised-May2015.pdf. (Accessed 20 December 2020).

Holland, T. (2008) *Millennium: The End of the World and the Forging of Christendom*. Little, Brown (London).

Horton, B. (2020) *What's the Arctic Death Spiral?* www.arcticdeathspiral.org. (Accessed 20 December 2020).

Horton, R. (2020) Offline: COVID-19 Is Not a Pandemic. *Lancet*, 396: 874, 26 September. https://doi.org/10.1016/S0140-6736(20)32000-6. (Accessed 20 December 2020).

Huxtable, S.-A., Fowler, C., Kefalas, C. and Slocombe, E. (eds) (2020) *Interim Report on the Connections between Colonialism and Properties Now in the Care of the National Trust, Including Links with Historic Slavery*. National Trust (Swindon). www.nationaltrust.org.uk/features/addressing-the-histories-of-slavery-and-colonialism-at-the-national-trust. (Accessed 20 December 2020).

IEA (2020) *Global Energy Review 2020: The Impacts of the Covid-19 Crisis on Global Energy Demand and CO_2 Emissions*. Revised edition July 2020. International Energy Agency (Paris). www.iea.org/reports/global-energy-review-2020/global-energy-and-co2-emissions-in-2020. (Accessed 20 December 2020).

IIED (2012) *TAMD: A Framework for Assessing Climate Adaptation and Development Effects, Briefing Paper*. International Institute for Environment and Development (London).

Iizumi, T., Hirata, R. and Matsuda, R. (eds) (2019) *Adaptation to Climate Change in Agriculture: Research and Practices*. Springer International (Singapore).

Indrawan, M., Luzar, J., Hanna, H. and Mayer, T. (eds) (2020) *Civic Engagement in Asia: Lessons from Transformative Learning in the Quest for a Sustainable Future*. Yayasan Pustaka Obor Indonesia (Jakarta).

Interfaith Statement on Climate Change (2015) Statement by religious, spiritual and faith-based leaders for the first meeting of the parties to the Paris Agreement during the twenty-second session of the Conference of the Parties, signed by 304 eminent faith leaders from 58 countries with representatives from Buddhist, Christian, Hindu, Jains, Quakers, Muslim, Sikh, Unitarian Universalists, as well as Indigenous and Spiritual leaders. https://unfccc.int/news/religious-leaders-call-for-effective-paris-agreement. (Accessed 20 December 2020).

IPBES (2019) *Global Assessment Report on Biodiversity and Ecosystem Services: Summary for Policymakers.* Intergovernmental Science-Policy Platform on Biodiversity and Ecosystem Services Secretariat (Bonn). https://ipbes.net/system/tdf/ipbes_global_assessment_report_summary_for_policymakers.pdf?file=1andtype=nodeandid=35329. (Accessed 20 December 2020).

IPCC (1992) *Climate Change: The 1990 and 1992 IPCC Assessments (IPCC First Assessment Report Overview and Policymaker Summaries and 1992 IPCC Supplement).* Intergovernmental Panel on Climate Change and World Meteorological Organization (Geneva), June 1992. www.ipcc.ch/report/climate-change-the-ipcc-1990-and-1992-assessments/. (Accessed 20 December 2020).

IPCC (1995) *IPCC Second Assessment: Climate Change 1995, a Report of the Intergovernmental Panel on Climate Change.* United Nations Environment Programme (Nairobi) and World Meteorological Organization (Geneva). www.ipcc.ch/site/assets/uploads/2018/05/2nd-assessment-en-1.pdf. (Accessed 20 December 2020).

IPCC (2001) *Climate Change 2001 Synthesis Report: Contribution of Working Groups I, II and III to the Third Assessment Report of the Intergovernmental Panel on Climate Change.* Cambridge University Press (Cambridge). www.ipcc.ch/report/ar3/syr/. (Accessed 20 December 2020).

IPCC (2007) *Climate Change 2007 Mitigation: Contribution of Working Group III to the Fourth Assessment Report of the Intergovernmental Panel on Climate Change.* Cambridge University Press (Cambridge). www.ipcc.ch/site/assets/uploads/2018/03/ar4_wg3_full_report-1.pdf. (Accessed 20 December 2020).

IPCC (2014) *Climate Change 2014: Synthesis Report. Contribution of Working Groups I, II and III to the Fifth Assessment Report of the Intergovernmental Panel on Climate Change* (core writing team, R.K. Pachauri and L.A. Meyer (eds)). IPCC (Geneva). www.ipcc.ch/site/assets/uploads/2018/02/SYR_AR5_FINAL_full.pdf. (Accessed 20 December 2020).

IPCC (2018) *Global Warming of 1.5°C: An IPCC Special Report on the Impacts of Global Warming of 1.5°C above Pre-industrial Levels and Related Global Greenhouse Gas Emission pathways, in the Context of Strengthening the Global Response to the Threat of Climate Change, Sustainable Development, and Efforts to Eradicate Poverty – Summary for Policymakers.* Intergovernmental Panel on Climate Change and World Meteorological Organization (Geneva). www.ipcc.ch/sr15/chapter/spm/. (Accessed 20 December 2020).

IPCC (2019a) *Special Report on the Ocean and Cryosphere in a Changing Climate: Summary for Policymakers*. Intergovernmental Panel on Climate Change and World Meteorological Organization (Geneva). www.ipcc.ch/srocc/chapter/summary-for-policymakers/. (Accessed 20 December 2020).

IPCC (2019b) *Climate Change and Land: An IPCC Special Report on Climate Change, Desertification, Land Degradation, Sustainable Land Management, Food Security, and Greenhouse Gas Fluxes in Terrestrial Ecosystems – Summary for Policymakers*. Intergovernmental Panel on Climate Change and World Meteorological Organization (Geneva). www.ipcc.ch/srccl/chapter/summary-for-policymakers/. (Accessed 20 December 2020).

Islamic Declaration on Global Climate Change (2015) Drafted by Fazlun Khalid, Fachruddin Mangunjaya, Othman Llewellyn, Azizan Baharuddin, Ibrahim Ozdemir and Abdelmajid Tribak; launched at the International Islamic Climate Change Symposium (Istanbul). www.ifees.org.uk/wp-content/uploads/2020/01/climate_declarationmmwb.pdf. (Accessed 20 December 2020).

Jarvie, J., Sutarto, R., Syam, D. and Jeffery, P. (2015) Lessons for Africa from urban climate change resilience building in Indonesia. *Current Opinion in Environmental Sustainability*, 13: 19–24. http://dx.doi.org/10.1016/j.cosust.2014.12.006. (Accessed 20 December 2020).

Jha, P. (2014) *Battles of the New Republic: A Contemporary History of Nepal*. Hurst (London).

Joosten, H. (2019) Permafrost peatlands: losing ground in a warming world, pages 28–51 in *Frontiers 2018/19: Emerging Issues of Environmental Concern*. United Nations Environment Programme (Nairobi).

Kabisch, N., Korn, H., Stadler, J. and Bonn, A. (eds) (2017) *Nature-Based Solutions to Climate Change Adaptation in Urban Areas: Linkages between Science, Policy and Practice*. Springer International (Berlin).

Kamphof, R. (2018a) *The European Union: Patterns of Collaboration with International Environmental and Climate Institutions*. UNU-CRIS Working Paper, W-2018/5. United Nations University Institute on Comparative Regional Integration Studies (Bruges). http://cris.unu.edu/european-union-patterns-collaboration-international-environmental-and-climate-institutions. (Accessed 20 December 2020).

Kamphof, R. (2018b) *EU and Member State Implementation of the UN Agenda 2030 and Sustainable Development Goals*. UNU-CRIS Working Paper, W-2018/1. United Nations University Institute on Comparative Regional Integration Studies (Bruges). http://cris.unu.edu/eu-and-member-state-implementation-un-agenda-2030-and-sustainable-development-goals. (Accessed 20 December 2020).

Kauffman, S. (1995) *At Home in the Universe: The Search for Laws of Complexity*. Oxford University Press (New York).

Keeling, C.D. (1986) *Atmospheric CO₂ Concentrations: Mauna Loa Observatory, Hawaii 1958–1986*. Environmental Sciences Division, Oak Ridge National Laboratory (Oak Ridge, TN). www.osti.gov/servlets/purl/537311. (Accessed 20 December 2020).

Kelman, I., Mercer, J. and Gaillard, J.C. (eds) (2017) *The Routledge Handbook of Disaster Risk Reduction Including Climate Change Adaptation*. Routledge (Abingdon and New York).

Khan, M., Robinson, S.-A., Weikmans, R., Ciplet, D. and Roberts, J.T. (2019) Twenty-five years of adaptation finance through a climate justice lens. *Climatic Change*. https://doi.org/10.1007/s10584-019-02563-x. (Accessed 20 December 2020).

Killian, B. (2008a) Do Elections Matter in Zanzibar? *Journal of African Elections*, 8 (2): 74–87.

Killian, B. (2008b) The state and identity politics in Zanzibar: challenges to democratic consolidation in Tanzania. *African Identities*, 6 (2): 99–125. https://doi.org/10.1080/14725840801933932. (Accessed 20 December 2020).

Killian, B. (2014) A proposed structure of the union in Tanzania: political parties at a crossroad. *African Review*, 41 (1): 116–138.

King, C. (2019) *The Reinvention of Humanity: A Story of Race, Sex, Gender and the Discovery of Culture*. The Bodley Head (London).

Klepp, S. and Chavez-Rodriguez, L. (eds) (2018) *A Critical Approach to Climate Change Adaptation: Discourses, Policies and Practices*. Routledge (Abingdon and New York).

Knight, T. (2020) *FFI Launches Global Campaign to Secure $500 Billion for Nature Conservation*. www.fauna-flora.org/news/ffi-launches-global-campaign-secure-500-billion-nature-conservation. (Accessed 20 December 2020).

Landi, S.J. (2020) *Two Lessons and a Common Way for International Recovery*. www.greendkinsea.com/post/two-lessons-and-a-common-way-for-international-recovery. (Accessed 20 December 2020).

Leaders' Pledge for Nature (2020) *Leaders' Pledge for Nature: United to Reverse Biodiversity Loss by 2030 for Sustainable Development*. www.leaderspledgefornature.org. (Accessed 20 December 2020).

Leagnavar, P., Bours, D. and McGinn, C. (2015) *Good Practice Study on Principles for Indicator Development, Selection, and Use in Climate Change Adaptation Monitoring and Evaluation*. Climate-Eval Community of Practice. www.eartheval.org/sites/ceval/files/studies/Good-Practice-Study.pdf. (Accessed 20 December 2020).

Leal Filho, W. (ed.) (2017) *Climate Change Adaptation in Pacific Countries: Fostering Resilience and Improving the Quality of Life*. Springer International (Cham, Switzerland).

Leal Filho, W. (ed.) (2018) *Climate Change Impacts and Adaptation Strategies for Coastal Communities*. Springer International (Cham, Switzerland).

Leal Filho, W., Adamson, K. Dunk, R.M., Azeiteiro, U.M., Illingworth, S. and Alves, F. (eds) (2016) *Implementing Climate Change Adaptation in Cities and Communities: Integrating Strategies and Educational Approaches*. Springer International (Cham, Switzerland).

Leal Filho, W., Azeiteiro, U.M. and Alves, F. (eds) (2016) *Climate Change and Health: Improving Resilience and Reducing Risks*. Springer International (Cham, Switzerland).

Leal Filho, W. and de Freitas, L.E. (eds) (2018) *Climate Change Adaptation in Latin America: Managing Vulnerability, Fostering Resilience*. Springer International (Cham, Switzerland).

Leal Filho, W. and Keenan, J.M. (eds) (2017) *Climate Change Adaptation in North America: Fostering Resilience and the Regional Capacity to Adapt*. Springer International (Cham, Switzerland).

Leal Filho, W., Manolas, E. Azul, A.M., Azeiteiro, U.M. and McGhie, H. (eds) (2018a) *Handbook of Climate Change Communication: Vol. 2 – Theory of Climate Change Communication*. Springer International (Cham, Switzerland).

Leal Filho, W., Manolas, E. Azul, A.M., Azeiteiro, U.M. and McGhie, H. (eds) (2018b) *Handbook of Climate Change Communication: Vol. 3 – Case Studies in Climate Change Communication*. Springer International (Cham, Switzerland).

Leal Filho, W. and Nalau, J. (eds) (2018) *Limits to Climate Change Adaptation*. Springer International (Cham, Switzerland).

Leal Filho, W., Oguge, N.O., Ayal, D., Adeleke, L. and da Silva, I. (eds) (2021) *African Handbook of Climate Change Adaptation*. Springer International (Cham, Switzerland).

Leal Filho, W., Simane, B., Kalangu, J., Wuta, M., Munishi, P. and Musiyiwa, K. (eds) (2017) *Climate Change Adaptation in Africa: Fostering Resilience and Capacity to Adapt*. Springer International (Cham, Switzerland).

Le Quéré, C. Moriarty, R., Andrew, R.M., Canadell, J.G., Sitch, S., Korsbakken, J.I., Friedlingstein, P., Peters, G.P., Andres, R.J., Boden, T.A., Houghton, R.A., House, J.I., Keeling, R.F., Tans, P., Arneth, A., Bakker, D.C.E., Barbero, L., Bopp, L., Chang, J., Chevallier, F., Chini, L.P., Ciais, P., Fader, M., Feely, R.A., Gkritzalis, T., Harris, I., Hauck, J., Ilyina, T., Jain, A.K., Kato, E., Kitidis, V., Klein Goldewijk, K., Koven, C., Landschützer, P., Lauvset, S.K., Lefèvre, N., Lenton, A., Lima, I.D., Metzl, N., Millero, F., Munro, D.R., Murata, A., Nabel,

J.E.M.S, Nakaoka, S., Nojiri, Y., O'Brien, K., Olsen, A., Ono, T., Pérez, F.F., Pfeil, B., Pierrot, D., Poulter, B., Rehder, G., Rödenbeck, C., Saito, S., Schuster, U., Schwinger, J., Séférian, R., Steinhoff, T., Stocker, B.D., Sutton, A.J., Takahashi, T., Tilbrook, B., van der Laan-Luijkx, I.T., van der Werf, G.R., van Heuven, S., Vandemark, D., Viovy, N., Wiltshire, A., Zaehle, S. and Zeng, N. (2015) Global carbon budget 2015. *Earth System Science Data*, 7: 349–396. https://doi.org/10.5194/essd-7-349-2015. (Accessed 20 December 2020).

Le Quéré, C., Andrew, R.M., Friedlingstein, P., Sitch, S., Hauck, J., Pongratz, J., Pickers, P.A., Korsbakken, J.I., Peters, G.P., Canadell, J.G., Arneth, A., Arora, V.K., Barbero, L., Bastos, A., Bopp, L., Chevallier, F., Chini, L.P., Ciais, P., Doney, S.C., Gkritzalis, T., Goll, D.S., Harris, I., Haverd, V., Hoffman, F.M., Hoppema, M., Houghton, R.A., Hurtt, G., Ilyina, T., Jain, A.K., Johannessen, T., Jones, C.D., Kato, E., Keeling, R.F., Goldewijk, K.K., Landschützer, P., Lefèvre, N., Lienert, S., Liu, Z., Lombardozzi, D., Metzl, N., Munro, D.R., Nabel, J.E.M.S., Nakaoka, S., Neill, C., Olsen, A., Ono, T., Patra, P., Peregon, A., Peters, W., Peylin, P., Pfeil, B., Pierrot, D., Poulter, B., Rehder, G., Resplandy, L., Robertson, E., Rocher, M., Rödenbeck, C., Schuster, U., Schwinger, J., Séférian, R., Skjelvan, I., Steinhoff, T., Sutton, A., Tans, P.P., Tian, H., Tilbrook, B., Tubiello, F.N., van der Laan-Luijkx, I.T., van der Werf, G.R., Viovy, N., Walker, A.P., Wiltshire, A.J., Wright, R., Zaehle, S. and Zheng, B. (2018) Global carbon budget 2018 *Earth System Science Data*, 10: 2141–2194. https://doi.org/10.5194/essd-10-2141-2018. (Accessed 20 December 2020).

Letslink UK (2020) *Theory of Complementary/Community Currencies*. UK Local Exchange Trading and Complementary Currencies Development Agency (London). www.letslinkuk.net/home/theory.htm. (Accessed 20 December 2020).

Liddell Hart, B.H. (1954) *Strategy, Revised Second Edition, 1967*. Faber & Faber (London).

Lipper, L., McCarthy, N. and Zilberman, D. (eds) (2017) *Climate Smart Agriculture: Building Resilience to Climate Change*. Springer International (Cham, Switzerland).

Lodge, A. (2014) Has air pollution made Kathmandu unliveable? *The Guardian*, 21 March 2014. www.theguardian.com/cities/2014/mar/21/air-pollution-kathmandu-nepal-liveable-smog-paris. (Accessed 20 December 2020).

Lovejoy, T.E. and Nobre, C. (2018) Amazon tipping point. *Science Advances*, 4 (2). https://doi.org/10.1126/sciadv.aat2340. (Accessed 20 December 2020).

Luintel, H., Bluffstone, R.A. and Scheller, R.M. (2018) The effects of the Nepal community forestry program on biodiversity conservation and carbon storage.

PLoS One, 13 (6): e0199526. https://journals.plos.org/plosone/article?id=10 .1371/journal.pone.0199526. (Accessed 20 December 2020).

Luttwak, E.N. (1976) *The Grand Strategy of the Roman Empire from the First Century A.D. to the Third*. Johns Hopkins University Press (Baltimore).

Lutz, E. and Caldecott, J.O. (eds) (1996) *Decentralization and Biodiversity Conservation*. The World Bank (Washington, DC).

Lynas, M. (2007) *Six Degrees*. Fourth Estate (London).

MA (2005a) *Living beyond Our Means: Natural Assets and Human Well-being. Millennium Ecosystem Assessment*. Island Press (Washington, DC).

MA (2005b) *Ecosystems and Human Well-Being: Current State and Trends. Millennium Ecosystem Assessment. Findings of the Condition and Trends Working Group*. Island Press (Washington, DC).

Magnusson, M. (1960) Introduction, pages 9–31 in *Njal's Saga* (translated by Magnus Magnusson and Herman Pálsson). Penguin (London).

Mallory, I.A. (1990) Conduct unbecoming: the collapse of the International Tin Agreement. *American University International Law Review*, 5 (3): 835–892.

MANRZ (2019) *Enhancing Climate Change Resilience of Coastal Communities of Zanzibar*. Proposal to the Adaptation Fund. Ministry of Agriculture, Natural Resources, Livestock and Fisheries, Revolutionary Government of Zanzibar (Zanzibar).

Marlier, M.E., DeFries, R.S., Kim, P.S., Koplitz, S.N., Jacob, D.J., Mickley, L.J. and Myers, S.S. (2015) Fire emissions and regional air quality impacts from fires in oil palm, timber, and logging concessions in Indonesia. *Environmental Research Letters*, 10. https://doi.org/10.1088/1748-9326/10/8/085005. (Accessed 20 December 2010).

Marselle, M.R., Stadler, J., Korn, H., Irvine, K.N. and Bonn, A. (eds) (2019) *Biodiversity and Health in the Face of Climate Change*. Springer International (Cham, Switzerland).

Matthews, H.D., Gillett, N.P., Stott, P.A. and Zickfeld, K. (2009) The proportionality of global warming to cumulative carbon emissions. *Nature*, 459: 829–832. www.nature.com/articles/nature08047. (Accessed 20 December 2020).

MCCN (2017) *Strategic Plan 2018–2020*. Mwambao Coastal Community Network (Stonetown, Zanzibar).

MCCN (2019) *SWIOfish projects with Department of Fisheries*. www.mwambao.or .tz/2503–2/. (Accessed 20 December 2020).

MCCN (2020) *The Sea Belongs to Everyone, It Is Our Responsibility to Look after It*. Mwambao Coastal Community Network (Stonetown, Zanzibar). www .mwambao.or.tz. (Accessed 20 December 2020).

McCormick, M. (2017) 'They lied': Bolivia's untouchable Amazon lands at risk once more. *The Guardian*, 11 September 2011. www.theguardian.com/environ

ment/2017/sep/11/they-lied-bolivia-untouchable-amazon-lands-tipnis-at-risk-once-more. (Accessed 20 December 2020).

McCurry, J. (2020) South Korea vows to go carbon neutral by 2050 to fight climate emergency. *The Guardian*, 28 October 2020. www.theguardian.com/world/2020/oct/28/south-korea-vows-to-go-carbon-neutral-by-2050-to-fight-climate-emergency. (Accessed 20 December 2020).

Meadows, D.H. (2008) *Thinking in Systems: A Primer*. Chelsea Green (White River Junction, VT).

Meyfroidt, P. and Lambin, E.F. (2011) Global forest transition: prospects for an end to deforestation. *Annual Review of Environment and Resources*, 36 (1): 343–371. https://doi.org/10.1146/annurev-environ-090710-143732. (Accessed 20 December 2020).

Miyaguchi, T. and Uitto, J.I. (2015) *A Realist Review of Climate Change Adaptation Programme Evaluations: Methodological Implications and Programmatic Findings*. Independent Evaluation Office, United Nations Development Programme (New York). http://web.undp.org/evaluation/documents/articles-papers/occasional_papers/Occasional_Paper_Climate_Change_Uitto_Miyaguchi.pdf. (Accessed 20 December 2020).

MoFSC (2014) *People and Forests: An SMF-Based Emission Reduction Program in Nepal's Terai Arc Landscape*. Emission Reduction Programme Idea Note (ER-PIN) prepared for the World Bank Forest Carbon Partnership Facility. REDD Forestry and Climate Change Cell, Ministry of Forests and Soil Conservation of Nepal (Singh Durbar, Kathmandu).

MoFSC (2017) *Forest Investment Program: Investment Plan for Nepal*. Ministry of Forests and Soil Conservation (Singh Durbar, Kathmandu).

Moorhead, B. (2019) Talanoa: dialogue for action. *The Interfaith Observer*, 15 February 2019. www.theinterfaithobserver.org/journal-articles/2018/12/13/talanoa-dialogue-for-action-mll5r. (Accessed 20 December 2020).

Mori, A.S. (2020) Next-generation meetings must be diverse and inclusive. *Nature Climate Change*, 10: 481. https://doi.org/10.1038/s41558-020-0795-z. (Accessed 20 December 2020).

Moser, S.C. and Boykoff, M.T. (eds) (2013a) *Successful Adaptation to Climate Change: Linking Science and Policy in a Rapidly Changing World*. Routledge (Abingdon and New York).

Moser, S.C. and Boykoff, M.T. (2013b) Preface, pages xxi–xxiv in *Successful Adaptation to Climate Change: Linking Science and Policy in a Rapidly Changing World* (edited by S.C. Moser and M.T. Boykoff). Routledge (Abingdon and New York).

Moser, S.C. and Boykoff, M.T. (2013c) Climate change and adaptation success: the scope of the challenge, pages 1–33 in *Successful Adaptation to Climate Change:*

Linking Science and Policy in a Rapidly Changing World (edited by S.C. Moser and M.T. Boykoff). Routledge (Abingdon and New York).

Moulton, P. and Cohen, M. (1998) *Promoting Electric Vehicles in the Developing World*. Global Resources Institute (Eugene, OR). www.grilink.org/ev.htm. (Accessed 20 December 2020).

Nakashima, D.J., Galloway McLean, K., Thulstrup, H.D., Ramos Castillo, A. and Rubis, J.T. (2012) *Weathering Uncertainty: Traditional Knowledge for Climate Change Assessment and Adaptation*. UNESCO (Paris) and United Nations University (Darwin, Australia). https://unesdoc.unesco.org/ark:/48223/pf0000216613. (Accessed 20 December 2020).

Nakashima, D., Krupnik, I. and Rubis, J.T. (eds) (2018) *Indigenous Knowledge for Climate Change Assessment and Adaptation*. UNESCO and Cambridge University Press (Cambridge).

Nepstad, D., Schwartzman, S., Bamberger, B., Santilli, M., Ray, D., Schlesinger, P., Lefebvre, P., Alencar, A., Prinz, E., Fiske, G. and Rolla, A. (2006) Inhibition of Amazon deforestation and fire by parks and Indigenous lands. *Conservation Biology*, 20 (1): 65–73.

NGFS (2019) *A Call for Action: Climate Change as a Source of Financial Risk*, April. Network for Greening the Financial System (Paris). www.ngfs.net/sites/default/files/medias/documents/ngfs_first_comprehensive_report_-_17042019_0.pdf. (Accessed 20 December 2020).

NGFS (2020) *Guide for Supervisors Integrating Climate-Related and Environmental Risks into Prudential Supervision*, May. Network for Greening the Financial System (Paris). www.ngfs.net/sites/default/files/medias/documents/ngfs_guide_for_supervisors.pdf. (Accessed 20 December 2020).

Nightingale, A.J. (2020) *The Power of Climate Change: Knowledge Politics, Adaptation Follies and Ontological Frictions*. Edinburgh Environment and Development Network (Edinburgh). www.ed.ac.uk/files/atoms/files/andrea_nightingale_the_power_of_climate_change_eedn_jan_2020.pdf. (Accessed 20 December 2020).

Nightingale, A.J., Eriksen, S., Taylor, M., Forsyth, T., Pelling, M., Newsham, A., Boyd, E., Brown, K., Harvey, B., Jones, L., Kerr, R.B., Mehta, L., Naess, L.O., Ockwell, D., Scoones, I., Tanner, T. and Whitfield, S. (2020) Beyond technical fixes: climate solutions and the great derangement. *Climate and Development*, 12 (4): 343–352. https://doi.org/10.1080/17565529.2019.1624495. (Accessed 20 December 2020).

OECD (2006) *Adaptation to Climate Change: Key Terms*. Organisation for Economic Co-operation and Development (Paris).

OECD (2011) *Handbook on the OECD-DAC Climate Markers*. Organisation for Economic Co-operation and Development (Paris).

OECD (2015) *Climate finance in 2013–14 and the USD 100 billion goal.* Organisation for Economic Co-operation and Development and Climate Policy Initiative (Paris). www.oecd.org/environment/cc/OECD-CPI-Climate-Finance-Report.htm. (Accessed 20 December 2020).

OECD (2018) *Aid Activities Targeting Global Environmental Objectives, 2018.* https://stats.oecd.org/Index.aspx?DataSetCode=RIOMARKERS. (Accessed 20 December 2020).

OECD (2019a) *Aligning Development Co-operation and Climate Action: The Only Way Forward.* The Development Dimension, OECD Publishing (Paris).

OECD (2019b) *Climate Finance Provided and Mobilised by Developed Countries in 2013–17.* OECD Publishing (Paris).

OECD/DAC (OECD Development Assistance Committee) (2019) *Development Aid at a Glance Statistics by Region, 4. Asia, 2019 edition.* www.oecd.org/dac/financing-sustainable-development/development-finance-data/Asia-Development-Aid-at-a-Glance-2019.pdf. (Accessed 20 December 2020).

OECD/DAC (undated) *OECD DAC Rio Markers for Climate: Handbook – Indicative Table to Guide Rio Marking by Sector/Sub-Sector.* www.oecd.org/dac/environment-development/Revised%20climate%20marker%20handbook_FINAL.pdf. (Accessed 20 December 2020).

Oldekop, J.A., Sims, K.R.E., Whittingham, M.J. and Arun Agrawal, A. (2018) An upside to globalization: international outmigration drives reforestation in Nepal. *Global Environmental Change*, 52: 66–74. https://doi.org/10.1016/j.gloenvcha.2018.06.004. (Accessed 20 December 2020).

Olivier, J., Leiter, T. and Linke, J. (2013) *Adaptation Made to Measure: A Guidebook to the Design and Results-Based Monitoring of Climate Change Adaptation Projects*, 2nd edition. Deutsche Gesellschaft für Internationale Zusammenarbeit (Eschborn, Germany).

Ostler, N. (2005) *Empires of the Word: A Language History of the World.* HarperCollins (London).

Our World in Data (2020a) *Atmospheric CO_2 Concentration: Global Average Long-Term Atmospheric Concentration of Carbon Dioxide (CO_2), Measured in Parts Per Million (ppm).* https://ourworldindata.org/grapher/co2-concentration-long-term. (Accessed 20 December 2020).

Our World in Data (2020b) *Global Greenhouse Gas Emissions and Warming Scenarios.* https://ourworldindata.org/co2-and-other-greenhouse-gas-emissions. (Accessed 20 December 2020).

Overdevest, C. and Zeitlin, J. (2011) Assembling an experimentalist regime: EU FLEGT and transnational governance interactions in the forest sector. *GR:EEN Working Paper Series*, 2: 1–38. United Nations University Institute on

Comparative Regional Integration Studies (Bruges, Belgium). http://cris.unu
.edu/assembling-experimentalist-regime-eu-flegt-and-transnational-govern
ance-interactions-forest-sector. (Accessed 20 December 2020).

Overdevest, C. and Zeitlin, J. (2014) Assembling an experimentalist regime: transnational governance interactions in the forest sector. *Regulation &*
Governance, 8 (1). https://doi.org/10.1111/j.1748-5991.2012.01133.x (Accessed
20 December 2020).

Oxfam (2020) *Climate Finance Shadow Report 2020: Assessing Progress towards*
the $100 Billion Commitment. T. Carty, J. Kowalzig and B. Zagema, Oxfam GB
(Cowley) for Oxfam International. https://oxfamibis.dk/sites/default/files/
media/climate_finance_shadow_report_-_english_-_embargoed_20_october_
2020.pdf. (Accessed 20 December 2020).

Palmer, M. and Finlay, V. (2003) *Faith in Conservation: New Approaches to*
Religions and the Environment. The World Bank (Washington, DC).

Parellada, A., Betancur, J., A.C., Aragón, M.A., Zurita, I.É. and Roca, C. (2010). *The*
Rights of Indigenous Peoples: Cooperation between Denmark and Bolivia
(2005–2009). The Danish Royal Embassy to Bolivia (La Paz) and the
International Work Group for Indigenous Affairs, IWGIA (Copenhagen). www
.iwgia.org/images/publications//0462_EB-DANIDA-BOLIVIA-ENGELSK.pdf.
(Accessed 20 December 2020).

Particip (2015) *Thematic Evaluation of the EU Support to Environment and*
Climate Change in Third Countries (2007–2013), Volume 1 – Main Report.
European Commission (Brussels).

Particip (2016) *Evaluation of the EU Support to Research and Innovation for*
Development in Partner Countries 2007–2013, Volume 1 – Main Report.
European Commission (Brussels).

Patiño, M. (2009) *Gestión Territorial Indígena*. Confederación de los Pueblos
Indígenas de Bolivia (CIDOB, Santa Cruz de la Sierra). www.slideshare.net/
FTIERRA2010/gestin-territorial-indgena. (Accessed 20 December 2020).

PEMconsult and ODI (2020) *Evaluation of Danish Support for Climate Change*
Adaptation in Developing Countries. PEMconsult and Overseas Development
Institute. Danida Ref. no. 2019-8044. Ministry for Foreign Affairs of Denmark
(Copenhagen).

Pendleton, S.L., Miller, G.H., Lifton, N., Lehman, S.J., Southon, J., Crump, S.E. and
Anderson, R.S. (2019) Rapidly receding Arctic Canada glaciers revealing land-
scapes continuously ice-covered for more than 40,000 years. *Nature*
Communications, 10, Article 445. https://doi.org/10.1038/s41467-019-08307-w.
(Accessed 20 December 2020).

Perez, C. (1985) Microelectronics, long waves and world structural change: new perspectives for developing countries, *World Development*, 13: 441–463. https://doi.org/10.1016/0305-750X(85)90140-8. (Accessed 20 December 2020).

Petzold, J. (2017) *Social Capital, Resilience and Adaptation on Small Islands: Climate Change on the Isles of Scilly*. Springer International (Cham, Switzerland).

Phillips, T. (2020) How Bolivia's left returned to power months after Morales was forced out. *The Guardian*, 23 October 2020. www.theguardian.com/world/2020/oct/23/bolivia-left-return-power-evo-morales-mas. (Accessed 20 December 2020).

Pichler, P.-P., Zwickel, T., Chavez, A., Kretschmer, T., Seddon, J. and Weisz, H. (2017) Reducing urban greenhouse gas footprints. *Scientific Reports*, 7: 14659. https://doi.org/10.1038/s41598-017-15303-x. (Accessed 20 December 2020).

Pielke, R.A., Jr (1998). Rethinking the role of adaptation in climate policy. *Global Environmental Change*, 8 (2): 159–170. www.sciencedirect.com/science/article/abs/pii/S0959378098000119?via%3Dihub. (Accessed 20 December 2020).

Piketty, T. (2014) *Capital in the Twenty-First Century*. Belknap, Harvard (Cambridge, MA).

Pope Francis (2015) *Laudato Si': On Care for Our Common Home*. Our Sunday Visitor (Huntington, IN).

Porter-Bolland, L., Ellis, E.A., Guariguata, M.R., Ruiz-Mallén, I., Negrete-Yankelevich, S. and Reyes-García, V. (2012) Community managed forests and forest protected areas: an assessment of their conservation effectiveness across the tropics. *Forest Ecology and Management*, 268: 6–17.

Portes, J. (2020) Pandemonics: the false choice between national wealth and national health. *Byline Times*, 13 April 2020. https://bylinetimes.com/2020/04/13/pandemonics-the-false-choice-between-national-wealth-and-national-health/. (Accessed 20 December 2020).

Pradhan, B.B., Dangol, P.M., Bhaunju, R.M. and Pradhan, S. (2012) *Rapid Urban Assessment of Air Quality for Kathmandu, Nepal*. International Centre for Integrated Mountain Development (ICIMOD, Kathmandu).

Preston, B.L., Rickards, L., Dessai, S. and Meyer, R. (2013) Water, seas, and wine: science for successful climate adaptation, pages 151–169 in *Successful Adaptation to Climate Change: Linking Science and Policy in a Rapidly Changing World* (edited by S.C. Moser and M.T. Boykoff). Routledge (Abingdon and New York).

Pringle, P. (2011) *AdaptME Toolkit: Adaptation Monitoring and Evaluation*. United Kingdom Climate Impacts Programme (Oxford).

PSC (2020a). *The Polar Science Center: 40+ Years of Polar Research*. http://psc.apl.uw.edu/chairs-welcome/. (Accessed 20 December 2020).

PSC (2020b) *Unified Sea Ice Thickness Climate Data Record*. http://psc.apl.uw
.edu/sea_ice_cdr/. (Accessed 20 December 2020).

Quevedo, A., Bird, N., Amsalu, A., Crick, F., Gargule, A. and Suji, O. (2019) *Country
Experiences with Decentralised Climate Finance: Early Outcomes*. BRACED
Working paper. Building Resilience and Adaptation to Climate Extremes and
Disasters Programme (London).

Read, R. (2018) *This Civilisation Is Finished: So What Is to Be Done?* IFLAS
Occasional Paper 32. Institute of Leadership and Sustainability, University of
Cumbria (Carlisle). http://lifeworth.com/IFLAS_OP_3_rr_whatistobedone.pdf.
(Accessed 20 December 2020).

Reed, S.O., Friend, R., Jarvie, J., Henceroth, J., Thinphanga, P., Singh, D., Tran, P.
and Sutarto, R. (2014) Resilience projects as experiments: implementing climate
change resilience in Asian cities. *Climate and Development*. http://dx.doi.org/
10.1080/17565529.2014.989190. (Accessed 20 December 2020).

Resilience Academy (2020) *About Resilience Academy*. https://resilienceacademy
.ac.tz. (Accessed 20 December 2020).

Ripple, W.J., Wolf, C., Newsome, T.M., Barnard, P. Moomaw, W.R. and 11,258
scientist signatories from 153 countries (2020) World scientists' warning of a
climate emergency. *BioScience*, 70 (1): 8–12. https://doi.org/10.1093/biosci/
biz088. (Accessed 20 December 2020).

Rittel, H.W.J. and Webber, M.M. (1973) Dilemmas in a general theory of planning.
Policy Sciences, 4 (2): 155–169. https://link.springer.com/article/10.1007/
BF01405730. (Accessed 20 December 2020).

Roberts, J.T. and Parks, B.C. (2007) *A Climate of Injustice: Global
Inequality, North–South Politics, and Climate Policy*. MIT Press (Cambridge,
MA).

Robinson, D. and Robinson, K. (2005) *Pacific Ways of Talk: Hui and Talanoa*.
Social and Civic Policy Institute (Wellington, New Zealand). www
.communityresearch.org.nz/wp-content/uploads/formidable/robinson4.pdf.
(Accessed 20 December 2020).

Rocker, R. (1937) *Nationalism and Culture* (translated from German by R.E.
Chase). Freedom Press (London).

Rose, M. (2018) Macron gathers world's top sovereign funds to send climate signal.
Reuters, 6 July 2018. https://uk.reuters.com/article/uk-france-climatechange/
macron-gathers-worlds-top-sovereign-funds-to-send-climate-signal-
idUKKBN1JW0II. (Accessed 20 December 2020).

Rossing, T., Ayers, J., Anderson, S. and Pradhan, S. (2012) *CARE Participatory
Monitoring, Evaluation, Reflection and Learning (PMERL) for Community-
Based Adaptation (CBA)*. CARE International (Chatelaine, Switzerland).

Rutz, C., Loretto, M.-C., Bates, A.E., Davidson, S.C., Duarte, C.M., Jetz, W., Johnson, M., Kato, A., Kays, R., Mueller, T., Primack, R.., Ropert-Coudert, Y., Tucker, M.A., Wikelski, M. and Cagnacci, F. (2020) COVID-19 lockdown allows researchers to quantify the effects of human activity on wildlife. *Nature Ecology & Evolution*, 4: 1156–1159. https://doi.org/10.1038/s41559-020-1237-z. (Accessed 20 December 2020).

Sabel, C. and Zeitlin, J. (2011) Experimentalist governance. *GR:EEN Working Paper Series*, 2: 1–25. United Nations University Institute on Comparative Regional Integration Studies (Brugge, Belgium). http://cris.unu.edu/experimentalist-governance. (Accessed 20 December 2020).

Sabel, C. and Zeitlin, J. (2012) Experimentalist governance, pages 169–183 in *The Oxford Handbook of Governance* (edited by D. Levi-Faur). Oxford University Press (Oxford).

Sahlins, M. (1995) *How 'Natives' Think, About Captain Cook, for Example.* University of Chicago Press (Chicago and London).

Sanneh, E.S. (2018) *Systems Thinking for Sustainable Development: Climate Change and the Environment.* Springer International (Berlin).

Sarkar, A., Sensarma, S.N. and vanLoon, G.W. (eds) (2019) *Sustainable Solutions for Food Security: Combating Climate Change by Adaptation.* Springer International (Cham, Switzerland).

Saud, B. and Paudel, G. (2018) The threat of ambient air pollution in Kathmandu, Nepal. *Journal of Environmental and Public Health*, Article ID 1504591. https://doi.org/10.1155/2018/1504591. (Accessed 20 December 2020).

Saunois, M., Bousquet, P., Poulter, B., Peregon, A., Ciais, O., Canadell, J.G., Dlugokencky, E.J., Etiope, G., Bastviken, D., Houweling, S., Janssens-Maenhout, G., Tubiello, F.N., Castaldi, S., Jackson, R.B., Alexe, M., Arora, V.K., Beerling, D.J., Bergamaschi, P., Blake, D.R., Brailsford, G., Brovkin, V., Bruhwiler, L., Crevoisier, C., Crill, P., Covey, K., Curry, C., Frankenberg, C., Gedney, N., Höglund-Isaksson, L., Ishizawa, M., Ito, A., Joos, F., Kim, H.-S, Kleinen, T., Krummel, P., Lamarque, J.-F., Langenfelds, R., Locatelli, R., Machida, T., Maksyutov, S., McDonald, K.C., Marshall, J., Melton, J.R., Morino, I., Naik, V., O'Doherty, S., Parmentier, F.-J.W., Patra, P.K., Peng, C., Peng, S., Peters, G.P., Pison, I., Prigent, C., Prinn, R., Ramonet, M., Riley, W.J., Saito, M., Santini, M., Schroeder, R., Simpson, I.J., Spahni, R., Steele, P., Takizawa, A., Thornton, B.F., Tian, H., Tohjima, Y., Viovy, N., Voulgarakis, A., van Weele, M., van der Werf, G.R., Weiss, R., Wiedinmyer, C., Wilton, D.J., Wiltshire, A., Worthy, D., Wunch, D., Xu, X., Yoshida, Y., Zhang, B., Zhang, Z. and Zhu, Q. (2016) The global methane budget 2000–2012. *Earth System Science Data*, 8: 697–751. https://doi.org/10.5194/essd-8-697-2016. (Accessed 20 December 2020).

Saunois, M., Stavert, A.R., Poulter, B., Bousquet, P., Canadell, J.G., Jackson, R.B., Raymond, P.A., Dlugokencky, E.J., Houweling, S., Patra, P.K., Ciais, P., Arora, V.K., Bastviken, D., Bergamaschi, P., Blake, D.R., Brailsford, G., Bruhwiler, L., Carlson, K.M., Carrol, M., Castaldi, S., Chandra, N., Crevoisier, C., Crill, P.M., Covey, K., Curry, C.L., Etiope, G., Frankenberg, C., Gedney, N., Hegglin, M.I., Höglund-Isaksson, L., Hugelius, G., Ishizawa, M., Ito, A., Janssens-Maenhout, G., Jensen, K.M., Joos, F., Kleinen, T., Krummel, P.B., Langenfelds, R.L., Laruelle, G.G., Liu, L., Machida, T., Maksyutov, S., McDonald, K.C., McNorton, J., Miller, P.A., Melton, J.R., Morino, I., Müller, J., Murguia-Flores, F., Naik, V., Niwa, Y., Noce, S., O'Doherty, S., Parker, R.J., Peng, C., Peng, S., Peters, G.P., Prigent, C., Prinn, R., Ramonet, M., Regnier, P., Riley, W.J., Rosentreter, J.A., Segers, A., Simpson, I.J., Shi, H., Smith, S.J., Steele, L.P., Thornton, B.F., Tian, H., Tohjima, Y., Tubiello, F.N., Tsuruta, A., Viovy, N., Voulgarakis, A., Weber, T.S., van Weele, M., van der Werf, G.R., Weiss, R.F., Worthy, D., Wunch, D., Yin, Y., Yoshida, Y., Zhang, W., Zhang, Z., Zhao, Y., Zheng, B., Zhu, Q., Zhu, Q. and Zhuang, Q. (2020) The global methane budget 2000–2017. *Earth System Science Data*, 12: 1561–1623. https://doi.org/10.5194/essd-12-1561-2020. (Accessed 20 December 2020).

Schleicher, J., Peres, C.A., Amano, T., Llactayo, W. and Leader-Williams, N. (2017) Conservation performance of different conservation governance regimes in the Peruvian Amazon. *Scientific Reports*, 7: 11318. https://doi.org/10.1038/s41598-017-10736-w. (Accessed 20 December 2020).

Schwensen, C., Mailloux, L., Ramos, J.M., Behrendt, A. and Smed, L.S. (2017) *Evaluation of Danish–Bolivian Cooperation (1994–2016), September 2017.* Nordic Consulting Group and Orbicon (Copenhagen) and Ministry for Foreign Affairs of Denmark (Copenhagen).

Scotland, P. (2020) *Statement by Commonwealth Secretary-General on the 2020 General Elections in the United Republic of Tanzania, 2 November 2020.* https://thecommonwealth.org/media/news/statement-commonwealth-secretary-general-2020-general-elections-united-republic-tanzania. (Accessed 20 December 2020).

Scotland's Climate Assembly (2020a) *How Should Scotland Change to Tackle the Climate Emergency in an Effective and Fair Way?* https://scotclimateca.dialogue-app.com/how-should-scotland-change-to-tackle-the-climate-emergency-in-an-effective-and-fair-way/home?sort_order=rated#idea-count-container. (Accessed 20 December 2020).

Scotland's Climate Assembly (2020b) *A Peace with Nature Constitution.* https://scotclimateca.dialogue-app.com/how-should-scotland-change-to-tackle-the-cli

mate-emergency-in-an-effective-and-fair-way/a-peace-with-nature-constitu tion. (Accessed 20 December 2020).

Scranton, R. (2015) *Learning to Die in the Anthropocene: Reflections on the End of a Civilization*. City Lights (San Francisco).

Shah, S. (2008) *Civil Society in Uncivil Places: Soft State and Regime Change in Nepal*. East-West Center Policy Studies No. 48. East-West Center in Washington (Washington, DC).

Siegel, F.R. (2019) *Adaptations of Coastal Cities to Global Warming, Sea Level Rise, Climate Change and Endemic Hazards*. Springer International (Cham, Switzerland).

Silvestrini, S., Bellino, I. and Väth, S. (2015) *Impact Evaluation Guidebook for Climate Change Adaptation Projects*. Deutsche Gesellschaft für Internationale Zusammenarbeit (Eschborn, Germany).

Smith, B., Cuccillato, E. and Anderson, S. (2014) *Monitoring and Evaluating Climate Adaptation: A review Of GCCA Experience*. International Institute for Environment and Development (London). http://pubs.iied.org/pdfs/ 17253IIED.pdf. (Accessed 20 December 2020).

Snyder, C.W. (2016) Evolution of global temperature over the past two million years. *Nature*, 538: 226–228. https://doi.org/10.1038/nature19798. (Accessed 20 December 2020).

Southworth, J., Nagendra, H. and Cassidy, L. (2012) Forest transition pathways in Asia: studies from Nepal, India, Thailand, and Cambodia. *Journal of Land Use Science*, 7 (1): 51–65. https://doi.org/10.1080/1747423X.2010.520342. (Accessed 20 December 2020).

Sparkes, R. (2002) *Socially Responsible Investment: A Global Revolution*. Wiley (Chichester).

Spearman, M. and McGray, H. (2011) *Making Adaptation Count: Concepts and Options for Monitoring and Evaluation of Climate Change Adaptation*. Deutsche Gesellschaft für Internationale Zusammenarbeit (Eschborn, Germany).

Sridhar, D. (2020) The evidence is clear: if countries act together, they can suppress Covid. *The Guardian*, 1 November 2020. www.theguardian.com/commentis free/2020/nov/01/suppress-covid-england-lockdown-east-asian-african. (Accessed 20 December 2020).

Steffen, W., Richardson, K., Rockström, J., Cornell, S.E., Fetzer, I., Bennett, E.M., Biggs, R., Carpenter, S.R., Vries, W., De Wit, C.A., De Folke, C., Gerten, D., Heinke, J., Mace, G.M., Persson, L.M., Ramanathan, V., Reyers, B. and Sörlin, S. (2015) Planetary boundaries: guiding human development on a changing planet.

Science, 347 (6223): 1–10. https://doi.org/10.1126/science.1259855. (Accessed 20 December 2020).

Steffen, W., Rockström, J., Richardson, K., Lenton, T.M., Folke, C., Liverman, D., Summerhayes, C.P., Barnosky, A.D., Cornell, S.E., Crucifix, M., Donges, J.F., Fetzer, I., Lade, S.J., Scheffer, M., Winkelmann, R. and Schellnhuber, H.J. (2018) Trajectories of the Earth system in the Anthropocene. *Proceedings of the National Academy of Sciences of the USA*, 115 (33): 8252–8259. www.pnas .org/cgi/doi/10.1073/pnas.1810141115. (Accessed 20 December 2020).

Stern, N. (2007) *Stern Review on the Economics of Climate Change*. H.M. Treasury and Cambridge University Press (Cambridge).

Stevens, A. (1993) *The Two Million-Year-Old Self*. Texas A&M University Press (College Station).

Stiglitz, J. (2013) *The Price of Inequality*. Penguin (London).

Stiglitz, J. (2017) *Globalization and Its Discontents Revisited*. Penguin (London).

Sturgeon, N. (2019) *Wellbeing Economy Governments (WEGo) Policy Labs: First Minister's Speech*. Scottish Government. www.gov.scot/publications/well being-economy-governments-wego-policy-labs/. (Accessed 20 December 2020).

Suiseeya, K.R.M. and Zanotti, L. (2019) Making influence visible: innovating ethnography at the Paris Climate Summit. *Global Environmental Politics*, 19(2): 38–60. https://doi.org/10.1162/glep_a_00507. (Accessed 20 December 2020).

Swinburn, B.A., Kraak, V.I., Allender, S., Atkins, V.J., Baker, P.I., Bogard, J.R., Brinsden, H., Calvillo, A., De Schutter, O., Devarajan, R., Ezzati, M., Friel, S., Goenka, S., Hammond, R.A., Hastings, G., Hawkes, C., Herrero, M., Hovmand, P.S., Howden, M., Jaacks, L.M., Kapetanaki, A.B., Kasman, M., Kuhnlein, H.V., Kumanyika, S.K., Larijani, B., Lobstein, T., Long, M.W., Matsudo, V.K.R., Mills, S.D.H., Morgan, G., Morshed, A., Nece, P.M., Pan, A., Patterson, D.W., Sacks, G., Shekar, M., Simmons, G.L., Smit, W., Tootee, A., Vandevijvere, S., Waterlander, W.E., Wolfenden, L. and Dietz, W.H. (2019) The global syndemic of obesity, undernutrition, and climate change: the Lancet Commission report. *Lancet*, 393 (10173): 791–846. https://doi.org/10.1016/S0140-6736(18)32822-8. (Accessed 20 December 2020).

SWIOFish (2018) *United Republic of Tanzania SWIOFish Implementation Support Mission (22 Oct-2 Nov 2018)*. South West Indian Ocean Fisheries Governance and Shared Growth Program. World Bank Group and Indian Ocean Commission.

SWIOFish (2019) *Mauritius Symposium on Addressing Fisheries Governance Challenges in the Western Indian Ocean*. South West Indian Ocean Fisheries Governance and Shared Growth Program. World Bank Group and Indian Ocean Commission.

Swiss Re (2020) *A Fifth of Countries Worldwide at Risk from Ecosystem Collapse as Biodiversity Declines, Reveals Pioneering Swiss Re Index.* Swiss Re Management Ltd/Swiss Re Institute (Zurich). www.swissre.com/media/news-releases/nr-20200923-biodiversity-and-ecosystems-services.html. (Accessed 20 December 2020).

Tangney, P. (2017) *Climate Adaptation Policy and Evidence: Understanding the Tensions between Politics and Expertise in Public Policy.* Routledge (Abingdon and New York).

Taylor, M. (2020) The evolution of Extinction Rebellion. *The Guardian*, 4 August 2020. www.theguardian.com/environment/2020/aug/04/evolution-of-extinction-rebellion-climate-emergency-protest-coronavirus-pandemic. (Accessed 20 December 2020).

TBUK (2020) *What Is Timebanking?* Timebanking UK (Stroud). www.timebanking .org/what-is-timebanking/what-is-timebanking/. (Accessed 20 December 2020).

TEEB (2010) *The Economics of Ecosystems and Biodiversity: Mainstreaming the Economics of Nature – A Synthesis of the Approach, Conclusions and Recommendations of TEEB.* United Nations Environment Programme (Nairobi).

Theilade, I. (2020) *Oprindelige Folks Rettigheder og Skovbevaring* (Indigenous Peoples' Rights and Forest Conservation). International Work Group for Indigenous Affairs (IWGIA) and University of Copenhagen (Copenhagen).

Theilade, I., Brofeldt, S., Turreira-García, N. and Argyriou, D. (2021) Community monitoring of illegal logging and forest resources using smartphones and the Prey Lang application in Cambodia, pages 266–281 in *Geographic Citizen Science Design: No One Left Behind* (edited by A. Skarlatidou and A. Haklay). UCL Press (London).

Thom, R. (1975) *Structural Stability and Morphogenesis.* W.A. Benjamin (Reading, MA).

Throup, D.W (2016) *The Political Crisis in Zanzibar.* Centre for Strategic and International Studies. www.csis.org/analysis/political-crisis-zanzibar. (Accessed 20 December 2020).

Thunberg, G. (2019) *No One Is Too Small to Make a Difference.* Penguin (London).

Topper, E. and Pallen, D. (2014) *Evaluation of the Intra ACP Global Climate Change Alliance (GCCA), World-Wide, Final Report.* EURONET Consortium for EC (Brussels).

Topper, E. and Pallen, D. (2015) *Evaluation of the Global Climate Change Alliance (GCCA) Global Programme, World-Wide, Final Report.* EURONET Consortium for EC (Brussels).

Tyler, S., Keller, M., Swanson, D., Bizikova, L., Hammill, A., Zamudio, A.N., Moench, M., Dixit, A., Flores, R.G., Heer, C., González, D., Sosa, A.R., Gough, A.M., Solórzano, J.L., Wilson, C., Hernandez, X. and Bushey, S. (2013) *Climate*

Resilience and Food Security: A Framework for Planning and Monitoring. International Institute for Sustainable Development (Winnipeg, Canada).

UCLG (2019) *Manifesto: The Future of Resilience.* United Cities and Local Governments Congress, World Summit of Local and Regional Leaders, Durban 11–15 November, 2019. www.uclg.org/sites/default/files/en_mani festo_resilience.pdf. (Accessed 20 December 2020).

UN (2015) *Transforming Our World: The 2030 Agenda for Sustainable Development A/RES/70/1.* https://sustainabledevelopment.un.org/content/documents/21252030%20Agenda%20for%20Sustainable%20Development%20web.pdf. (Accessed 20 December 2020).

UN (2016) *Leaving No One Behind: The Imperative of Inclusive Development. Report on the World Social Situation 2016.* United Nations (New York).

UN (2020) *About the Sustainable Development Goals.* www.un.org/sustainablede velopment/sustainable-development-goals/. (Accessed 20 December 2020).

UNDESA (2018) *Sustainable Development Goals.* Sustainable Development Knowledge Platform. United Nations Department of Economic and Social Affairs (New York). https://sustainabledevelopment.un.org/sdgs. (Accessed 20 December 2020).

UNDP (2007a) *Human Development Report 2007/2008 Fighting Climate Change: Human Solidarity in a Divided World.* United Nations Development Programme (New York).

UNDP (2007b) *UNDP Monitoring and Evaluation Framework for Adaptation to Climate Change.* United Nations Development Programme (New York).

UNDP (2014a) *Community Based Resilience Assessment (CoBRA): Conceptual Framework and Methodology.* United Nations Development Programme (New York).

UNDP (2014b) *Community Based Resilience Analysis (CoBRA): Implementation Guidelines.* United Nations Development Programme (New York).

UNDP (2014c) *Understanding Community Resilience: Findings from Community-Based Resilience Analysis (CoBRA Assessments).* United Nations Development Programme (New York).

UNDP (2018) *Enhancing Climate Change Resilience in Zanzibar*, pipeline project, July 2018. https://info.undp.org/docs/pdc/Documents/TZA/ENHANCING%20CLIMATE%20CHANGE%20RESILIENCE%20IN%20ZANZIBAR%20-%20SIGNED%20PRODOC.pdf. (Accessed 20 December 2020).

UNEP (2012) *Fifth Global Environment Outlook (GEO-5): Chapter 5 – Biodiversity.* United Nations Environment Programme (Nairobi).

UNEP (2014) *The Adaptation Gap Report 2014: A Preliminary Assessment.* United Nations Environment Programme (Nairobi).

UNEP (2017) *The Adaptation Gap Report 2017: Towards Global Assessment.* United Nations Environment Programme (Nairobi).

UNEP (2018) *The Adaptation Gap Report 2018: Health Report.* United Nations Environment Programme (Nairobi).

UNEP (2019a) *Sixth Global Environment Outlook (GEO-6): Healthy Planet, Healthy People.* United Nations Environment Programme. Cambridge University Press (Cambridge).

UNEP (2019b) *Thawing Arctic Peatlands Risk Unlocking Huge Amounts of Carbon.* www.unenvironment.org/news-and-stories/story/thawing-arctic-peatlands-risk-unlocking-huge-amounts-carbon. (Accessed 20 December 2020).

UNEP (2019c) *Emissions Gap Report 2019.* United Nations Environment Programme (Nairobi). https://wedocs.unep.org/bitstream/handle/20.500.11822/30797/EGR2019.pdf?sequence=1andisAllowed=y. (Accessed 20 December 2020).

UNEP (2019d) *Costa Rica Named 'UN Champion of the Earth' for Pioneering Role in Fighting Climate Change.* United Nations Environment Programme, press release, 20 September 2019. www.unenvironment.org/news-and-stories/press-release/costa-rica-named-un-champion-earth-pioneering-role-fighting-climate. (Accessed 20 December 2020).

UNEP (2020) *Cities and Climate Change.* www.unenvironment.org/explore-topics/resource-efficiency/what-we-do/cities/city-activities. (Accessed 20 December 2020).

UNEP-DTIE (2015) *Awareness and Preparedness for Emergencies at Local Level: A Process for Improving Community Awareness and Preparedness for Technological Hazards and Environmental Emergencies,* 2nd edition. UNEP Division of Technology, Industry and Economics (Paris). www.preventionweb.net/publications/view/45469. (Accessed 20 December 2020).

UNFCCC (2015) *Synthesis Report on the Aggregate Effect of the Intended Nationally Determined Contributions.* FCCC/CP/2015/7. Note by the secretariat, Conference of the Parties Twenty-first session Paris, 30 November to 11 December 2015. UNFCCC Secretariat (Bonn). https://unfccc.int/resource/docs/2015/cop21/eng/07.pdf. (Accessed 20 December 2020).

UNFCCC (2016a) *Summary of the Paris Agreement.* United Nations Framework Convention on Climate Change eHandbook. UNFCCC Secretariat (Bonn). http://bigpicture.unfccc.int/#content-the-paris-agreemen. (Accessed 20 December 2020).

UNFCCC (2016b) *Aggregate Effect of the Intended Nationally Determined Contributions: An Update.* FCCC/CP/2016/2. Synthesis report by the secretariat, Conference of the Parties Twenty-second session Marrakech, 7–18 November 2016. UNFCCC Secretariat (Bonn). https://unfccc.int/resource/docs/2016/cop22/eng/02.pdf. (Accessed 20 December 2020).

UNFCCC (2020a) *What Do Adaptation to Climate Change and Climate Resilience Mean?* UNFCCC Secretariat (Bonn). https://unfccc.int/topics/adaptation-and-resilience/the-big-picture/what-do-adaptation-to-climate-change-and-climate-resilience-mean#eq-1. (Accessed 20 December 2020).

UNFCCC (2020b) *Understanding INDCs, NDCs and Long-Term Strategies.* UNFCCC Secretariat (Bonn). https://unfccc.int/resource/bigpicture/index .html#content-indcs-and-ndcs. (Accessed 20 December 2020).

UNFCCC (2020c) *2018 Talanoa Dialogue Platform.* unfccc.int/process-and-meet ings/the-paris-agreement/the-paris-agreement/2018-talanoa-dialogue-platform. (Accessed 20 December 2020).

UNFCCC (2020d) *Statistics on Participation and In-Session Engagement.* UNFCCC Secretariat (Bonn). https://unfccc.int/process-and-meetings/parties-non-party-stakeholders/non-party-stakeholders/statistics-on-non-party-stakeholders/statis tics-on-participation-and-in-session-engagement. (Accessed 20 December 2020).

US Embassy in Nepal (2020) *Air Quality Monitor.* https://np.usembassy.gov/ embassy/air-quality-monitor/. (Accessed 20 December 2020).

Verdens Skove (2020) *The Civil Society Fund Programme Completion Report.* Verdens Skove/Forests of the World for CISU (Aarhus and Copenhagen).

Villanueva, P.S. (2011) *Learning to ADAPT: Monitoring and Evaluation Approaches in Climate Change Adaptation and Disaster Risk Reduction – Challenges, Gaps and Ways Forward.* SCR Discussion Paper 9. Institute of Development Studies (Brighton).

Voiland, A. (2019) Smoke Blankets Borneo. *NASA Earth Observatory,* 18 September 2019. https://earthobservatory.nasa.gov/images/145614/smoke-blankets-borneo. (Accessed 20 December 2020).

Wadhams, P. (2016) *A Farewell to Ice.* Allen Lane (London).

Wadhams, P. (2017) *What is the 'Arctic Death Spiral'.* Oxford Academic (Oxford University Press). www.youtube.com/watch?v=Rl1xleN-Zp4. (Accessed 20 December 2020).

Walker, T.W. and Wade, C.J. (2011) *Nicaragua: Living in the Shadow of the Eagle.* Westview Press (Boulder, CO).

Waters, C.N., Zalasiewicz, J., Summerhayes, C., Barnosky, A.D., Poirier, C., Gałuszka, A., Cearreta, A., Edgeworth, M., Ellis, E.C., Ellis, M., Jeandel, C., Leinfelder, R., McNeill, J.R., Richter, D. deB., Steffen, W., Syvitski, J., Vidas, D., Wagreich, M., Williams, M., An, Z.-S., Grinevald, J., Odada, E., Oreskes, N. and Wolfe, A.P. (2016) The Anthropocene is functionally and stratigraphically distinct from the Holocene. *Science,* 351 (6269): 138–141. https://doi.org/10.1126/ science.aad2622. (Accessed 20 December 2020).

Watkiss, P., Pye, S., Hendriksen, G., Maclean, A., Bonjean, M., Shaghude, Y., Jiddawi, N., Sheikh, M.A. and Khamis, Z. (2012). *The Economics of Climate Change in Zanzibar*. Global Climate Adaptation Partnership and DFID Study Report for the Climate Change Committee of the Revolutionary Government of Zanzibar (Zanzibar).

WEAll (2020) *How Would Things Be Different in a Wellbeing Economy? New WEAll Resource Explores the Differences*. https://wellbeingeconomy.org/how-would-things-be-different-in-a-wellbeing-economy-new-website-section-explores-the-differences. (Accessed 20 December 2020).

Weart, S.R. (2004) *The Discovery of Global Warming*. Harvard University Press (Cambridge, MA).

WEF (2020a) *Nature Risk Rising: Why the Crisis Engulfing Nature Matters for Business and the Economy*. World Economic Forum in collaboration with PwC (Geneva). www3.weforum.org/docs/WEF_New_Nature_Economy_Report_2020.pdf. (Accessed 20 December 2020).

WEF (2020b) *The Future of Nature and Business*. World Economic Forum in collaboration with AlphaBeta (Geneva). www3.weforum.org/docs/WEF_The_Future_Of_Nature_And_Business_2020.pdf. (Accessed 20 December 2020).

WEF (2020c) *The Global Risks Report 2020*. World Economic Forum in partnership with Marsh and McLennan and Zurich Insurance Group (Geneva). www3.weforum.org/docs/WEF_Global_Risk_Report_2020.pdf. (Accessed 20 December 2020).

Weikmans, R., Timmons Roberts, J., Baum, J., Bustos, M.C. and Durand, A. (2017) Assessing the credibility of how climate adaptation aid projects are categorised. *Development in Practice*, 27 (4): 458–471. https://doi.org/10.1080/09614524.2017.1307325. (Accessed 20 December 2020).

Weinberg, S. (1993) *The First Three Minutes: A Modern View of the Origin of the Universe*, 2nd edition. Basic Books (New York).

Weisbrot, M. (2020) Silence reigns on the US-backed coup against Evo Morales in Bolivia. *The Guardian*, 18 September 2020. www.theguardian.com/commentisfree/2020/sep/18/silence-us-backed-coup-evo-morales-bolivia-american-states. (Accessed 20 December 2020).

Welch, C. (2019) Arctic permafrost is thawing fast: that affects us all. *National Geographic*, September 2019. www.nationalgeographic.com/environment/2019/08/arctic-permafrost-is-thawing-it-could-speed-up-climate-change-feature/. (Accessed 20 December 2020).

Whelpton, J. (2005) *A History of Nepal*. Cambridge University Press (Cambridge).

Willsher, K. (2020) Emmanuel Macron pledges €15bn to tackle climate crisis. *The Guardian*, 29 June 2020. www.theguardian.com/world/2020/jun/29/emmanuel-macron-pledges-15bn-to-tackle-climate-crisis. (Accessed 20 December 2020).

Wilson, P.H. (2016) *The Holy Roman Empire: A Thousand Years of Europe's History*. Allen Lane (London).

Wilson, S.J., Schelhas, J., Grau, R., Nanni, A.S. and Sloan, S. (2017) Forest ecosystem-service transitions: the ecological dimensions of the forest transition. *Ecology and Society*, 22 (4): Article 38. https://doi.org/10.5751/ES-09615-220438. (Accessed 20 December 2020).

World Bank (2015) *Financing Climate-Resilient Growth in Tanzania. Environment and Natural Resources Global Practice Policy Note*. World Bank Group Report Number ACS11581. The World Bank (Washington, DC).

World Bank (2017) *Tanzania Economic Update: Managing Water Wisely – The Urgent Need to Improve Water Resources Management in Tanzania*. World Bank (Washington, DC).

World Bank (2019) *Tanzania: Country Environmental Analysis – Environmental Trends and Threats, and Pathways to Improved Sustainability*. World Bank (Washington, DC).

World Bank (2020) *Mapping Zanzibar Using Low-Cost Drones*. https://olc.worldbank.org/content/mapping-zanzibar-using-low-cost-drones. (Accessed 20 December 2020).

WRI (2020) *INDCS: Post-2020 Climate Action Commitments*. www.wri.org/our-work/topics/indcs. (Accessed 20 December 2020).

WWF (2012) *Living Planet Report 2012: Biodiversity, Biocapacity and Better Choices*. Edited by M. Grooten, R.E.A. Almond and R. McLellan. WWF International (Gland).

WWF (2014) *Living Planet Report 2014: Species and Spaces, People and Places*. Edited by R. McLellan, L. Iyengar, B. Jeffries and N. Oerlemans. WWF International (Gland).

WWF (2016) *Living Planet Report 2016: Risk and Resilience in a New Era*. Edited by N. Oerlemans, H. Strand, A. Winkelhagen, M. Barrett and M. Grooten. WWF International (Gland).

WWF (2018) *Living Planet Report 2018: Aiming Higher*. Edited by M. Grooten and R.E.A. Almond. WWF International (Gland).

WWF (2020) *Living Planet Report 2020: Bending the Curve of Biodiversity Loss*. Edited by R.E.A. Almond, M. Grooten and T. Petersen. WWF International (Gland).

XR (Extinction Rebellion) (2019) *This Is Not a Drill: An Extinction Rebellion Handbook*. Penguin (London).

Xu, C., Kohler, T.A., Lenton, T.M., Svenningg , J.-C. and Scheffer, M. (2020) Future of the human climate niche. www.pnas.org/content/pnas/early/2020/04/28/1910114117.full.pdf. (Accessed 20 December 2020).

Yaro, J.A. and Hesselberg, J. (eds) (2016) *Adaptation to Climate Change and Variability in Rural West Africa.* Springer International (Cham, Switzerland).

Yokomatsu, M. and Hochrainer-Stigler, S. (eds) (2020) *Disaster Risk Reduction and Resilience.* Springer International (Cham, Switzerland).

ZAN-SDI (2015) *Project Document: National Spatial Data Infrastructure for Integrated Coastal and Marine Spatial Planning in Zanzibar (ZAN-SDI).* Submitted by the Finnish Environment Institute and National Land Survey of Finland in cooperation with the Zanzibar Commission for Lands. MFA Finland, Helsinki (26 November 2015).

ZanSea (2020) *Zanzibar Social Environmental Atlas for Coastal and Marine Areas.* www.suza.ac.tz/zansea-website/index.php. (Accessed 20 December 2020).

Zeitlin, J. and Sabel, C. (2013) Experimentalism in Transnational Governance: Emergent Pathways and Diffusion Mechanisms. *GR:EEN Working Paper Series*, 3: 1–9. United Nations University Institute on Comparative Regional Integration Studies (Brugge, Belgium). http://cris.unu.edu/experimentalism-transnational-governance-emergent-pathways-and-diffusion-mechanisms. (Accessed 20 December 2020).

ZMI (2020) *Zanzibar Mapping Initiative.* www.zanzibarmapping.org. (Accessed 20 December 2020).

Index